Principles of

Power Engineering Analysis

Principles of
Power Engineering Analysis

Robert C. Degeneff | M. Harry Hesse

CRC Press
Taylor & Francis Group
Boca Raton   London   New York

CRC Press is an imprint of the
Taylor & Francis Group, an **informa** business

CRC Press
Taylor & Francis Group
6000 Broken Sound Parkway NW, Suite 300
Boca Raton, FL 33487-2742

First issued in paperback 2017

© 2012 by Taylor & Francis Group, LLC
CRC Press is an imprint of Taylor & Francis Group, an Informa business

No claim to original U.S. Government works

ISBN 13: 978-1-4398-9231-2 (hbk)
ISBN 13: 978-1-138-07506-1 (pbk)

Visit the Taylor & Francis Web site at
http://www.taylorandfrancis.com

and the CRC Press Web site at
http://www.crcpress.com

Contents

List of Figures

Preface

In the late 1800s as electrical engineering programs were taking shape, they were structured to emphasize power generation, transmission, and its utilization. However, by the middle of the 20th century in recognition of the vast advances in controls, electronics, and computers these programs were being drastically restructured as they moved away from the traditional core. This transition was so swift and complete that within a decade few electrical engineering programs offered more than a class or two in electric power. Utilities and manufacturers of heavy electrical equipment, still in need of competent practitioners, found it difficult to find engineers with the desired skills in heavy three-phase electrical power. Recognizing this situation Dr. Eric T. B. Gross, with the financial support of American Electric Power, formed the Department of Electric Power Engineering at Rensselaer Polytechnic Institute (RPI). The primary purpose of this department was to educate power engineers to fill this void. A unique characteristic of this department from its onset was its focus on the master's degree rather than the bachelor's or doctorate. Additionally, the student was encouraged to complete the program in a calender year. For the following four decades this program was one of the very few that offered graduate work in electric power engineering. In recognition of its successfully achieving its goal, students were consistently attracted to it from around the world.

To facilitate graduation in a year, the program required the completion of 10 three credit hour classes without a dissertation. It was felt, I think with substantial justification, that a thesis at the master's degree level was of less value to an engineer than several well-taught classes. This class work was subdivided into two groups. The first was comprised of four required classes: Principles of Power Engineering Analysis, Computer Methods, Mechanical Aspects of Electric Power Equipment, and Electric and Magnetic Fields. Since the faculty was small, focused and communicated daily, coordination between the courses was highly effective. The faculty felt that these four integrated classes would provide the electric power engineering graduate student with the basic tools necessary to be technically competent as a practicing power engineer. The remaining six classes were selected by mutual agreement between the student and his or her advisor. They were composed of additional power engineering classes (relaying, transients, machines, economics, etc.), other appropriate engineering classes (heat transfer, fluids, controls), mathematics, or possibly a well-chosen business class. The pendulum of interest that swung away from electric power engineering a half century ago is now swinging back. This book fills the need for a graduate level text that addresses three-phase power system analysis. As long as electric power is transmitted with lines and cables and voltages are varied with transformers, the material presented in this text will be of interest to the power engineer.

This text introduces, at the graduate level, the concepts needed to understand and apply three-phase electric power transmission. The material in this text can be comfortably presented in a one-semester class. Chapter 1 introduces the transmission line and cable characteristics by developing from basic principles the expressions that predict how the transmission systems will perform. Chapter 2 in a similar fashion develops expressions for the performance of single-phase transformers. Chapter 3 extends the single-phase concepts presented in the first two chapters to that of balanced three-phase systems. Chapter 4 expands the discussion further to unbalanced systems. Chapter 5 introduces the basic concepts of symmetrical components analysis of transmission systems and

Chapter 6 extends the discussion of symmetrical components to three-phase transformers. Chapter 7 introduces the concepts of symmetrical component analysis to faulted systems. The presentation of symmetrical component analysis contained in Chapters 5, 6, and 7 are at a detail not readily available to the practicing engineer. Chapter 8 discussed the design of untransposed transmission lines and Chapter 9 introduces other analysis component systems. Synchronous machines are not covered in this volume since it was expected that they would be addressed in another required course in the master's degree program.

During the last forty years the electrical industry has undergone tremendous change, e.g., deregulation, soaring energy costs, the transition to computer controlled systems, power electronics, the aging of the installed equipment base, and major changes in the industry supplying equipment. However, the basic need for power engineering education has not changed nor has the need for a fundamental understanding of power system analysis. My first exposure to power engineering education was at RPI where I was fortunate first to be a student (one of Dr. Gross's boys) and for the last two decades as faculty member and part of this small, focused department. Drs. Gross, Greenwood, Nelson, Salon, Torrey, and Hesse made this work a rewarding time in my life. Dr. Hesse taught the Power Engineering Analysis class from 1970 until his retirement in 1994. This book is an outgrowth of the notes he developed and used while teaching Principles of Power Engineering Analysis. On numerous occasions I would encounter him at his computer adding or revising the material and I would often ask him when the notes were going to be completed and published. His answer was almost always the same–when I retire. Dr. Hesse retired in 1994 and regrettably passed away in 1995. Upon Dr. Hesse's retirement it was my good fortune to be asked to teach the class. What I quickly found was that it had now become my responsibility to respond to the frequently asked question of when were the notes going to be published. Often my answer was "soon." To make this "soon" possible Col. J. Sullivan helped in transitioning these notes and figures to LaTeX.

It is my hope that Dr. Hesse would have been pleased with the final form of what he started so many years ago.

Robert C. Degeneff

Niskayuna, New York

Authors

M. Harry Hesse Dr. Hesse received his bachelor's degree in Electrical Engineering from Marquette University in 1951 and a master's degree in Electrical Power Systems Engineering from the Illinois Institute of Technology in 1953. As a Fulbright Fellow he earned his doctoral degree in Electrical Power System Engineering from the Rheinisch-Wesfalische Technische Hochschule in Aachen, Germany. His dissertation was titled, "Short Circuit Calculation of Idealized Meshed Networks." He was a member of Sigma Xi, Pi Mu Epsilon, Tau Beta Pi, and Eta Kappa Nu. He received the Westinghouse Power Systems Fellowship from 1952 to 1953.

Dr. Hesse was a design engineer in GE's Large Power Transformer Department in Pittsfield, MA from 1956 - 1958. He was a senior analytic engineer with the Electric Utility Engineering Operation at GE from 1958 - 1970. Dr. Hesse was a lecturer for the Advanced Engineering Course at GE from 1957 until 1960. He was an instructor for the company's power system engineering course from 1960 - 1970.

From 1965 - 1970, Dr. Hesse was an adjunct professor for Electric Power Engineering at the Rensselaer Polytechnic Institute. He was named professor in 1970 and granted tenure in 1976. He retired in 1994 with 30 years of service and was later to receive the status of professor emeritus.

Dr. Hesse was named a Fellow of the Institute of Electrical Engineers (IEE) in London in 1976. He was named a Fellow in the IEEE in 1990 and a Life Fellow in 1993. He received the Power Engineering Educator Award from the Edison Electric Institute in 1980. He was also a registered professional engineer in New York state. He was published in numerous technical journals.

He was a Navy veteran of World War II.

Dr. Hesse passed away on August 24, 1995.

Robert C. Degeneff Dr. Degeneff received his bachelor's degree in Mechanical Engineering from Kettering Institute (formerly General Motors Institute) in 1966 and his master's in Electric Power Engineering from Rensselaer Polytechnic Institute (RPI) in 1967. In 1967 he entered the USAF and served as a technical intelligence officer, leaving the service as a Captain in 1970. He returned to RPI and was graduated with a Doctorate in Electric Power Engineering in 1973. His dissertation was titled, "Transient Interaction of Transformers and Transmission Lines." He is a member of Alpha Tau Iota, Tau Beta Pi, Eta Kappa Nu, and Sigma Xi. He received a National Science Fellowship from 1972-1973.

In 1973 he joined GE's Large Power Transformer Department as a senior development engineer. From 1978 - 1981 he managed the Department's Electrical and Analytical Development Unit developing computer tools and standard practices to insure transformer design integrity. From 1981 - 1985 he managed GE's HVDC Transmission Engineering Subsection and was responsible for the studies and research necessary to design an HVDC substation. From 1985 - 1989 he was responsible for managing GE's Software Services Section, which constructed computer programs to assist utilities in performing generation utilization planning.

In 1989 he joined the faculty of RPI as a professor of electric power engineering and was granted tenure in 1995. During his tenure at RPI he taught power engineering analysis and electrical transients in power systems. He was also actively involved in undergraduate education. The department was also involved in research into the design and performance of utility and industrial power apparatus. He retired from RPI in 2006 and was granted emeritus status in 2008.

In 1991 Dr. Degeneff founded and is currently the president of Utility Systems Technologies, Inc. which builds electronic voltage regulators and power quality mitigation equipment and provides consulting to the utility industry.

Dr. Degeneff is a PE in New York, a Fellow in the IEEE, and the 2008 recipient of the IEEE Herman Halprin Award. He has published over seven dozen papers (two IEEE prize papers), chapters in five books, and holds eight patents. He has been very active in the IEEE Transformer Committee and is currently the chair of the IEEE working group that wrote the C57.142 guide, *IEEE Guide to Describe the Occurrence and Mitigation of Switching Transients Induced by Transformers, Switching Devices, and System Interaction.*

Dr. Degeneff's research interests are computing the transient response of electrical equipment, power quality, and utility system planning.

Chapter 1

Transmission Line Characteristics

The popularity of the electric energy conversion process lies, amongst other things, upon the ease with which large blocks of energy can be transmitted from a source of generation to a remote point of utilization. Overhead transmission lines, cables, bus-bars, etc., which are used to carry this electrical energy, have electromagnetic characteristics that influence and limit the electrical energy, which they are able to transport effectively. The power transfer capability limit and degree of voltage variation with load changes are generally expressed in terms of inductive reactance, shunt capacitance, series resistance, and shunt conductance.

In many studies these four characteristics can be satisfactorily reflected by a relatively simple lumped-parameter network model. For example see References [1], [2], [3], [4], [5], [7], [8], and [9].

In studies dealing with switching surges or overvoltages due to lightning, a more complex distributed parameter model is required. And finally, the electric and magnetic fields associated with transmission lines and cables can have a significant impact upon nearby communication lines, fences, metal vehicles, etc., in which case relatively complex mathematical or network models may be required.

With the advent of modern computers, the arithmetic evaluation of the traditional transmission line parameters has become a relatively trivial matter. Consequently we are now free to devote more attention to an in-depth study of the physical nature of transmission line electromagnetic field characteristics and expand our horizon to permit greater flexibility in selecting suitable representations for a greater variety of studies.

1.1 The Magnetic Field

Exactly what is a magnetic field? Since a magnetic field is, per se, invisible, yet exists by virtue of numerous observations of its effects, we can at best propose a model that appears to satisfy all of these observations. We visualize a magnetic field in terms of a mathematical model of vector fields. Whether such vector fields actually exist is not critically important. What is important, though, is the fact that such a vector field model has successfully permitted predictions of observable magnetic phenomena for over a century and therefore leads us to continue to accept this as a valid model for analytical purposes.

One component of the magnetic vector field is defined by Ampere's law:

$$curl\vec{H} = \nabla \times \vec{H} = \vec{J} \qquad [A/m^2] \qquad (1.1.1)$$

Equation 1.1.1 states that whenever there exists a conduction current density vector $\vec{J}$, a magnetic vector $\vec{H}$ field is created. Conversely, in order to establish an $\vec{H}$ field, we must have a current density

1

$\vec{J}$ vector field. Even the magnetic field about permanent magnets is attributable to a net alignment of electron spins in the magnet.

We recall from vector algebra:

$$\vec{C} = \vec{A} \times \vec{B} \qquad (1.1.2)$$

that $\vec{C}$ lies perpendicular to the plane formed by vectors $\vec{A}$ and $\vec{B}$. On this basis we recognize that the vector $\vec{H}$ field lies in a plane perpendicular to the current density vector $\vec{J}$.

Via Stoke's theorem, we may convert Equation 1.1.1 into the more convenient integral form:

$$\oint \vec{H} \cdot d\vec{\ell} = i \qquad [A] \qquad (1.1.3)$$

which is known as Ampere's circuital law, and which is much easier to deal with in practical engineering problems than Equation 1.1.1. Equation 1.1.3 states that a line integral, taken around a closed path in the plane of a vector $\vec{H}$ field, is numerically equal to the electric current enclosed within the path of integration (i.e., equal to the current passing through the area whose perimeter is the closed path of integration).

$\vec{H}$ (amperes/meter) may be considered to be the strength of the force (excitation) available to create a magnetic flux field according to the relationship:

$$\vec{B} = \mu\vec{H} \qquad [Wb/m^2] \qquad (1.1.4)$$

whereby the magnitude of the resultant flux density $\vec{B}$ $[Wb/m^2]$ is also a function of the magnetic permeability μ of the medium within which the magnetic field resides.

It is common practice to write:

$$\mu = \mu_r\mu_o \qquad [H/m] \qquad (1.1.5)$$

where:

$$\mu_o = 4\pi(10^{-7}) \qquad [H/m] \qquad (free\ space) \qquad (1.1.6)$$
$$\mu_r = relative\ permeability \qquad (1.1.7)$$

For all nonferromagnetic materials such as air, copper, aluminum, earth, etc., the relative magnetic permeability will be taken equal to unity.

Maxwell's equation:

$$div\vec{B} = \nabla \cdot \vec{B} = 0 \qquad (1.1.8)$$

states that the magnetic flux density $\vec{B}$ field has no sources or sinks. Visualized in terms of lines of flux, flux lines are continuous and close upon themselves (unlike electrostatic flux lines that terminate on point charges). It is of interest also to note (since $div\ curl\ \vec{A} \equiv 0$) that Equation 1.1.1 leads to the conclusion that vector current density lines are continuous and must close upon themselves (i.e., since $div\ curl\ \vec{H} = 0 = div\vec{J}$). We shall shortly consider a single current-carrying conductor in free space, which violates the above constraint, and therefore observe some unrealistic consequences. Maxwell's equation:

$$\oint \vec{E} \cdot d\vec{\ell} = -\frac{d\left[\int\int \vec{B} \cdot d\vec{S}\right]}{dt} \qquad [V] \qquad (1.1.9)$$

is a generalized expression reflecting Faraday's observations that a voltage is induced in an electrical circuit when changes occur in the amount of magnetic flux linking (encircling) the circuit.

Equation 1.1.9 is actually quite profound and the reader is not encouraged to employ it directly unless he or she has exceptional expertise in vector calculus. On the surface one might interpret Equation 1.1.9 as follows:

$$e_i = -\frac{d\left[\int\int \vec{B} \cdot d\vec{S}\right]}{dt} = -\frac{d\Phi}{dt} \qquad [V] \qquad (1.1.10)$$

which would be valid for a single loop circuit, and also for N turns linked by a common flux Φ if Φ is properly identified as volts per turn. However, as we shall see shortly, the effective *flux-linkages* as required by Faraday's law inside a current-carrying conductor are not at all easily identifiable per Equation 1.1.10. We prefer to write Faraday's law in the form:

$$v = -e_i = \frac{d\lambda}{dt} \qquad [V] \qquad (1.1.11)$$

and:

$$\lambda = \int v \cdot dt \qquad [V \cdot s] \qquad (1.1.12)$$

The term *flux-linkages* (λ, volt-seconds) is introduced here to emphasize the fact that magnetic flux (Φ, Webers) and λ are not always directly interchangeable. The effectiveness with which an element $d\Phi$ causes an induced voltage according to Faraday's law determines $d\lambda$. The distinction in character between magnetic flux and magnetic flux-linkages is further amplified in Appendix B.

Equation 1.1.11 is presented in the two forms most commonly found in the literature. Those who prefer e_i refer to this as "induced voltage" or "back EMF." In view of the fact that:

$$v = L\frac{di}{dt} = \frac{d\lambda}{dt} \qquad [V] \qquad (1.1.13)$$

is almost always employed in conventional circuit analysis, whereby it is understood that v is a voltage drop (similar to an Ri drop), we shall consider the term e_i to be superfluous.

While the above approach is valid for singly-excited magnetic circuits, it does not resolve the polarity question for mutually coupled circuits. In the latter instances we merely invoke Lenz's law, which states that when the amount of flux linking a closed circuit is changed, a current will flow in that circuit in a direction as to oppose the change in flux-linkages. This uniquely establishes the polarity of the voltage across the open-circuit terminals.

1.1.1 Single Conductor in Free Space

Consider a portion of a current-carrying conductor in free space, shown in Figure 1.1.1. A magnetic field $\vec{H}$ exists and lies in a plane perpendicular to $\vec{J}$. By virtue of geometric symmetry, lines of constant $\vec{H}$ are circles. Ampere's circuital law is therefore easy to evaluate:

$$\oint \vec{H} \cdot d\vec{\ell} = H\int d\ell = H(2\pi r) = i \qquad [A] \qquad (1.1.14)$$

and

$$H_e = \frac{i}{2\pi r} \qquad [A/m] \qquad (1.1.15)$$

which is valid only for the magnetic field *external* to the conductor, in which case the path of the closed line integral encircles (links) the entire current i.

For the magnetic field *inside* the current-carrying conductor, when the current density is uniform across the cross section of the conductor, we must take:

$$\oint \vec{H} \cdot d\vec{\ell} = H(2\pi r) = \left(\frac{r}{r_c}\right)^2 i \qquad [A] \qquad (1.1.16)$$

since the path of integration encloses (links) only a portion of the total current i. Then:

$$H_i = \frac{i}{2\pi r}\left(\frac{r}{r_c}\right)^2 \qquad [A/m] \tag{1.1.17}$$

For the magnetic flux density we have:

$$\vec{B} = \mu_o \vec{H} \qquad [Wb/m^2] \tag{1.1.18}$$

and an associated tube of magnetic flux is defined by:

$$d\Phi = \vec{B} \cdot d\vec{S} = B dr \Delta Z \qquad [Wb] \tag{1.1.19}$$

For the magnetic field *external* to the conductor, each tube of flux links (encircles) the entire current i so that:

$$d\lambda_e = d\Phi \qquad [V \cdot s] \tag{1.1.20}$$

and:

$$\frac{\lambda_e}{\Delta Z} = \int_{r_c}^{R} \lim_{R\to\infty} \frac{\mu_o i}{2\pi r} dr = \lim_{R\to\infty} \frac{\mu_o i}{2\pi} \ln \frac{R}{r_c} \qquad [V \cdot s/m] \tag{1.1.21}$$

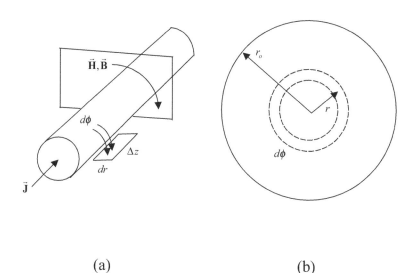

(a)   (b)

Figure 1.1.1: Magnetic Field about and within a Current-Carrying Conductor (Free Space)

Since the magnetic flux-linkages are directly proportional to the exciting current, it becomes convenient to write:

$$\frac{\lambda}{\Delta Z} = Li \qquad [V \cdot s/m] \tag{1.1.22}$$

where L is called the inductance per unit length with dimensions of Henrys per meter. Thus, for Equation 1.1.21, we have:

$$L_e = \lim_{R\to\infty} \frac{\mu_o}{2\pi} \ln \frac{R}{r_c} \qquad [H/m] \tag{1.1.23}$$

We note that L_e becomes infinitely large (since the flux-linkages extend to infinity), a fact that we intuitively sense to be unrealistic. This dilemma stems from the fact that we have assumed only a single current-carrying conductor in free space which, in view of the continuity of electric current, is unrealistic. We shall see in the next section that L_e remains of finite value when a return current is properly taken into account.

For the magnetic field *inside* a current-carrying conductor, the tube of flux $d\Phi$ does not encircle *all* of the current I. Thus, the effective flux-linkages are:

$$d\lambda_i = \left(\frac{r}{r_c}\right)^2 d\Phi \qquad [V \cdot s] \tag{1.1.24}$$

and:

$$\frac{\lambda_i}{\Delta Z} = \int_0^{r_c} \frac{\mu_o i}{2\pi r} \left(\frac{r}{r_c}\right)^4 dr = \frac{\mu_o i}{8\pi} \qquad [V \cdot s/m] \tag{1.1.25}$$

and:

$$L_i = \frac{\mu_o}{8\pi} \qquad [H/m] \tag{1.1.26}$$

The reader who may find it difficult to accept the premise for Equation 1.1.24 may find the alternative analysis in Appendix B more substantive.

The *total* flux linking a unit length of single current-carrying conductor in free space is therefore:

$$\frac{\lambda}{\Delta Z} = \lim_{R \to \infty} \frac{\mu_o i}{2\pi} \left(\ln \frac{R}{r_c} + \frac{1}{4}\right) \qquad [V \cdot s/m] \tag{1.1.27}$$

The term (1/4) in Equation 1.1.27 is valid only for the conductors of circular cross-section and in which the current density is uniform across the cross-section of the conductor. It has become customary to write:

$$\frac{1}{4} = \ln e^{\frac{1}{4}} = \ln 1.284025 = \ln \frac{1}{0.7788} \tag{1.1.28}$$

and to express Equation 1.1.27 in the form:

$$\frac{\lambda}{\Delta Z} = \lim_{R \to \infty} \frac{\mu_o i}{2\pi} \ln \frac{R}{0.7788 r_c} \qquad [V \cdot s/m] \tag{1.1.29}$$

whereby:

$$0.7788 r_c = GMR = Geometric\, Mean\, Radius \qquad [m] \tag{1.1.30}$$

While this example of a single current-carrying conductor in free space is admittedly physically unrealistic, it serves well to establish the basic procedure for evaluating the net magnetic flux-linkages about the current-carrying conductor. We shall make use of these results in the following analysis of some physically realistic transmission line configurations.

1.1.2 Parallel Conductors in Free Space

Consider a set of parallel current-carrying conductors in free space as shown in Figure 1.1.2. A conductor located at (x_1, y_1), for example, carries a current i_1. Also let there be a parallel line filament p located at (x_p, y_p), with a vanishingly small radius and carrying no current. We wish to determine the net flux linking the line filament p.

We might, perhaps, consider the contributions of each current-carrying conductor to the $\vec{H}$ field at p on the basis of Equation 1.1.15. While this approach is perfectly possible, it is, however, awkward since the field is a vector field $\vec{H}$ involving both magnitude and direction.

On the other hand, observe that Φ (or λ) is the result of a vector dot product $\vec{H} \cdot d\vec{S}$ and is therefore a *scalar* quantity. Consequently the contributions of the various current-carrying conductors to the net flux linking p can be obtained directly by an arithmetic superposition. We shall prefer this simpler approach.

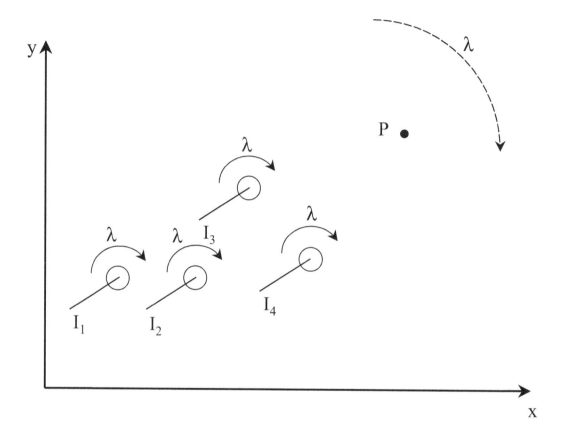

Figure 1.1.2: Magnetic Field Due to Parallel Conductors in Free Space

Consider first the contribution due to i_1. Let the center-to-center distance between conductor #1 and the filament p be given by:

$$d_{p1} = \sqrt{(x_p - x_1)^2 + (y_p - y_1)^2} \qquad [m] \qquad (1.1.31)$$

None of the flux-linkages *internal* to conductor #1 as given by Equation 1.1.25 link with p. Of the flux-linkages *external* to conductor #1 as given by Equation 1.1.21, none of the flux between r_1 and d_{p1} links filament p. Therefore, the only flux linking filament p due to current i_1 is given by:

$$\frac{\lambda_{p1}}{\Delta Z} = \lim_{R \to \infty} \int_{d_{p1}}^{R} \frac{\mu_o i_1}{2\pi r} dr = \lim_{R \to \infty} \frac{\mu_o i_1}{2\pi} \ln \frac{R}{d_{p1}} \qquad [V \cdot s/m] \qquad (1.1.32)$$

In general, then, the contribution to the flux linking filament p due to any current i_k is:

$$\frac{\lambda_{pk}}{\Delta Z} = \lim_{R \to \infty} \frac{\mu_o i_k}{2\pi} \ln \frac{R}{d_{pk}} \qquad [V \cdot s/m] \qquad (1.1.33)$$

and the total flux linking filament p due to all of the current-carrying conductors is:

$$\frac{\lambda_{pk}}{\Delta Z} = \lim_{R \to \infty} \frac{\mu_o}{2\pi} \sum_{k=1}^{n} \left(\ln \frac{R}{d_{pk}} \right) i_k \qquad [V \cdot s/m] \qquad (1.1.34)$$

We may write Equation 1.1.34 in expanded form:

$$\frac{\lambda_p}{\Delta Z} = \frac{\mu_o}{2\pi} \sum_{k=1}^{n} \left(\ln \frac{1}{d_{pk}} \right) i_k + \frac{\mu_o}{2\pi} \left[\lim_{R \to \infty} \ln R \right] \sum_{k=1}^{n} i_k \qquad [V \cdot s/m] \qquad (1.1.35)$$

If we consider a *complete* system of current-carrying conductors for which:

$$\sum_{k=1}^{n} i_k = 0 \qquad [A] \qquad (1.1.36)$$

then the second term in Equation 1.1.35 becomes zero.

For every set of straight parallel current-carrying conductors in free space, for which Equation 1.1.36 is satisfied, and not considering effects at the ends of the conductors, the total flux linking any parallel line filament p is given by:

$$\frac{\lambda_p}{\Delta Z} = \frac{\mu_o}{2\pi} \sum_{k=1}^{n} (\ln \frac{1}{d_{pk}}) i_k \qquad [V \cdot s/m] \qquad (1.1.37)$$

While Equations 1.1.34 and 1.1.37 are mathematically and physically equivalent, the terms in 1.1.34 clearly reflect the flux lying between the filament p and some very remote distance R.

Equation 1.1.37 may be employed directly in calculating the total flux-linkages about any non-current-carrying conductors (fences, communication lines, ground wires, etc.) due to a set of parallel current-carrying conductors in free space. Equation 1.1.37 may also be used as a basis for evaluating the net flux-linkages about any of the actual current-carrying conductors, thereby defining practical self- and mutual inductances.

For example, the total flux per unit length linking conductor i is obtained by taking line filament p to lie along the centerline of conductor i:

$$\frac{\lambda_i}{\Delta Z} = \frac{\mu_o}{2\pi} \sum_{\substack{k=1 \\ i \neq 1}}^{n} \left(\ln \frac{1}{d_{ik}} \right) i_k + \frac{\mu_o}{2\pi} \left(\ln \frac{1}{GMR_i} \right) i_i \qquad [V \cdot s/m] \qquad (1.1.38)$$

We define:

$$L_{ii} = \frac{\mu_o}{2\pi} \ln \frac{1}{GMR_i} \qquad [H/m] \qquad (1.1.39)$$

$$L_{ij} = \frac{\mu_o}{2\pi} \ln \frac{1}{d_{ij}} \qquad [H/m] \qquad (1.1.40)$$

(Free Space)

whereby the numerators in the ln terms are to be interpreted as 1 meter spacing. For the sake of typographical convenience, we shall hereafter take $\lambda_i/\Delta Z = \Delta\lambda_i$ and write:

$$\Delta\lambda_i = \sum_{j=1}^{n} L_{ij}i_j \qquad [V \cdot s/m] \qquad (1.1.41)$$

For a four-conductor system (Figure 1.1.3, for example) we have four simultaneous equations, which may be written in matrix format as:

$$\begin{bmatrix} \Delta\lambda_1 \\ \Delta\lambda_2 \\ \Delta\lambda_3 \\ \Delta\lambda_4 \end{bmatrix} = \begin{bmatrix} L_{11} & L_{12} & L_{13} & L_{14} \\ L_{21} & L_{22} & L_{23} & L_{24} \\ L_{31} & L_{32} & L_{33} & L_{34} \\ L_{41} & L_{42} & L_{43} & L_{44} \end{bmatrix} \begin{bmatrix} i_1 \\ i_2 \\ i_3 \\ i_4 \end{bmatrix} \qquad [V \cdot s/m] \qquad (1.1.42)$$

whereby the elements of the [L] matrix are defined by Equations 1.1.39 and 1.1.40. For a general n-conductor system we may define the associated system of flux-linkage equations in terms of symbolic matrix notation by:

$$[\Delta\lambda] = [L][i] \qquad [V \cdot s/m] \qquad (1.1.43)$$

When the conductor currents are purely sinusoidal at $\omega = 2\pi f$, we have by Faraday's law (in phasor notation and rms quantities):

$$\left[\Delta\tilde{V}\right] = j\omega\left[\Delta\tilde{\Lambda}\right] = j(2\pi f)\left[\Delta\tilde{\Lambda}\right] \qquad [V \cdot s/m] \qquad (1.1.44)$$

where $\Delta\tilde{V}$ is interpreted as a (positive) voltage drop along the conductor in the direction of (positive) $\tilde{I}$. Then:

$$\begin{bmatrix} \Delta\tilde{V}_1 \\ \Delta\tilde{V}_2 \\ \Delta\tilde{V}_3 \\ \Delta\tilde{V}_4 \end{bmatrix} = (j)\begin{bmatrix} X_{11} & X_{12} & X_{13} & X_{14} \\ X_{21} & X_{22} & X_{23} & X_{24} \\ X_{31} & X_{32} & X_{33} & X_{34} \\ X_{41} & X_{42} & X_{43} & X_{44} \end{bmatrix} \begin{bmatrix} \tilde{I}_1 \\ \tilde{I}_2 \\ \tilde{I}_3 \\ \tilde{I}_4 \end{bmatrix} \qquad [V \cdot s/m] \qquad (1.1.45)$$

whereby the elements of the [X] matrix are the inductive self- and mutual *reactances* of the conductors.

Example 1.1.1

Consider the configuration shown in Figure 1.1.3(a), and assume that $(\tilde{I}_1 + \tilde{I}_2 + \tilde{I}_3 + \tilde{I}_4) = 0$ so that Equations 1.1.39 and 1.1.40 are applicable. Then:

$$\begin{bmatrix} \Delta\tilde{\Lambda}_1 \\ \Delta\tilde{\Lambda}_2 \\ \Delta\tilde{\Lambda}_3 \\ \Delta\tilde{\Lambda}_4 \end{bmatrix} = \begin{bmatrix} 0.9210 & -0.4605 & -0.3442 & -0.4398 \\ -0.4605 & 0.9210 & -0.4398 & -0.3442 \\ -0.3442 & -0.4398 & 1.1983 & -0.3219 \\ -0.4398 & -0.3442 & -0.3219 & 1.1983 \end{bmatrix} \begin{bmatrix} \tilde{I}_1 \\ \tilde{I}_2 \\ \tilde{I}_3 \\ \tilde{I}_4 \end{bmatrix}$$

$$[mV \cdot s/km]$$

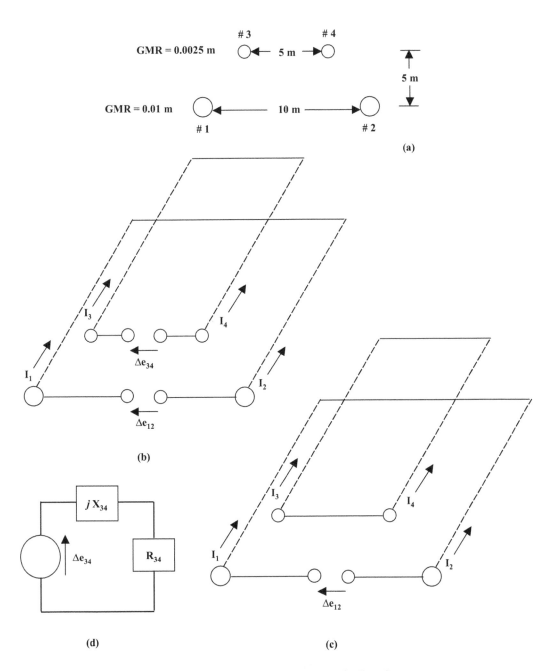

Figure 1.1.3: Four Parallel Conductors in Free Space

and for $f = 60$ Hz:

$$
\begin{bmatrix} \Delta \tilde{V}_1 \\ \Delta \tilde{V}_2 \\ \Delta \tilde{V}_3 \\ \Delta \tilde{V}_4 \end{bmatrix} = (j) \begin{bmatrix} 347.21 & -173.60 & -129.76 & -165.80 \\ -173.60 & 347.21 & -165.80 & -129.76 \\ -129.76 & -165.80 & 451.75 & -121.35 \\ -165.80 & -129.76 & -121.35 & 451.75 \end{bmatrix} \begin{bmatrix} \tilde{I}_1 \\ \tilde{I}_2 \\ \tilde{I}_3 \\ \tilde{I}_4 \end{bmatrix}
$$

$$[mV/km]$$

Now let $\tilde{I}_3 = \tilde{I}_4 = 0$ with $\tilde{I}_1 = -\tilde{I}_2 = 1000$ amperes. Then:

$$
\begin{bmatrix} \Delta \tilde{\Lambda}_1 \\ \Delta \tilde{\Lambda}_2 \\ \Delta \tilde{\Lambda}_3 \\ \Delta \tilde{\Lambda}_4 \end{bmatrix} = \begin{bmatrix} 1.3815 \\ -1.3815 \\ 0.0956 \\ -0.0956 \end{bmatrix} [V \cdot s/km]
\qquad
\begin{bmatrix} \Delta \tilde{V}_1 \\ \Delta \tilde{V}_2 \\ \Delta \tilde{V}_3 \\ \Delta \tilde{V}_4 \end{bmatrix} = (j) \begin{bmatrix} 520.81 \\ -520.81 \\ 36.04 \\ -36.04 \end{bmatrix}
$$

$$[V/km]$$

Now consider two loops formed by the conductors as shown in Figure 1.1.3(b). Since we have assumed $\tilde{I}_3 = \tilde{I}_4 = 0$, the voltage around the upper loop is:

$$\Delta \tilde{V}_{3-4} = \Delta \tilde{V}_3 - \Delta \tilde{V}_4 = j72.08 \qquad [V/km]$$

For the lower loop we have:

$$\Delta \tilde{V}_{1-2} = \Delta \tilde{V}_1 - \Delta \tilde{V}_2 = j1041.62 \qquad [V/km]$$

In practical terms, this voltage (neglecting conductor resistance) is represented by:

$$\Delta \tilde{V}_{1-2} = j X_{1-2} \tilde{I}_1 \qquad [V/km]$$

For $\tilde{I}_1 = 1000$ amperes rms, we have, therefore:

$$X_{1-2} = 1.0416 \qquad [ohms/km]$$

for the lower loop, and:

$$X_1 = X_2 = 0.5208 \quad [ohms/km] \qquad = 0.8381 \quad [ohms/mile]$$

per conductor.

Example 1.1.2
Consider now the configuration shown in Figure 1.1.3(c) whereby the upper loop is closed, thereby permitting a current $\tilde{I}_3 = -\tilde{I}_4$ to circulate. We wish to determine the magnitude of $\tilde{I}_3$. First we shall assume the upper loop to be lossless ($R_3 = R_4 = 0$). We must now recalculate all flux-linkages to include the additional effect of $\tilde{I}_3 = -\tilde{I}_4$. Now clearly, since we have assumed the upper loop to be lossless, if $\Delta \tilde{\Lambda}_{3-4} \neq 0$, then $\Delta \tilde{V}_{3-4} \neq 0$, and $\tilde{I}_3 = -\tilde{I}_4$ would become infinitely large. Since we intuitively feel that this is not reasonable, we reflect and recognize that a constraint to our problem is $\Delta \tilde{\Lambda}_{3-4} = 0$. The voltage around the upper loop is given by:

$$
\begin{aligned}
\Delta \tilde{\Lambda}_{3-4} &= \Delta \tilde{\Lambda}_3 - \Delta \tilde{\Lambda}_4 \\
&= (L_{31} - L_{41})\tilde{I}_1 + (L_{32} - L_{42})\tilde{I}_2 + (L_{33} - L_{43})\tilde{I}_3 + (L_{34} - L_{44})\tilde{I}_4
\end{aligned}
$$

If we neglect the resistance of the conductors in the upper loop, the net voltage around the upper loop must be zero according to Kirchhoff's voltage law; i.e., the voltage induced in the upper loop is equal to the reactance drop around the upper loop . Thus, we have:

$$2\left[(L_{31} - L_{41})\tilde{I}_1 + (L_{33} - L_{43})\tilde{I}_3\right] = 0$$

and:

$$\tilde{I}_3 = -\frac{(L_{31} - L_{41})}{(L_{33} - L_{43})}\tilde{I}_1 = -62.89 \qquad [A]$$

We observe that our solution also satisfies Lenz's law. If we wish to include the effect of losses in the upper loop, we may do this quite simply on the basis of the Thevenin-Helmholtz equivalent network shown in Figure 1.1.3(d) where ΔV_{3-4} is the open-loop voltage:

$$\Delta V_{3-4} = 72.08 \qquad [V/km]$$

and on the basis of the closed-loop current:

$$X_{3-4} = \frac{72.08}{62.89} = 1.1461 \qquad [\Omega/km]$$

In retrospect, we could have written the loop equations directly as:

$$\begin{bmatrix} \Delta\tilde{\Lambda}_{1-2} \\ \Delta\tilde{\Lambda}_{3-4} \end{bmatrix} = (2)\begin{bmatrix} (L_{11} - L_{21}) & (L_{13} - L_{23}) \\ (L_{31} - L_{41}) & (L_{33} - L_{43}) \end{bmatrix}\begin{bmatrix} \tilde{I}_1 \\ \tilde{I}_2 \end{bmatrix}$$

$$= \begin{bmatrix} 2.7630 & 0.1912 \\ 0.1912 & 3.0404 \end{bmatrix}\begin{bmatrix} \tilde{I}_1 \\ \tilde{I}_2 \end{bmatrix} \qquad [mV.s/km]$$

1.1.3   Single Conductor above Idealized Conducting Plane

The analysis in Section 1.1.2 does not take into account the influence of conducting surfaces adjacent to parallel current-carrying conductors. We shall first consider a single conductor parallel to a perfectly conducting (zero resistance) plane surface. Consider a pair of parallel conductors in free space, as illustrated in Figure 1.1.4, of equal radii, separated by $2h$ meters, and carrying currents $\tilde{I}_2 = -\tilde{I}_1$. According to Section 1.1.2, we may write:

$$\begin{bmatrix} \Delta\tilde{\Lambda}_1 \\ \Delta\tilde{\Lambda}_2 \end{bmatrix} = \begin{bmatrix} L_{11} & L_{12} \\ L_{21} & L_{22} \end{bmatrix}\begin{bmatrix} \tilde{I}_1 \\ \tilde{I}_2 \end{bmatrix} \qquad [V \cdot s/m] \qquad (1.1.46)$$

so that:

$$\Delta\tilde{\Lambda}_1 = (L_{11} - L_{12})\tilde{I}_1 = \frac{\mu_o}{2\pi}\left(\ln\frac{2h}{GMR}\right)\tilde{I}_1 \qquad [V \cdot s/m] \qquad (1.1.47)$$

$$\Delta\tilde{\Lambda}_2 = (L_{22} - L_{21})\tilde{I}_2 = \frac{\mu_o}{2\pi}\left(\ln\frac{2h}{GMR}\right)\tilde{I}_2 \qquad [V \cdot s/m] \qquad (1.1.48)$$

Because of the geometric symmetry indicated in Figure 1.1.4, an infinitely thin, perfectly conducting, nonmagnetic sheet may be inserted at the midplane between the conductors without affecting the external magnetic field distribution. In Appendix C, we show that this thin sheet may also have an appropriate surface current distribution without affecting the external magnetic field distribution.

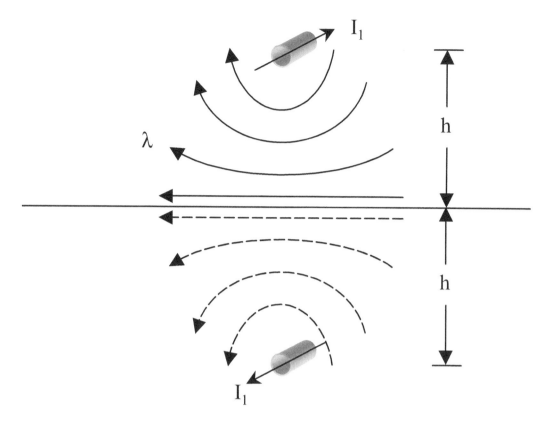

Figure 1.1.4: Single Current-Carrying Conductor above Perfectly Conducting Plane

We may now remove the lower conductor and the lower half of the thin sheet and thereby obtain a solution for a current-carrying conductor above a perfectly conducting plane with return current in the plane.

What we have done is to establish the basis for the so called "*image method*." Given a current-carrying conductor above a perfectly conducting nonmagnetic plane, with return current in the plane, we inject an identical current-carrying conductor at the mirror image below the surface of the plane with equal and opposite current. The flux linking the current-carrying conductor is then given by Equation 1.1.47.

1.1.4   Parallel Conductors above Ideal Conducting Plane

The practical approach for determining the flux-linkages associated with a set of conductors, parallel to one another, and parallel to an ideal conducting plane is based upon an extension of the "*method of images*" developed in Section 1.4. Thus, a set of "*image*" conductors is located in positions of the mirror image about the plane surface as shown in Figure 1.1.5. The image conductors carry the negative of the currents in their counterparts.

For the configuration shown in Figure 1.1.5, we have for the conductor 1:

$$\Delta\tilde{\Lambda}_1 = L_{11}\tilde{I}_1 + L_{12}\tilde{I}_2 + L_{13}\tilde{I}_3 - L_{11'}\tilde{I}_1 - L_{12'}\tilde{I}_2 - L_{13'}\tilde{I}_3 \qquad [V \cdot s/m] \qquad (1.1.49)$$

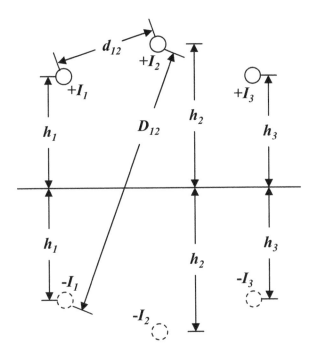

Figure 1.1.5: Parallel Current-Carrying Conductors above Perfectly Conducting Plane

and in detail:

$$
\begin{aligned}
\Delta\tilde{\Lambda}_1 &= \frac{\mu_o}{2\pi}\left(\ln\frac{1}{GMR_1} + \ln 2h_1\right)\tilde{I}_1 \\
&\quad + \frac{\mu_o}{2\pi}\left(\ln\frac{1}{d_{12}} + \ln D_{12}\right)\tilde{I}_2 \\
&\quad + \frac{\mu_o}{2\pi}\left(\ln\frac{1}{d_{13}} + \ln D_{13}\right)\tilde{I}_3 \qquad [V\cdot s/m]
\end{aligned}
\qquad (1.1.50)
$$

For conductors above an *ideal conducting* plane we have the general forms:

$$
L_{ii-p} = \frac{\mu_o}{2\pi}\left[\ln\frac{1}{GMR_i} + \ln 2h_i\right] \qquad [H/m] \qquad (1.1.51)
$$

$$
L_{ij-p} = \frac{\mu_o}{2\pi}\left[\ln\frac{1}{d_{ij}} + \ln D_{ij}\right] \qquad [H/m] \qquad (1.1.52)
$$

Ideal Plane

with:
GMR_i = geometric mean radius of conductor i, meters
h_i = height of conductor i above ideal conducting plane, meters
d_{ij} = separation distance between conductors i and j, meters
D_{ij} = distance between conductor i and the image of conductor j, meters

1.1.5   Parallel Conductors above Resistive Earth

As shown in Appendix C, time-varying flux lines do not penetrate a perfectly conducting plane. Induced voltages and the resultant surface layer currents prevent this. The layer of current is infinitesimally thin; i.e., the "skin depth" approaches zero.

For a resistive plane (lossy material), the skin depth of the induced currents becomes greater and greater; i.e., it diffuses into the semi-infinite conductor with a plane surface. Also, the flux lines penetrate deeper and deeper into the semi-infinite conductor.

For a conductor h meters above the surface of the earth, the effective image conductor appears to be considerably greater than h meters below the surface of the earth. The next section describes the manner in which finite earth resistivity may be handled in evaluating overhead line inductances.

1.1.6   Carson's Earth-Return Correction

In 1926, Dr. John R. Carson published results of his analysis of the influence of a homogeneous resistive earth. The details of portions of his analysis are presented in Appendix D. The results for 60 Hz and practical transmission line configurations take the approximate form:

$$L_{ii-e} = \frac{\mu_o}{2\pi} \ln \frac{1}{GMR_i} + \frac{L_e}{3} \quad [H/m] \qquad (1.1.53)$$

$$L_{ij-e} = \frac{\mu_o}{2\pi} \ln \frac{1}{d_{ij}} + \frac{L_e}{3} \quad [H/m] \qquad (1.1.54)$$

with:

$$\begin{aligned}
\frac{L_e}{3} &= \frac{\mu_o}{2\pi} \left[-0.0772 + \ln \frac{2}{\sqrt{\alpha}} \right] \\
&= \frac{\mu_o}{2\pi} \ln \frac{1.8514}{\sqrt{\alpha}} \\
&= \frac{\mu_o}{2\pi} \ln 85.06 \sqrt{\rho_e} \quad [H/m]
\end{aligned} \qquad (1.1.55)$$

and for $\rho_e = 100$ [ohm.m]:

$$\frac{L_e}{3} = 0.2 \ln 850.6 = 1.3492 \quad [mH/km] \qquad (1.1.56)$$

In order to obtain an appreciation for the penetration of the magnetic field into the earth, recall that in Equation 1.1.52, d_{ij} and D_{ij} represent the distance between the actual conductors (above an ideal conducting plane) and the distance between an actual conductor and the image of the other. In this case:

$$D_e = 85.06 \sqrt{\rho_e} = 850.6 \quad [m] \quad for \ \rho_e = 100 \quad [\Omega \cdot m] \qquad (1.1.57)$$

represents an effective separation distance between the actual conductors and their images.

It is sometimes erroneously assumed that the earth-return current is, in fact, concentrated at the location of the "effective" image conductor. This is, of course, not true. Recall in Section 1.4 that the return current is concentrated at the surface of the perfectly conducting plane. In the case of resistive earth, current density diffuses below the surface of the earth. At a $D_e = 850.6$ m, however, the current density is essentially zero.

1.1.7 Three-Phase Single-Circuit Lines above Earth

With an understanding of the physical principles developed so far, writing the flux-linkage equations for a system of parallel overhead lines becomes a rather simple mechanical procedure. For most practical three-phase single-circuit lines above earth, the Carson's correction factor is independent of the height of the conductors above earth. Therefore, the flux-linkage equations take the form:

$$
\begin{bmatrix} \Delta\tilde{\Lambda}_a \\ \Delta\tilde{\Lambda}_b \\ \Delta\tilde{\Lambda}_c \end{bmatrix} = \begin{bmatrix} L_{aa} & L_{ab} & L_{ac} \\ L_{ba} & L_{bb} & L_{bc} \\ L_{ca} & L_{cb} & L_{cc} \end{bmatrix} \begin{bmatrix} \tilde{I}_a \\ \tilde{I}_b \\ \tilde{I}_c \end{bmatrix} + (\frac{L_e}{3}) \begin{bmatrix} 1 & 1 & 1 \\ 1 & 1 & 1 \\ 1 & 1 & 1 \end{bmatrix} \begin{bmatrix} \tilde{I}_a \\ \tilde{I}_b \\ \tilde{I}_c \end{bmatrix}
$$
$$[V \cdot s/m] \qquad\qquad (1.1.58)$$

with:

$$
L_{ii} = \frac{\mu_o}{2\pi} \ln \frac{1}{GMR_i} \qquad [H/m] \qquad (1.1.59)
$$

$$
L_{ij} = \frac{\mu_o}{2\pi} \ln \frac{1}{d_{ij}} \qquad [H/m] \qquad (1.1.60)
$$

where L_{ii} is commonly referred to as the inductance to one-meter spacing [compare these equations with Equations 1.1.39 and 1.1.40, 1.1.51, and 1.1.52]. The practicality of this format lies in the fact that L_{ii} depends *only* upon the characteristics of the conductor itself, and L_{ij} depends *only* upon the spacing between conductors.

Observe [in Equation 1.1.58] that when the three-phase currents are perfectly *balanced*, there is *no* influence due to earth. However, when the currents are *unbalanced*, the effect of earth return must be taken into account.

Untransposed Lines

Most conductors in overhead transmission lines are disposed horizontally, or vertically, but rarely in an equilateral triangle. Figures 1.1.8 and 1.1.9 illustrate some typical single-circuit line configurations. Most often, then, $L_{ab} \neq L_{bc} \neq L_{ca}$. Even for a *balanced* three-phase current, where *earth has no influence*, we have:

$$
\begin{bmatrix} \Delta\tilde{\Lambda}_a \\ \Delta\tilde{\Lambda}_b \\ \Delta\tilde{\Lambda}_c \end{bmatrix} = (\tilde{I}_a) \begin{bmatrix} L_{aa} + a^2 L_{ab} + a L_{ac} \\ L_{ba} + a^2 L_{bb} + a L_{bc} \\ L_{ca} + a^2 L_{cb} + a L_{cc} \end{bmatrix} \qquad [V \cdot s/m] \qquad (1.1.61)
$$

Although the three-phase conductors will generally be of the same conductor material and size so that $L_{aa} = L_{bb} = L_{cc}$, it is apparent that there will be an unbalance in flux-linkages. On the basis of Faraday's law, this translates into an unbalanced inductive reactance voltage drop along the line. This aspect is treated in greater detail in Chapters 5 and 8.

Example 1.1.3

Consider the horizontal line configuration shown in Figure 1.1.6 for which we have:

$$
\begin{bmatrix} \Delta\tilde{\Lambda}_a \\ \Delta\tilde{\Lambda}_b \\ \Delta\tilde{\Lambda}_c \end{bmatrix} = \begin{bmatrix} 2.2702 & 0.8887 & 0.8887 \\ 0.8887 & 2.2702 & 0.8887 \\ 0.8887 & 0.7500 & 2.2702 \end{bmatrix} \begin{bmatrix} \tilde{I}_a \\ \tilde{I}_b \\ \tilde{I}_c \end{bmatrix} \qquad [mV \cdot /km]
$$

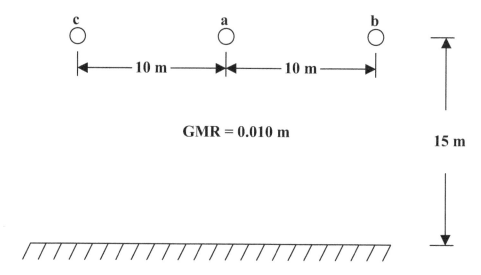

Figure 1.1.6: Single-Circuit Three-Phase Line

and for balanced three-phase currents:

$$\begin{bmatrix} \Delta\tilde{\Lambda}_a \\ \Delta\tilde{\Lambda}_b \\ \Delta\tilde{\Lambda}_c \end{bmatrix} = (\tilde{I}_a) \begin{bmatrix} (L_{aa} - L_{ab}) \\ a^2(L_{aa} - L_{ab}) \\ a(L_{aa} - L_{ab}) \end{bmatrix} \quad [V \cdot s/m]$$

Thus, $\mid \Delta\tilde{\Lambda}_b \mid = \mid \Delta\tilde{\Lambda}_c \mid = 1.0538 \mid \Delta\tilde{\Lambda}_a \mid$, and $\mid \Delta\tilde{\Lambda}_b \mid$ and $\mid \Delta\tilde{\Lambda}_c \mid$ are advanced and retarded, respectively, in phase angle by 4.73 degrees.

When the single-circuit line, energized by a balanced source voltage, terminates in a balanced load impedance, the voltage at the load will be slightly unbalanced due to the unbalanced voltage drop along the line. However, the voltage drop along the line is very much smaller than the source voltage so that the effect of the unbalanced induced voltages is generally negligible.

Transposed Lines

In order to eliminate the unbalanced induced voltages described in the preceding section, the line may be transposed according to Figure 1.1.7. Phase *rotation*, shown in Figure 1.1.7(b), is *always* effective if the rotations are performed at the one-third points along the length of the line section. Phase *transpositions*, illustrated in Figure 1.1.7(a), at the one-third points are completely effective only for *some* line configurations; a symmetrically spaced horizontal line configuration is one of these.

For the completely transposed line:

$$\begin{aligned} L_{aa} &= L_{bb} = L_{cc} \\ L_{ab} &= L_{bc} = L_{ca} \end{aligned} \tag{1.1.62}$$

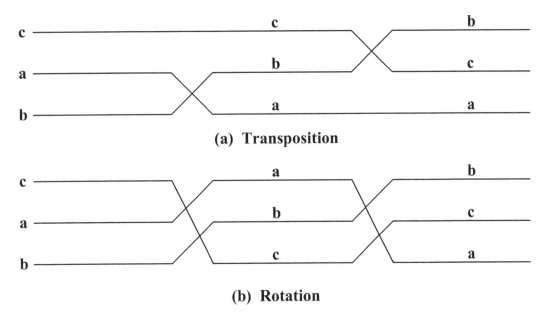

(a) Transposition

(b) Rotation

Figure 1.1.7: Line Transposition

and for balanced three-phase currents:

$$
\begin{bmatrix} \Delta\tilde{\Lambda}_a \\ \Delta\tilde{\Lambda}_b \\ \Delta\tilde{\Lambda}_c \end{bmatrix} = (\tilde{I}_a) \begin{bmatrix} (L_{aa} - L_{ab}) \\ a^2(L_{aa} - L_{ab}) \\ a(L_{aa} - L_{ab}) \end{bmatrix} \qquad [V \cdot s/m] \qquad (1.1.63)
$$

and the *physically mutually-coupled* three-phase circuits may be viewed in terms of three *equivalent independent* (mutually uncoupled) single-phase lines.

In order to actually accomplish phase rotation or transposition at the one-third points along the line requires special, more expensive, steel tower structures. Historically, there is also ample evidence that transposition points in the line are more vulnerable to faults. It is therefore important to be able to evaluate the gain in performance versus the increased expense and possible lessening of reliability. Present-day consensus of opinion is that, for single-circuit lines, line transposition cannot be justified. However, every opportunity to perform phase rotation or transposition at existing substations, where this may be done with little effort, should be taken to advantage.

Influence of Ground Wires

Often ground wires (shield wires) are located above the phase conductors to intercept lightning strokes. These ground wires will generally be conductively connected to the steel towers which, in turn, are assumed to be effectively grounded. Only in the most exacting type of analysis will it be necessary to make allowance for the fact that this grounding takes place at discrete span intervals. For power frequency analysis, it is common practice to assume that ground wires are continuously grounded; i.e., the ground wire is assumed to be continuously at zero potential with respect to earth.

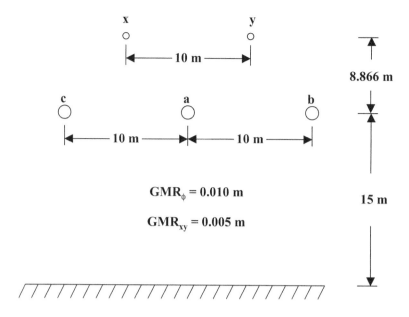

Figure 1.1.8: Single-Circuit Three-Phase Line with Ground Wires

For the configuration shown in Figure 1.1.8, we have:

$$
\begin{bmatrix}
\Delta\tilde{\Lambda}_a \\
\Delta\tilde{\Lambda}_b \\
\Delta\tilde{\Lambda}_c \\
0 \\
0
\end{bmatrix}
=
\begin{bmatrix}
L_{aa} & L_{ab} & L_{ac} & L_{ax} & L_{ay} \\
L_{ba} & L_{bb} & L_{bc} & L_{bx} & L_{by} \\
L_{ca} & L_{cb} & L_{cc} & L_{cx} & L_{cy} \\
L_{xa} & L_{xb} & L_{xc} & L_{xx} & L_{xy} \\
L_{ya} & L_{yb} & L_{yc} & L_{yx} & L_{yy}
\end{bmatrix}
\begin{bmatrix}
\tilde{I}_a \\
\tilde{I}_b \\
\tilde{I}_c \\
\tilde{I}_x \\
\tilde{I}_y
\end{bmatrix}
$$

$$
+\left(\frac{L_e}{3}\right)
\begin{bmatrix}
1 & 1 & 1 & 1 & 1 \\
1 & 1 & 1 & 1 & 1 \\
1 & 1 & 1 & 1 & 1 \\
1 & 1 & 1 & 1 & 1 \\
1 & 1 & 1 & 1 & 1
\end{bmatrix}
\begin{bmatrix}
\tilde{I}_a \\
\tilde{I}_b \\
\tilde{I}_c \\
\tilde{I}_x \\
\tilde{I}_y
\end{bmatrix}
$$

$$[V \cdot s/m] \qquad (1.1.64)$$

For an untransposed line section, while we may assume balanced three-phase currents, we do not know, a priori, that $\tilde{I}_x + \tilde{I}_y = 0$. We have stated previously that in the real world, and for a complete set of current-carrying conductors, the summation of currents is zero. Because of the manner in which we have included the effects of earth return, the *earth-return current* does not appear explicitly in Equation 1.1.64, but is indeed implied. Therefore, we actually have:

$$\tilde{I}_a + \tilde{I}_b + \tilde{I}_c + \tilde{I}_x + \tilde{I}_y + \tilde{I}_e = 0 \qquad (1.1.65)$$

Therefore we must include $L_e/3$ in all of the L_{ii} and L_{ij} elements. For the configuration shown in

Figure 1.1.8, we have, with all inductances expressed in terms of mH/km:

$$
\begin{bmatrix} \Delta\tilde{\Lambda}_a \\ \Delta\tilde{\Lambda}_b \\ \Delta\tilde{\Lambda}_c \\ 0 \\ 0 \end{bmatrix} =
\begin{bmatrix}
2.2702 & 0.8887 & 0.8887 & 0.8851 & 0.8851 \\
0.8887 & 2.2702 & 0.7500 & 0.7776 & 0.8851 \\
0.8887 & 0.7500 & 2.2702 & 0.8851 & 0.7776 \\
0.8851 & 0.7776 & 0.8851 & 2.4089 & 0.8887 \\
0.8851 & 0.8851 & 0.7776 & 0.8887 & 2.4089
\end{bmatrix}
\begin{bmatrix} \tilde{I}_a \\ \tilde{I}_b \\ \tilde{I}_c \\ \tilde{I}_x \\ \tilde{I}_y \end{bmatrix}
$$
$$[mV \cdot s/km] \qquad (1.1.66)$$

We may write this system of equations symbolically as:

$$
\begin{bmatrix} \Delta\tilde{\Lambda}_\Phi \\ 0 \end{bmatrix} =
\begin{bmatrix} A & B \\ B^T & C \end{bmatrix}
\begin{bmatrix} \tilde{I}_\Phi \\ \tilde{I}_{gw} \end{bmatrix}
\qquad (1.1.67)
$$

Then:

$$[\tilde{I}_{gw}] = -[C]^{-1}[B]^T[\tilde{I}_\Phi] \qquad (1.1.68)$$

and:

$$[\Delta\tilde{\Lambda}_\Phi] = \{[A] - [B][C]^{-1}[B]^T\}[\tilde{I}_\Phi] \qquad (1.1.69)$$

The influence upon the three-phase single-circuit line, due to the ground wires, is given by $[B][C]^{-1}[B]^T$, and for Equation 1.1.66 is numerically:

$$
\begin{aligned}
[B][C]^{-1}[B]^T &=
\begin{bmatrix}
0.4752 & 0.4463 & 0.4463 \\
0.4463 & 0.4230 & 0.4154 \\
0.4463 & 0.4154 & 0.4230
\end{bmatrix} \\
&= (0.4375)
\begin{bmatrix}
1.086 & 1.020 & 1.020 \\
1.020 & 0.967 & 0.949 \\
1.020 & 0.949 & 0.967
\end{bmatrix} [mH/km]
\end{aligned}
$$
$$(1.1.70)$$

For most practical purposes we may write (approximately):

$$
[B][C]^{-1}[B]^T \approx (0.4375)
\begin{bmatrix}
1 & 1 & 1 \\
1 & 1 & 1 \\
1 & 1 & 1
\end{bmatrix} = (0.4375)[\xi] \qquad [mH/km] \qquad (1.1.71)
$$

Thus, we have (approximately) for balanced three-phase currents:

$$
\begin{bmatrix} \Delta\tilde{\Lambda}_a \\ \Delta\tilde{\Lambda}_b \\ \Delta\tilde{\Lambda}_c \end{bmatrix} = (\tilde{I}_a) \left\{
\begin{bmatrix}
2.2702 & 0.8887 & 0.8887 \\
0.8887 & 2.2702 & 0.7500 \\
0.8887 & 0.7500 & 2.2702
\end{bmatrix} - (0.4375)[\xi] \right\}
\begin{bmatrix} 1 \\ a^2 \\ a \end{bmatrix}
$$
$$[mV \cdot s/km] \qquad (1.1.72)$$

Thus, the influence of ground wires, under balanced three-phase operation, is generally negligibly small (even though not *exactly* zero). The more accurate result, based upon Equation 1.1.70 is:

$$
\begin{bmatrix} \Delta\tilde{\Lambda}_a \\ \Delta\tilde{\Lambda}_b \\ \Delta\tilde{\Lambda}_c \end{bmatrix} = (\tilde{I}_a)
\begin{bmatrix}
1.7950 & 0.4424 & 0.4424 \\
0.4424 & 1.8472 & 0.3346 \\
0.4424 & 0.3346 & 1.8472
\end{bmatrix}
\begin{bmatrix} 1 \\ a^2 \\ a \end{bmatrix} \qquad [mV \cdot s/km] \qquad (1.1.73)
$$

It is left for the reader to verify that if the three-phase conductors are completely transposed (by phase rotation), that the influence of ground wires, under *balanced* three-phase currents, is *identically* null. This is not true for unbalanced three-phase currents.

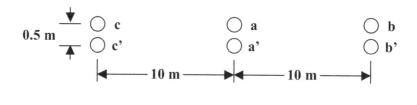

$$\text{GMR} = 0.010 \text{ m}$$

Figure 1.1.9: Single-Circuit, Bundled Conductor Three-Phase Line

Bundled Conductors

At lower voltages, transmission lines generally have only a single conductor per phase. However, at 345 kV and above, it is generally necessary to employ two or more conductors per phase due to corona and radio noise considerations. While it may appear, on the surface, that employing two conductors per phase would also present one-half the inductance per phase, this is not exactly true.

Consider the configuration shown in Figure 1.1.9, with two subconductors spaced at 0.5 meters. For this configuration we obtain:

$$
\begin{bmatrix}
\Delta\tilde{\Lambda}_a \\
\Delta\tilde{\Lambda}_b \\
\Delta\tilde{\Lambda}_c \\
\Delta\tilde{\Lambda}_{a'} \\
\Delta\tilde{\Lambda}_{b'} \\
\Delta\tilde{\Lambda}_{c'}
\end{bmatrix}
=
\begin{bmatrix}
L_{aa} & L_{ab} & L_{ac} & L_{aa'} & L_{ab'} & L_{ac'} \\
L_{ba} & L_{bb} & L_{bc} & L_{ba'} & L_{bb'} & L_{bc'} \\
L_{ca} & L_{cb} & L_{cc} & L_{ca'} & L_{cb'} & L_{cc'} \\
L_{a'a} & L_{a'b} & L_{a'c} & L_{a'a'} & L_{a'b'} & L_{a'c'} \\
L_{b'a} & L_{b'b} & L_{b'c} & L_{b'a'} & L_{b'b'} & L_{b'c'} \\
L_{c'a} & L_{c'b} & L_{c'c} & L_{c'a'} & L_{c'b'} & L_{c'c'}
\end{bmatrix}
\begin{bmatrix}
\tilde{I}_a \\
\tilde{I}_b \\
\tilde{I}_c \\
\tilde{I}_{a'} \\
\tilde{I}_{b'} \\
\tilde{I}_{c'}
\end{bmatrix}
\qquad (1.1.74)
$$

and numerically:

$$
\begin{bmatrix}
\Delta\tilde{\Lambda}_a \\
\Delta\tilde{\Lambda}_b \\
\Delta\tilde{\Lambda}_c \\
\Delta\tilde{\Lambda}_{a'} \\
\Delta\tilde{\Lambda}_{b'} \\
\Delta\tilde{\Lambda}_{c'}
\end{bmatrix}
=
\begin{bmatrix}
2.2702 & 0.8887 & 0.8887 & 1.4878 & 0.8884 & 0.8884 \\
0.8887 & 2.2702 & 0.7500 & 0.8884 & 1.4878 & 0.7500 \\
0.8887 & 0.7500 & 2.2702 & 0.8884 & 0.7500 & 1.4878 \\
1.4878 & 0.8884 & 0.8884 & 2.2702 & 0.8887 & 0.8887 \\
0.8884 & 1.4878 & 0.7500 & 0.8887 & 2.2702 & 0.7500 \\
0.8884 & 0.7500 & 1.4878 & 0.8887 & 0.7500 & 2.2702
\end{bmatrix}
\begin{bmatrix}
\tilde{I}_a \\
\tilde{I}_b \\
\tilde{I}_c \\
\tilde{I}_{a'} \\
\tilde{I}_{b'} \\
\tilde{I}_{c'}
\end{bmatrix}
\qquad (1.1.75)
$$

with inductance in terms of mH/km.

In symbolic matrix notation we write:

$$
\begin{bmatrix}
\Delta\tilde{\Lambda}_\Phi \\
\Delta\tilde{\Lambda}_{\phi'}
\end{bmatrix}
=
\begin{bmatrix}
A & B \\
B^T & C
\end{bmatrix}
\begin{bmatrix}
\tilde{I}_\Phi \\
\tilde{I}_{\Phi'}
\end{bmatrix}
\qquad (1.1.76)
$$

Since all subconductors are electrically in parallel, per phase, and neglecting resistance losses, we have:

$$
\left[\Delta\tilde{\Lambda}_{\Phi'} \right] = \left[\Delta\tilde{\Lambda}_\Phi \right]
\qquad (1.1.77)
$$

By an appropriate subtraction of corresponding rows we obtain:

$$\begin{bmatrix} \Delta\tilde{\Lambda}_\Phi \\ 0 \end{bmatrix} = \begin{bmatrix} 1 & 0 \\ -1 & 1 \end{bmatrix}\begin{bmatrix} \Delta\tilde{\Lambda}_\Phi \\ \Delta\tilde{\Lambda}_{\phi'} \end{bmatrix} = \begin{bmatrix} 1 & 0 \\ -1 & 1 \end{bmatrix}\begin{bmatrix} A & B \\ B^T & C \end{bmatrix}\begin{bmatrix} \tilde{I}_\Phi \\ \tilde{I}_{\Phi'} \end{bmatrix}$$

$$= \begin{bmatrix} A & B \\ (B^T - A) & (C - B) \end{bmatrix}\begin{bmatrix} \tilde{I}_\Phi \\ \tilde{I}_{\Phi'} \end{bmatrix} \tag{1.1.78}$$

and by an appropriate subtraction of corresponding columns we obtain:

$$\begin{bmatrix} \Delta\tilde{\Lambda}_\Phi \\ 0 \end{bmatrix} = \begin{bmatrix} A & B \\ (B^T - A) & (C - B) \end{bmatrix}\begin{bmatrix} 1 & -1 \\ 0 & 1 \end{bmatrix}\begin{bmatrix} 1 & 1 \\ 0 & 1 \end{bmatrix}\begin{bmatrix} \tilde{I}_\Phi \\ \tilde{I}_{\Phi'} \end{bmatrix}$$

$$= \begin{bmatrix} A & (B - A) \\ (B^T - A) & (A + C - B - B^T) \end{bmatrix}\begin{bmatrix} \tilde{I}_\Phi + \tilde{I}_{\Phi'} \\ \tilde{I}_{\Phi'} \end{bmatrix} \tag{1.1.79}$$

This last step, subtraction of columns, is not necessary if we are only interested in an analysis based purely upon numerical computation. However, if the analysis is to be carried out in terms of analog or model representation, then the next step is necessary in order to preserve a *symmetrical* inductance matrix.

According to Equation 1.1.75, we have, for the *phasing selected in our special case*:

$$[C] = [A]$$
$$[B]^T = [B] \tag{1.1.80}$$

so that we have:

$$\begin{bmatrix} \Delta\tilde{\Lambda}_\Phi \\ 0 \end{bmatrix} = \begin{bmatrix} A & (B - A) \\ (B - A) & 2(A - B) \end{bmatrix}\begin{bmatrix} \tilde{I}_\Phi + \tilde{I}_{\Phi'} \\ \tilde{I}_{\Phi'} \end{bmatrix} \tag{1.1.81}$$

Then, by the same technique employed in the ground wire reduction method, we obtain:

$$\left[\Delta\tilde{\Lambda}_\Phi\right] = \{[A] - \frac{1}{2}[B - A][A - B]^{-1}[B - A]\}[\tilde{I}_\Phi + \tilde{I}_{\Phi'}] = \frac{1}{2}[A + B][\tilde{I}_\Phi + \tilde{I}_{\Phi'}] \tag{1.1.82}$$

In this manner, a single-circuit line with two subconductors per phase may be expressed as a single-circuit line with *an effective single equivalent conductor per phase*. Numerically we obtain:

$$\begin{bmatrix} \Delta\tilde{\Lambda}_a \\ \Delta\tilde{\Lambda}_b \\ \Delta\tilde{\Lambda}_c \end{bmatrix} = \begin{bmatrix} 1.8790 & 0.8885 & 0.8885 \\ 0.8885 & 1.8790 & 0.7500 \\ 0.8885 & 0.7500 & 1.8790 \end{bmatrix}\begin{bmatrix} \tilde{I}_a + \tilde{I}_{a'} \\ \tilde{I}_b + \tilde{I}_{b'} \\ \tilde{I}_c + \tilde{I}_{c'} \end{bmatrix} \quad [mV \cdot s/km] \tag{1.1.83}$$

If we assume the three-phase currents to be balanced and equal in related subconductors, we obtain:

$$\begin{bmatrix} \Delta\tilde{\Lambda}_a \\ \Delta\tilde{\Lambda}_b \\ \Delta\tilde{\Lambda}_c \end{bmatrix} = (2\tilde{I}_a)\begin{bmatrix} 0.9905 \\ a^2(0.9905 - 0.1385a^2) \\ a(0.9905 - 0.1385a) \end{bmatrix} \quad [mV \cdot s/km] \tag{1.1.84}$$

Thus, a twin-subconductor line has about 72% of the inductance to balanced three-phase currents as does the single-conductor line [compared to Example 1.1.3]. The influence of ground wires upon bundled-conductor lines is essentially the same as for single-conductor lines.

1.1.8 Three-Phase Multicircuit Lines above Earth

When more than a single circuit is involved, whether the other circuits are either adjacent, or adjacent and electrically in parallel, the expansion to a larger system of equations is a simple mechanical

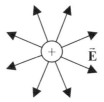

Figure 1.2.1: Electric Field about a Point Charge

procedure. Inevitably, a digital computer program is desirable for performing the numerical evaluations. While there are some important properties associated with electrically-parallel multi-circuit lines, their identification in terms of physical parameters (phase quantities) tends to become cumbersome and obscure. A much more enlightening analysis is possible in terms of symmetrical components as shown in Chapter 5.

1.1.9   Cables

For a rather detailed description of the electromagnetic characteristics of cables, see Chapter 3 of Reference [2]. Cable characteristics will also be found in References [5] and [4].

1.2   The Electric Field

Based upon experimental evidence, the force between two separated spherical (point) electric charges is given by Coulomb's law as:

$$F = \frac{q_1 q_2}{4\pi\varepsilon r^2} \qquad [N] \tag{1.2.1}$$

For analytical purposes it is more convenient to deal with single charges rather than pairs of charges as required by Equation 1.2.1. It has therefore become accepted practice to associate (define, hypothesize), the existence of a force field about a single charge according to:

$$\vec{E} = \frac{q}{4\pi\varepsilon r^2} \frac{\vec{r}}{r} \qquad [N/C] = [V/m] \tag{1.2.2}$$

which defines the force field associated with a single charge as illustrated in Figure 1.2.1. We may visualize an electric field in terms of a mathematical model of vector fields. One component of the electric vector field is defined by Maxwell's equation:

$$\nabla \cdot \vec{D} = div\vec{D} = \varrho \qquad [C/m^3] \tag{1.2.3}$$

Equation 1.2.3 states that whenever there exists an electric charge density distribution, there also exists an electric flux density vector $\vec{D}$ field with positive charges as sources and negative charges as sinks. Thus, electric flux lines emanate from positive charges and terminate on negative charges. This indeed reflects the fact of nature that atoms are basically neutral, and energy is required to separate electric charges.

Via Gauss's theorem, we may convert Equation 1.2.3 into the integral form:

$$\iiint \rho dv = \iint \vec{D} \cdot d\vec{S} = q \qquad [C] \tag{1.2.4}$$

which is known as Gauss's law and is much easier to deal with in practical engineering problems than Equation 1.2.3. Equation 1.2.4 states that the component of vector electric flux density $vecD$

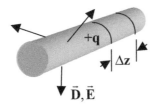

Figure 1.2.2: Electric Field about Single Conductor in Free Space

normal to a surface $\vec{S}$ enclosing a volume of charge density distribution, is equal numerically to the total charge enclosed within the surface $\vec{S}$.

The associated vector electric force field is related to $\vec{D}$ according to:

$$\vec{E} = \frac{\vec{D}}{\varepsilon} \qquad [N/C] = [V/m] \tag{1.2.5}$$

where:

$$\varepsilon = \varepsilon_r \varepsilon_o \qquad [F/m] \tag{1.2.6}$$

$$\varepsilon_o \approx \frac{10^{-9}}{36\pi} \qquad [F/m] \tag{1.2.7}$$

$$\varepsilon_r = relative\,permittivity \tag{1.2.8}$$

We note, in passing, that:

$$\nu = \frac{1}{sqrt\mu_o\varepsilon_o} \approx 300(10^6) \qquad [m/s] \quad (free\,space) \tag{1.2.9}$$

represents the velocity of light in free space.

A characteristic of the vector $\vec{E}$ field is given by Maxwell's equation:

$$curl\vec{E} = \nabla \times \vec{E} = 0 \tag{1.2.10}$$

and in integral form:

$$\oint \vec{E} \cdot d\ell = 0 \qquad [V] \tag{1.2.11}$$

whereby Equations 1.2.10 and 1.2.11 apply only to quasistationary electric fields in which there is no time-varying magnetic field [see also Equation 1.1.9].

1.2.1 Single Conductor in Free Space

Consider the hypothetical case of a single-charged conductor in free space as illustrated in Figure 1.2.2. Because like charges repel, all charges within the conductor reside on the surface of the conductor. Consequently there is neither a $\vec{D}$ nor $\vec{E}$ field *within* the conductor. External to the conductor we have:

$$\vec{D} = \frac{\Delta q}{2\pi r}\frac{\vec{r}}{r} \qquad [C/m^2] \tag{1.2.12}$$

and:

$$\vec{E} = \frac{\Delta q}{2\pi\varepsilon r}\frac{\vec{r}}{r} \qquad [N/C] = [V/m] \tag{1.2.13}$$

with Δq = charge per unit length of conductor, [C/m].

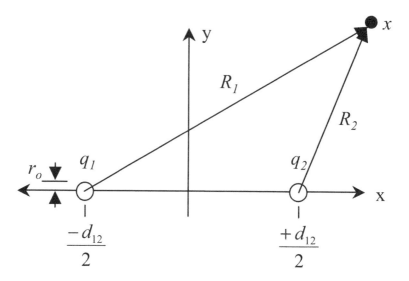

Figure 1.2.3: Parallel-Charged Conductors in Free Space

We associate a scalar potential (energy level) field with the vector force field defined by:

$$dv = -\vec{E} \cdot d\vec{r} \qquad [J/C] = [V] \tag{1.2.14}$$

interpreted as the energy per Coulomb required to move a unit positive (point) charge against this $\vec{E}$ field. The negative sign indicates that a positive unit (point) charge will have work done upon it in moving in the direction of the vector $\vec{E}$ field. For the charged conductor in Figure 1.2.2, we have:

$$v_R - v_C = -\int_{r_i}^{R} \frac{\Delta q}{2\pi\varepsilon r} dr = -\frac{\Delta q}{2\pi\varepsilon_o} \ln\frac{R}{r_i} \qquad [V] \tag{1.2.15}$$

as the energy per Coulomb expended in moving a unit positive (point) charge from the surface of the conductor r_i, to some position R remote from the centerline of the conductor.

1.2.2   Parallel Conductors in Free Space

Consider two parallel conductors in free space shown in Figure 1.2.3. The contribution of Δq_1 to the scalar potential at some point x is:

$$v_{x1} = -\frac{\Delta q_1}{2\pi\varepsilon} \ln\frac{d_{x1}}{r_1} + v_1 \qquad [V] \tag{1.2.16}$$

and due to Δq_2:

$$v_{x2} = -\frac{\Delta q_2}{2\pi\varepsilon} \ln\frac{d_{x2}}{r_2} + v_2 \qquad [V] \tag{1.2.17}$$

The total potential at x is:

$$v_x = v_1 + v_2 - \frac{\Delta q_1}{2\pi\varepsilon_o} \ln\frac{d_{x1}}{r_1} - \frac{\Delta q_2}{2\pi\varepsilon_o} \ln\frac{d_{x2}}{r_2} \qquad [V] \tag{1.2.18}$$

In a real-world situation, considering a complete system of electrically charged conductors, we must have:

$$\sum_{i=1}^{n} \Delta q_i = 0 \qquad [C] \tag{1.2.19}$$

Therefore, let $\Delta q_2 = -\Delta q_1$. Then:

$$v_x = v_1 + v_2 - \frac{\Delta q_1}{2\pi\varepsilon} \ln \frac{d_{x1} r_2}{r_1 d_{x2}} \qquad [V] \tag{1.2.20}$$

(a) Let dx1 = r1 and dx2 = d12 so that and:

$$v_1 = v_1 + v_2 - \frac{\Delta q_1}{2\pi\varepsilon} \ln \frac{r_2}{d_{12}} \qquad [V] \tag{1.2.21}$$

and:

$$v_2 = -\frac{\Delta q_1}{2\pi\varepsilon} \ln \frac{d_{12}}{r_2} \qquad [V] \tag{1.2.22}$$

(b) Let dx1 = d12 and dx2 = r2 so that and:

$$v_2 = v_1 + v_2 - \frac{\Delta q_1}{2\pi\varepsilon} \ln \frac{d_{12}}{r_1} \qquad [V] \tag{1.2.23}$$

and:

$$v_1 = \frac{\Delta q_1}{2\pi\varepsilon} \ln \frac{d_{12}}{r_1} \qquad [V] \tag{1.2.24}$$

(c) At the midplane, let dx1 = dx2 and r1 = r2 so that:

$$v_{xmp} = v_1 + v_2 = 0 \tag{1.2.25}$$

Thus, the scalar potential of conductor 1 with respect to the midplane is:

$$v_1 = \frac{\Delta q_1}{2\pi\varepsilon} \ln \frac{d_{12}}{r_1} \qquad [V] \tag{1.2.26}$$

and:

$$v_2 = -v_1 \tag{1.2.27}$$

1.2.3 Single Conductor above Ideal Conducting Plane

For the pair of parallel, equal but oppositely charged conductors, considered in the previous section, the electric field takes the form shown in Figure 1.2.4. In particular, the equipotential surfaces are circular cylinders eccentric to the charged line.

A perfectly conducting, infinitesimally thin, conducting cylinder of the same shape as an equipotential surface may be appropriately inserted into the field without affecting it. We may therefore insert a thin, perfectly conducting plane into the field at midplane. Charge separation takes place in the conducting sheet:

$$\sigma(x) = \frac{-h_i}{\pi(h_i^2 + x^2)} \Delta q_i = -D_n(x) \qquad [C/m^2] \tag{1.2.28}$$

analogous to the current density distribution for the magnetic field case (see Appendix C).

Thus, we have directly the case of a charged conductor above, and parallel to a perfectly conducting plane with:

$$v_i = \frac{\Delta q_i}{2\pi\varepsilon_o} \ln \frac{2h_i}{r_i} \qquad [V] \tag{1.2.29}$$

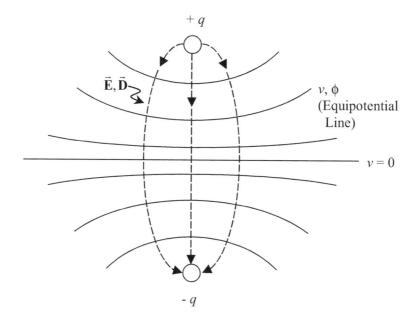

Figure 1.2.4: Single-Charged Conductor above Ideal Conducting Plane

1.2.4   Parallel Conductors above Ideal Conducting Plane

For parallel charged conductors (Figure 1.2.5) above a perfectly conducting plane, we employ the method of images. Extending the method of Section 1.2.2, we have:

$$v_x = v_1 + v_2 - (v_1 + v_2) - \frac{\Delta q_1}{2\pi\varepsilon_o} \ln \frac{d_{1x}r_1}{r_1 D_{1x}} - \frac{\Delta q_2}{2\pi\varepsilon_o} \ln \frac{d_{2x}r_2}{r_2 D_{2x}} \qquad [V] \qquad (1.2.30)$$

(a) For d1x = r1, D1x = 2h1, d2x = d12, D2x = D12, we have and:

$$v_1 = \frac{\Delta q_1}{2\pi\varepsilon_o} \ln \frac{2h_1}{r_1} + \frac{\Delta q_2}{2\pi\varepsilon_o} \ln \frac{D_{12}}{d_{12}} \qquad [V] \qquad (1.2.31)$$

(b) For d1x = d12, D1x = D12, d2x = r2, D2x = 2h2, we have and:

$$v_2 = \frac{\Delta q_2}{2\pi\varepsilon_o} \ln \frac{2h_2}{r_2} + \frac{\Delta q_1}{2\pi\varepsilon_o} \ln \frac{D_{12}}{d_{12}} \qquad [V] \qquad (1.2.32)$$

Thus, for a system of parallel-charged conductors above a perfectly conducting plane we have:

$$[V] = [P][\Delta q] \qquad [V] \qquad (1.2.33)$$

with:

$$P_{ii} = \frac{1}{2\pi\varepsilon_o} \ln \frac{2h_i}{r_i} \qquad [m/F] \qquad (1.2.34)$$

$$P_{ij} = \frac{1}{2\pi\varepsilon_o} \ln \frac{D_{ij}}{d_{ij}} \qquad [m/F] \qquad (1.2.35)$$

The coefficients P_{ii} and P_{ij} are called self- and mutual *potential coefficients*.

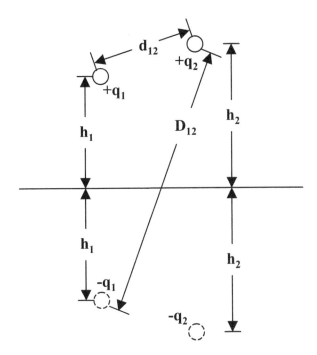

Figure 1.2.5: Parallel Conductors above Ideal Conducting Plane

1.2.5 Parallel Conductors above Earth

For moist, homogeneous earth, the surface charge distribution defined by Equation 1.2.23 is easily attained. In this case, earth behaves like a perfect conductor and the surface of the earth may be taken as the zeropotential equipotential surface.

Where the surface layer is rocky, or dry sand, the effective zero-potential equipotential surface may lie several meters below the actual surface of the earth. In addition, shrubs and trees on and adjacent to the right-of-way of transmission lines also influence the electric field.

Thus, the evaluation of the potential coefficients is subject to some inaccuracy, and the actual potential coefficients will vary with weather conditions. For most practical purposes, Equations 1.2.29 and 1.2.30 offer sufficient accuracy.

1.2.6 Three-Phase Single-Circuit Lines above Earth

With an understanding of the physical principles developed so far, writing the voltage equations for a system of parallel overhead lines becomes a rather simple mechanical procedure. For most practical three-phase single-circuit lines above earth, earth behaves like a zero-potential equipotential surface. Therefore, the voltage equations take the form:

$$
\begin{bmatrix} v_a \\ v_b \\ v_c \end{bmatrix} = \begin{bmatrix} P_{aa} & P_{ab} & P_{ac} \\ P_{ba} & P_{bb} & P_{bc} \\ P_{ca} & P_{cb} & P_{cc} \end{bmatrix} \begin{bmatrix} \Delta q_a \\ \Delta q_b \\ \Delta q_c \end{bmatrix} \quad [V] \quad (1.2.36)
$$

where:

$$P_{ii} = \frac{1}{2\pi\varepsilon_o} \ln \frac{2h_i}{r_i} \qquad [m/F] \qquad (1.2.37)$$

$$P_{ij} = \frac{1}{2\pi\varepsilon_o} \ln \frac{D_{ij}}{d_{ij}} \qquad [m/F] \qquad (1.2.38)$$

with Δq = charge per unit length of conductor, [C/m].

When the line voltages are purely sinusoidal at $\omega = 2\pi f$, electric power engineers prefer to think in terms of *line-charging current* rather than line charge. We have:

$$\Delta q = \int \Delta i \, dt \qquad (1.2.39)$$

and for sinusoidal voltages (charges) we have, in phasor notation and rms quantities:

$$\begin{bmatrix} \tilde{V}_a \\ \tilde{V}_b \\ \tilde{V}_c \end{bmatrix} = \left(\frac{1}{j\omega}\right) \begin{bmatrix} P_{aa} & P_{ab} & P_{ac} \\ P_{ba} & P_{bb} & P_{bc} \\ P_{ca} & P_{cb} & P_{cc} \end{bmatrix} \begin{bmatrix} \Delta \tilde{I}_a \\ \Delta \tilde{I}_b \\ \Delta \tilde{I}_c \end{bmatrix} \qquad [V] \qquad (1.2.40)$$

where $\Delta \tilde{I}$ represents the line-charging current in the direction from $+\tilde{V}$ to 0.

Untransposed Lines

Most conductors in overhead lines are disposed horizontally, or vertically, but rarely in the form of an equilateral triangle. Most often, then, $P_{ab} \neq P_{bc} \neq P_{ca}$. Observe also that the P_{ii} depend upon conductor height above earth so that for vertical arrangements, $P_{aa} \neq P_{bb} \neq Pcc$.

Normally we assume that the voltages $\tilde{V}_a$, $\tilde{V}_b$, and $\tilde{V}_c$ are *balanced* and that, as a consequence, the $\Delta \tilde{I}_a$, $\Delta \tilde{I}_b$, and $\Delta \tilde{I}_c$ are unbalanced. To observe this rigorously would require that we determine $[C] = [P]^{-1}$. However, we shall circumvent this complexity by assuming $\Delta \tilde{I}_a$, $\Delta \tilde{I}_b$, and $\Delta \tilde{I}_c$ to be balanced, with a resultant unbalance in $\tilde{V}_a$, $\tilde{V}_b$, and $\tilde{V}_c$, in which case we have:

$$\begin{bmatrix} \tilde{V}_a \\ \tilde{V}_b \\ \tilde{V}_c \end{bmatrix} = \left(\frac{\Delta \tilde{I}_a}{j\omega}\right) \begin{bmatrix} P_{aa} + a^2 P_{ab} + a P_{ac} \\ P_{ba} + a^2 P_{bb} + a P_{bc} \\ P_{ca} + a^2 P_{cb} + a P_{cc} \end{bmatrix} \qquad [V] \qquad (1.2.41)$$

Example 1.2.1

Consider the horizontal line configuration shown in Figure 1.1.6 (where we now let $r_a = r_b = r_c = 0.010m$) for which we obtain:

$$\begin{bmatrix} \tilde{V}_a \\ \tilde{V}_b \\ \tilde{V}_c \end{bmatrix} = \left(\frac{1}{j\omega}\right) \begin{bmatrix} 144.1 & 20.7 & 20.7 \\ 20.7 & 144.1 & 10.6 \\ 20.7 & 10.6 & 144.1 \end{bmatrix} \begin{bmatrix} \Delta \tilde{I}_a \\ \Delta \tilde{I}_b \\ \Delta \tilde{I}_c \end{bmatrix} \qquad [V]$$

and for balanced line charging currents:

$$\begin{bmatrix} \tilde{V}_a \\ \tilde{V}_b \\ \tilde{V}_c \end{bmatrix} = \left(\frac{\Delta \tilde{I}_a}{j\omega}\right) \begin{bmatrix} 123.4 \\ a^2(123.4 - 10.1a^2) \\ a(123.4 - 10.1a) \end{bmatrix} \qquad [V]$$

where the P coefficients are expressed in terms of km/μF. Thus, $\mid \tilde{V}_b \mid = \mid \tilde{V}_c \mid = \mid \tilde{V}_a \mid$, and $\tilde{V}_b$ and $\tilde{V}_c$ are advanced and retarded, respectively, in phase angle by 3.9 degrees.

While we have chosen to show the voltage unbalance due to balanced line charges (for the sake of avoiding the inversion of the [P] matrix), in reality the phase voltages will be balanced, and the line charge (charging current for ac) will have an unbalance similar to that shown above. For effectively neutral grounded systems, with lines of moderate length, this charging unbalance is sufficiently small and of no concern. However, where Peterson coil (resonant grounding, ground fault neutralizer) grounding is employed, this unbalance may be troublesome. This aspect is better treated in terms of symmetrical components as will be described in Chapters 5 and 8.

Transposed Lines

For a completely transposed line (see Section 1.1.7) we have:

$$P_{aa} = P_{bb} = P_{cc}$$
$$P_{ab} = P_{bc} = P_{ca} \tag{1.2.42}$$

and for balanced three-phase line charges:

$$\begin{bmatrix} \tilde{V}_a \\ \tilde{V}_b \\ \tilde{V}_c \end{bmatrix} = (\frac{\Delta \tilde{I}_a}{j\omega}) \begin{bmatrix} (P_{aa} - P_{ab}) \\ a^2(P_{aa} - P_{ab}) \\ a(P_{aa} - P_{ab}) \end{bmatrix} \quad [V] \tag{1.2.43}$$

and the physically mutually-coupled three-phase circuit may be viewed in terms of three equivalent independent (mutually uncoupled) single-phase lines.

Influence of Ground Wires

When the line is protected by ground wires (see Section 1.1.7), the voltage equations take the form:

$$\begin{bmatrix} \tilde{V}_\Phi \\ 0 \end{bmatrix} = (\frac{1}{j\omega}) \begin{bmatrix} P_{\Phi\Phi} & P_{\Phi x} \\ P_{x\Phi} & P_{xx} \end{bmatrix} \begin{bmatrix} \Delta \tilde{I}_\Phi \\ \Delta \tilde{I}_x \end{bmatrix} \tag{1.2.44}$$

For the configuration shown in Figure 1.1.8 (with $r_\Phi = 0.010$ m and $r_x = 0.005$ m) we have, with the potential coefficients expressed in terms of $km/\mu F$:

$$\begin{bmatrix} \tilde{V}_a \\ \tilde{V}_b \\ \tilde{V}_c \\ 0 \\ 0 \end{bmatrix} = (\frac{1}{j\omega}) \begin{bmatrix} 144.1 & 20.7 & 20.7 & 24.3 & 24.3 \\ 20.7 & 144.1 & 10.6 & 15.7 & 24.3 \\ 20.7 & 10.6 & 144.1 & 24.3 & 15.7 \\ 24.3 & 15.7 & 24.3 & 165.0 & 28.5 \\ 24.3 & 24.3 & 15.7 & 28.5 & 165.0 \end{bmatrix} \begin{bmatrix} \Delta \tilde{I}_a \\ \Delta \tilde{I}_b \\ \Delta \tilde{I}_c \\ \Delta \tilde{I}_x \\ \Delta \tilde{I}_y \end{bmatrix} \quad [V] \tag{1.2.45}$$

We may write this system of equations symbolically as:

$$\begin{bmatrix} \tilde{V}_\Phi \\ 0 \end{bmatrix} = (\frac{1}{j\omega}) \begin{bmatrix} A & B \\ B^T & C \end{bmatrix} \begin{bmatrix} \Delta \tilde{I}_\Phi \\ \Delta \tilde{I}_x \end{bmatrix} \quad [V] \tag{1.2.46}$$

Then:

$$\begin{bmatrix} \Delta \tilde{I}_x \end{bmatrix} = -[C]^{-1}[B^T] \begin{bmatrix} \Delta \tilde{I}_\Phi \end{bmatrix} \quad [A/m] \tag{1.2.47}$$

and:

$$\left[\tilde{V}_\Phi\right] = (\frac{1}{j\omega})\{[A] - [B][C]^{-1}[B^T]\}\left[\Delta\tilde{I}_\Phi\right] \qquad [V] \qquad (1.2.48)$$

The influence upon the three-phase single-circuit line, by the ground wires, is given by $[B][C]^{-1}[B^T]$, and numerically:

$$[B][C]^{-1}[B^T] = \begin{bmatrix} 6.1 & 5.0 & 5.0 \\ 5.0 & 4.4 & 3.9 \\ 5.0 & 3.9 & 4.4 \end{bmatrix} = (4.75)\begin{bmatrix} 1.28 & 1.05 & 1.05 \\ 1.05 & 0.93 & 0.82 \\ 1.05 & 0.82 & 0.93 \end{bmatrix} \qquad [km/F] \quad (1.2.49)$$

Since all the elements in the matrix in Equation 1.2.49 are close to unity, for all practical purposes we may write:

$$[B][c]^{-1}[B^T] \approx (4.75)[\xi] \qquad (1.2.50)$$

Comparing Equations 1.2.49 and 1.1.70, we observe that the electrostatic unbalance amongst the elements is somewhat greater than the electromagnetic unbalance. However, the overall electrostatic effect of the ground wires is sufficiently smaller to justify Equation 1.2.50. Thus, we have, approximately, for balanced three-phase line charges:

$$\begin{bmatrix} \tilde{V}_a \\ \tilde{V}_b \\ \tilde{V}_c \end{bmatrix} \approx (\frac{\Delta\tilde{I}_a}{j\omega})\left\{ \begin{bmatrix} 144.1 & 20.7 & 20.7 \\ 20.7 & 144.1 & 10.6 \\ 20.7 & 10.6 & 144.1 \end{bmatrix} - (4.75)[\xi] \right\} \begin{bmatrix} 1 \\ a^2 \\ a \end{bmatrix} \qquad [V] \qquad (1.2.51)$$

The influence of ground wires, under balanced three-phase operation, is generally negligibly small. The correct, and accurate, numerical values are:

$$\begin{bmatrix} \tilde{V}_a \\ \tilde{V}_b \\ \tilde{V}_c \end{bmatrix} \approx (\frac{\Delta\tilde{I}_a}{j\omega}) \begin{bmatrix} 138.0 & 15.7 & 15.7 \\ 15.7 & 139.7 & 6.8 \\ 15.7 & 6.8 & 139.7 \end{bmatrix} \begin{bmatrix} 1 \\ a^2 \\ a \end{bmatrix} \qquad [V] \qquad (1.2.52)$$

Bundled Conductors

Consider the configuration shown in Figure 1.1.9, with two subconductors spaced at 0.5 meters. For this configuration we obtain:

$$\begin{bmatrix} \tilde{V}_a \\ \tilde{V}_b \\ \tilde{V}_c \\ \tilde{V}_{a'} \\ \tilde{V}_{b'} \\ \tilde{V}_{c'} \end{bmatrix} = (\frac{1}{j\omega}) \begin{bmatrix} 144.1 & 20.7 & 20.7 & 73.4 & 20.4 & 20.4 \\ 20.7 & 144.1 & 10.6 & 20.4 & 73.4 & 10.4 \\ 20.7 & 10.6 & 144.1 & 20.4 & 10.4 & 73.4 \\ 73.4 & 20.4 & 20.4 & 143.5 & 20.2 & 20.2 \\ 20.4 & 73.4 & 10.4 & 20.2 & 143.5 & 10.2 \\ 20.4 & 10.4 & 73.4 & 20.2 & 10.2 & 143.5 \end{bmatrix} \begin{bmatrix} \Delta\tilde{I}_a \\ \Delta\tilde{I}_b \\ \Delta\tilde{I}_c \\ \Delta\tilde{I}_{a'} \\ \Delta\tilde{I}_{b'} \\ \Delta\tilde{I}_{c'} \end{bmatrix} \quad (1.2.53)$$

with potential coefficients in terms of $km/\mu F$. In symbolic matrix notation we write:

$$\begin{bmatrix} \tilde{V}_\Phi \\ \tilde{V}_{\Phi'} \end{bmatrix} = (\frac{1}{j\omega}) \begin{bmatrix} A & B \\ B^T & C \end{bmatrix} \begin{bmatrix} \Delta\tilde{I}_\Phi \\ \Delta\tilde{I}_{\Phi'} \end{bmatrix} \qquad (1.2.54)$$

and proceeding in the manner described in Section 1.1.7, we obtain for the equivalent single-circuit line:

$$\left[\tilde{V}_\Phi\right] = (\frac{1}{j\omega})\{[A] - [B-A][A+C-B-B^T]^{-1}[B^T-A]\}\left[\Delta\tilde{I}_\Phi + \Delta\tilde{I}_{\Phi'}\right] \qquad (1.2.55)$$

and for the *phasing selected in our special case* where:

$$[C] \approx [A]$$
$$[B^T] = [B] \tag{1.2.56}$$

we have:

$$\left[\tilde{V}_\Phi \right] \approx (\frac{1}{j2\omega})\{[A] + [B]\} \left[\tilde{I}_\Phi + \tilde{I}_{\Phi'} \right] \tag{1.2.57}$$

In this manner, a single-circuit line with two subconductors per phase may be expressed as a single-circuit line with an effective single equivalent conductor per phase. Numerically we obtain:

$$\begin{bmatrix} \tilde{V}_a \\ \tilde{V}_b \\ \tilde{V}_c \end{bmatrix} \approx (\frac{1}{j\omega}) \begin{bmatrix} 108.75 & 20.55 & 20.55 \\ 20.55 & 108.75 & 10.50 \\ 20.55 & 10.50 & 108.75 \end{bmatrix} \begin{bmatrix} \tilde{I}_a + \tilde{I}_{a'} \\ \tilde{I}_b + \tilde{I}_{b'} \\ \tilde{I}_c + \tilde{I}_{c'} \end{bmatrix} \tag{1.2.58}$$

If we assume the three-phase line charges to be balanced we obtain:

$$\begin{bmatrix} \tilde{V}_a \\ \tilde{V}_b \\ \tilde{V}_c \end{bmatrix} \approx (\frac{\tilde{I}_a}{j\omega}) \begin{bmatrix} 88.2 \\ a^2(88.2 - 10.05a^2) \\ a(88.2 - 10.05a) \end{bmatrix} \tag{1.2.59}$$

with potential coefficients in terms of $km/\mu F$.

Thus, a twin-subconductor line has about 72% of the capacitive reactance to ground, under balanced line charges, as does the single-conductor line [compared to Equation 1.2.43. The influence of ground wires upon bundled-conductor lines is essentially the same as for single-conductor lines.

Equivalent Capacitance Model

For a system of two parallel conductors above earth we have:

$$\begin{bmatrix} \tilde{V}_1 \\ \tilde{V}_2 \end{bmatrix} = \frac{1}{j\omega} \begin{bmatrix} P_{11} & P_{12} \\ P_{21} & P_{22} \end{bmatrix} \begin{bmatrix} \tilde{I}_1 \\ \tilde{I}_2 \end{bmatrix} \tag{1.2.60}$$

and therefore:

$$\begin{aligned} \begin{bmatrix} \tilde{I}_1 \\ \tilde{I}_2 \end{bmatrix} &= \frac{j\omega}{P_{11}P_{22} - P_{12}P_{21}} \begin{bmatrix} P_{22} & -P_{12} \\ -P_{21} & P_{11} \end{bmatrix} \begin{bmatrix} \tilde{V}_1 \\ \tilde{V}_2 \end{bmatrix} \\ &= j\omega \begin{bmatrix} C_{11} & C_{12} \\ C_{21} & C_{22} \end{bmatrix} \begin{bmatrix} \tilde{V}_1 \\ \tilde{V}_2 \end{bmatrix} \qquad [A/m] \end{aligned} \tag{1.2.61}$$

Several observations are to be made. For elementary two-electrode configurations we always have:

$$C = 1/P \qquad [F/m] \tag{1.2.62}$$

For configurations consisting of more than two electrodes, a matrix inversion is required to determine the appropriate capacitance coefficients. Furthermore, whereas the P_{ii} and P_{ij} coefficients are individually and directly related to the geometry of the conductor configuration, the $C_{(ii)}$ and C_{ij} are interrelated functions of the entire conductor geometry.

While the capacitance coefficients in Equation 1.2.61 are mathematically correct, the system of equations cannot be represented by lumped capacitors equal to the capacitance coefficients. In order to be able to represent the system of equations by a practical, physically realizable model, another set of capacitance coefficients has been introduced. Figure 1.2.6 illustrates a physically realizable model for a two-conductor above earth arrangement.

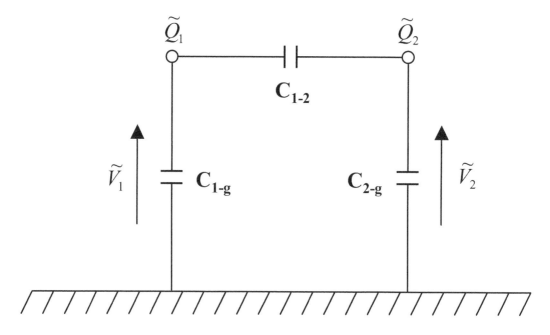

Figure 1.2.6: Equivalent Capacitance Model

According to Figure 1.2.6, we have:

$$\begin{aligned}
\Delta \tilde{I}_1 &= j\omega \left[C_{1-g}\tilde{V}_1 + C_{1-2}(\tilde{V}_1 - \tilde{V}_2) \right] \\
&= j\omega \left[(C_{1-g} + C_{1-2})\tilde{V}_1 - C_{1-2}\tilde{V}_2 \right] \qquad [A/m]
\end{aligned} \qquad (1.2.63)$$

and:

$$\begin{aligned}
\Delta \tilde{I}_2 &= j\omega \left[C_{2-g}\tilde{V}_2 + C_{1-2}(\tilde{V}_2 - \tilde{V}_1) \right] \\
&= j\omega \left[(C_{2-g} + C_{1-2})\tilde{V}_2 - C_{1-2}\tilde{V}_1 \right] \qquad [A/m]
\end{aligned} \qquad (1.2.64)$$

But according to Equation 1.2.61 we have:

$$\Delta \tilde{I}_1 = j\omega \left[C_{11}\tilde{V}_1 + C_{12}\tilde{V}_2 \right] \qquad [A/m] \qquad (1.2.65)$$

and:

$$\Delta \tilde{I}_2 = j\omega \left[C_{21}\tilde{V}_1 + C_{22}\tilde{V}_2 \right] \qquad [A/m] \qquad (1.2.66)$$

Thus:

$$\begin{aligned}
C_{ii} &= C_{1-g} + C_{1-2} & (1.2.67) \\
C_{22} &= C_{2-g} + C_{1-2} & (1.2.68) \\
C_{12} &= -C_{1-2} & (1.2.69)
\end{aligned}$$

and conversely:

$$\begin{aligned}
C_{1-g} &= C_{11} + C_{12} & (1.2.70) \\
C_{2-g} &= C_{22} + C_{21} & (1.2.71) \\
C_{1-2} &= -C_{12} & (1.2.72)
\end{aligned}$$

Observe that C_{1-2} is now a positive number (where C_{12} was a negative number).

Equation 1.2.61 may therefore be written in the form:

$$\begin{bmatrix} \tilde{I}_1 \\ \tilde{I}_2 \end{bmatrix} = j\omega \begin{bmatrix} (C_{1-g} + C_{1-2}) & -C_{1-2} \\ -C_{1-2} & (C_{2-g} + C_{1-2}) \end{bmatrix} \begin{bmatrix} \tilde{V}_1 \\ \tilde{V}_2 \end{bmatrix} \tag{1.2.73}$$

This equation can easily be extended to a larger system of conductors by recognizing the similarity to the nodal admittance equations. Thus:

$$C_{ii} = C_{i-g} + \sum_{\substack{j=1 \\ j\neq i}}^{n} C_{i-j} \qquad C_{ij} = -C_{i-j} \tag{1.2.74}$$

and conversely:

$$C_{i-g} = C_{ii} + \sum_{\substack{j=1 \\ j\neq i}}^{n} C_{ii} \qquad C_{ij} = -C_{i-j} \tag{1.2.75}$$

1.2.7   Three-Phase Multicircuit Lines above Earth

When more than a single circuit is involved, whether the other circuits are either adjacent, or adjacent and electrically in parallel, the expansion to a larger system of equations is a simple mechanical procedure. Inevitably, a digital computer program is required for performing the numerical evaluations. While there are some important properties associated with electrically-parallel multicircuit lines, their identification in terms of physical parameters (phase quantities) tends to become cumbersome and obscure. A much more enlightening analysis is possible in terms of symmetrical components as shown in Chapter 5.

1.2.8   Cables

For a rather detailed description of the electrostatic characteristics of cables, see Chapter 3 of Reference [2]. Cable characteristics will also be found in References [5] and [4].

1.2.9   Conductor Corona

See [2], Appendix B, pp. 524–528. Also see [6].

1.2.10   Radio Noise

See [6].

1.3   Induced Voltages

While the primary objective in the preceding sections has been to understand the magnetic and electric performance characteristics of transmission lines, the analysis has been presented in a manner that permits direct extension to the determination of the impact that these electromagnetic fields can have upon adjacent circuits and objects. It is the latter that is becoming of increased interest to environmentalists, and therefore to electric power engineers who must be in a position to properly evaluate such impacts. Also of increasing interest is the manner in which voltages can be induced in very low power control circuits whose lines are often adjacent to power lines or cables.

Given a conductor (fence, control wire, metal roof, etc.), parallel to a three-phase transmission line, and "floating" (i.e., not connected to earth or to any source voltage). A voltage is induced on the conductor in two ways:

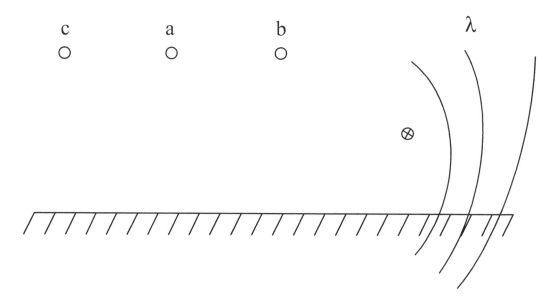

Figure 1.3.1: Magnetic Flux-Linkage

(a) Magnetically

(b) Electrically

as illustrated in Figures 1.3.1 and 1.3.2.

For the magnetically induced voltage we recognize the net flux (due to the overhead line currents) linking (encircling, enclosing) conductor x. We have:

$$\Delta\lambda_x = L_{xa}i_a + L_{xb}I_b + L_{xc}i_c \qquad [V \cdot s/m] \tag{1.3.1}$$

The *magnetically* induced voltage, according to Faraday's law, is given by:

$$\Delta V_x = \frac{d(\Delta\lambda_x)}{dt} \qquad [V/m] \tag{1.3.2}$$

We can determine that this voltage must act longitudinally (i.e., along the wire) because Lenz's law states that if current were permitted to flow along conductor x, its direction would be in such a direction as to oppose the current (overhead line currents) causing the flux. This is illustrated in Figure 1.3.3(a).

For the electrically induced voltage we recognize that an uncharged conductor x will assume the potential of the equipotential surface (due to the overhead lines) at the location of the conductor x. We have:

$$V_x = P_{xa}\Delta q_a + P_{xb}\Delta q_b + P_{xc}\Delta q_c \qquad [V] \tag{1.3.3}$$

In this case, every element of conductor x will have the same voltage with respect to earth, as illustrated in Figure 1.3.3(b).

While Equation 1.3.1 clearly shows the direct relationship between the overhead line currents and the flux linking conductor x, we observe that Equation 1.3.3 needs some help. Ordinarily the overhead line voltages are known, and only through the intermediate step of:

$$[\Delta q_\Phi] = [P_\Phi]^{-1}[V_\phi] \tag{1.3.4}$$

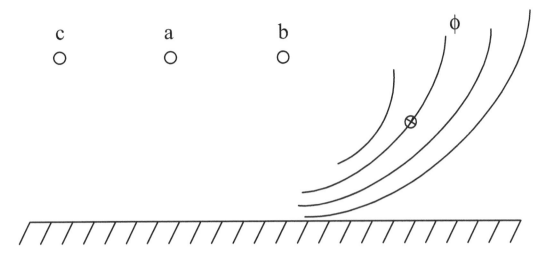

Figure 1.3.2: Electric Scalar Potential

do we recognize that Equation 1.3.3 depends upon the overhead line voltages.

In summary, the electric power engineer will often be called upon to ascertain the magnitudes of voltages and currents induced upon adjacent circuits. While this can be done quite readily in most cases, the impact or significance of these magnitudes often must be decided by others. Where light electronic equipment is involved, the designer of this equipment must decide whether these induced voltages and currents are hazardous; suitable protective devices might be required. Whether these induced voltages and currents are fatal or injurious to animals and humans in contact with them generally requires the knowledge of a physician.

In the case of humans coming in contact with induced voltages, while the electric current may not be of a level to be injurious to an average healthy human, the "surprise" element of even a minor shock may be responsible for physical injury (falling off a roof or ladder).

1.4   Conductor Resistance

Even the best electrical conductors transmit electrical energy imperfectly, some of the energy being converted to heat. For overhead lines, the dc resistance may be considered to be a constant at a given temperature (except for steel ground wires, etc.) and can be measured and predicted quite accurately. Of course the influence of spirality, stranding, and current distribution among strands of different materials must be duly taken into account. Stranded conductors are widely used when solid conductors are not flexible enough for the application at hand. When a conductor is stranded, the effective dc resistance will be slightly larger than is calculated for the total conductor length and cross section because the length of the outside strands will be slightly greater on account of the strand spiraling.

The ac resistance is increased above the dc resistance by skin effect and proximity effects. Skin effect accounts for the fact that ac current is not uniformly distributed over the conductor cross section but rather the current density will be greater near the conductor surface. Skin effect is computed under the assumption that the nonuniformity of current density in a conductor can be attributed due to the voltage induced by flux due to current in the conductor alone. In a parallel wire line, there is an additional nonuniformity caused by the flux due to the current in the adjacent strands. This phenomenon is known as the *proximity effect*. It will produce an increase in the conductor resistance and a decrease in the conductor inductance. For steel ground wires these

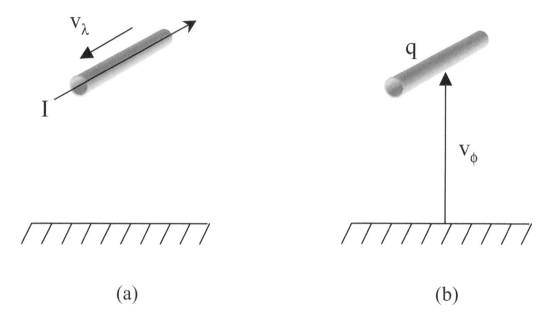

Figure 1.3.3: Nature of Induced Voltages

effects vary with the magnitude of the current in the ground wire.

For a detailed analysis of the concepts itemized in this subsection the reader is referred to Reference [3], Chapter 2. Ordinarily the electric power engineer need not concern himself with the evaluation of the resistance of a conductor, such data being easily available from the conductor manufacturer. References [1] to [5] contain such conductor tables.

For copper, aluminum, and ACSR (aluminum conductor with steel reinforcing), the conductor resistance varies only little as a function of conductor temperature, and is essentially independent of the current. On the other hand, the resistance (and inductance) of steel ground wires varies significantly as a function of current density. This is shown in Reference [2]. Table B.22, Part C.

In addition to the resistance of the conductor itself, allowance must be made for the resistive losses in earth when earth-return currents exist. We shall consider this aspect in Section 1.6.

1.5   Conductance (Leakage)

By conductance we mean the leakage of current radially from a conductor. For overhead lines, at normal power system frequencies and voltages, conductance is negligible. Under similar normal operating conditions, the leakage may have an appreciable effect in limiting the current rating of insulated cables (See Reference [3], Chapter 3).

For EHV (Extra High Voltage) and UHV (Ultra High Voltage) lines (and for lower voltage lines during system overvoltages) the corona starting voltage may be exceeded and the resultant corona loss must be taken into account. This subject is treated rather thoroughly in Referenece [3].

Relatively high leakage currents have also been observed immediately preceding a direct lightning stroke to the line.

Conductance from a conductor to earth is a desirable attribute of ground rods and counterpoises.

1.6 Transmission Line Performance Models

In the preceding sections we have studied the physical principles that relate to the electromagnetic performance characteristics of transmission lines. We now consider the interpretation of these characteristics into useful network models, suitable for practical analytical purposes.

1.6.1 Series Impedance

In Section 1.3, we studied the nature of the magnetic field and the consequent induced voltages for time-varying fields. This voltage drop acts along the length of the current-carrying conductor similar to the resistance drop. In Appendix D, we also showed that the influence of the earth can be accommodated by Carson's correction factors $L_e/3$ and $R_e/3$. The physical characteristics of the magnetic field and the ohmic losses can be reflected algebraically by the series impedance as defined by Equations D.4.7 and D.4.8. Observe that the mutual impedance for lines above earth also contain $R_e/3$ terms.

Figure 1.6.1 illustrates several network models that suitably reflect the series impedance of overhead lines.

1.6.2 Shunt Admittance

In Section 1.3, we studied the nature of the electric field and the consequent induced voltages for time-varying fields. This voltage acts radially from a charged (energized with voltage) line. For a time-varying field we therefore have displacement (capacitive, line-charging) currents, which are similar in nature to the leakage (conductance) currents.

For overhead lines we may generally ignore conductance and represent the shunt admittance characteristics entirely in terms of the effective line capacitance. The applicable potential coefficients are defined by Equations 1.2.29 and 1.2.30. For a system of overhead lines we have:

$$[V] = [P][\Delta q] \qquad [V] \tag{1.6.1}$$

and for a sinusoidal voltage and in terms of line-charging current we have:

$$[\tilde{V}] = \frac{1}{j\omega}[P][\Delta \tilde{I}] \qquad [V] \tag{1.6.2}$$

and taking the inverse:

$$[\Delta \tilde{I}] = j\omega[P]^{-1}[\tilde{V}] = j\omega[C][\tilde{V}] \qquad [A/m] \tag{1.6.3}$$

While Equation 1.6.3 may be employed in computerized analysis, we pointed out in Section 1.2.6 that the elements in [C] were not suitable for physical modeling purposes. The network models for shunt admittance illustrated in Figure 1.6.2 are therefore in terms of the equivalent capacitance model elements.

1.6.3 Long Line Equations

For a single-conductor line above earth, operating at sinusoidal power frequencies, we have for the series voltage drop (in phasor format):

$$-\frac{\delta \tilde{V}}{\delta x} = \tilde{z}\tilde{I} \qquad [V/m] \tag{1.6.4}$$

and for the shunt-charging current:

$$-\frac{\delta \tilde{I}}{\delta x} = \tilde{y}\tilde{V} \qquad [A/m] \tag{1.6.5}$$

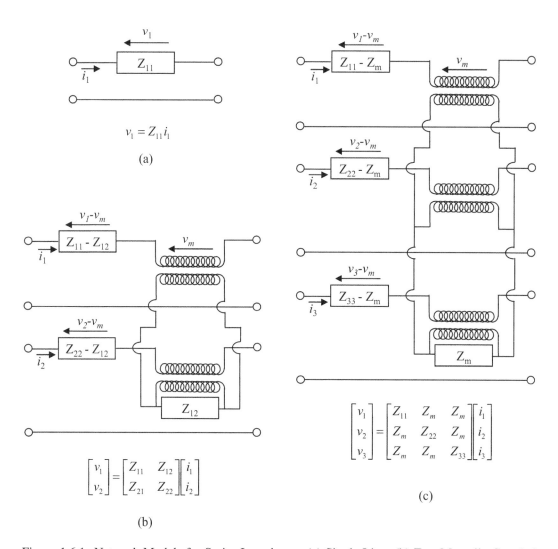

Figure 1.6.1: Network Models for Series Impedance: (a) Single Line; (b) Two Mutually Coupled Lines; (c) Three Mutually Coupled Lines with Equal Mutual Impedance

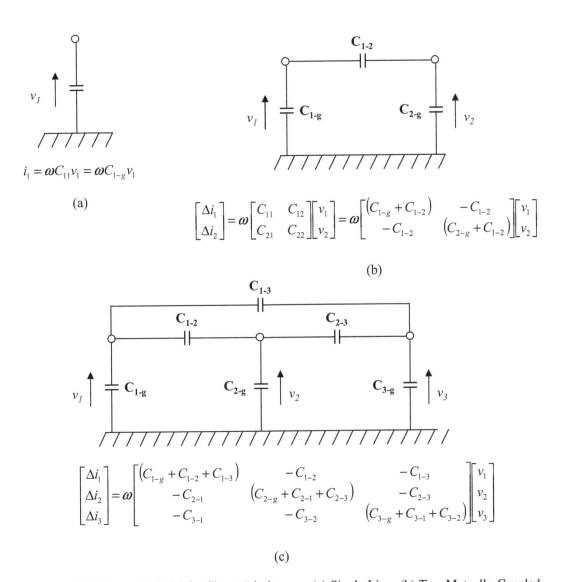

$$i_1 = \omega C_{11} v_1 = \omega C_{1-g} v_1$$

(a)

$$\begin{bmatrix} \Delta i_1 \\ \Delta i_2 \end{bmatrix} = \omega \begin{bmatrix} C_{11} & C_{12} \\ C_{21} & C_{22} \end{bmatrix} \begin{bmatrix} v_1 \\ v_2 \end{bmatrix} = \omega \begin{bmatrix} \left(C_{1-g} + C_{1-2} \right) & -C_{1-2} \\ -C_{1-2} & \left(C_{2-g} + C_{1-2} \right) \end{bmatrix} \begin{bmatrix} v_1 \\ v_2 \end{bmatrix}$$

(b)

$$\begin{bmatrix} \Delta i_1 \\ \Delta i_2 \\ \Delta i_3 \end{bmatrix} = \omega \begin{bmatrix} \left(C_{1-g} + C_{1-2} + C_{1-3} \right) & -C_{1-2} & -C_{1-3} \\ -C_{2-1} & \left(C_{2-g} + C_{2-1} + C_{2-3} \right) & -C_{2-3} \\ -C_{3-1} & -C_{3-2} & \left(C_{3-g} + C_{3-1} + C_{3-2} \right) \end{bmatrix} \begin{bmatrix} v_1 \\ v_2 \\ v_3 \end{bmatrix}$$

(c)

Figure 1.6.2: Network Model for Shunt Admittance: (a) Single Line; (b) Two Mutually Coupled Lines; (c) Three Mutually Coupled Lines

A further differentiation of Equations 1.6.4 and 1.6.5 leads to the forms:

$$\frac{\delta^2 \tilde{V}}{\delta x^2} = \tilde{z}\tilde{y}\tilde{V} \quad [V/m^2] \qquad \frac{\delta^2 \tilde{I}}{\delta x^2} = \tilde{y}\tilde{z}\tilde{I} \quad [A/m^2] \tag{1.6.6}$$

A solution of this system of simultaneous equations can be expressed in the form:

$$\begin{bmatrix} \tilde{V}_s \\ \tilde{I}_s \end{bmatrix} = \begin{bmatrix} \cosh\sqrt{\tilde{z}\tilde{y}}\ell & \sqrt{\dfrac{\tilde{z}}{\tilde{y}}}\sinh\sqrt{\tilde{z}\tilde{y}}\ell \\[2ex] \sqrt{\dfrac{\tilde{y}}{\tilde{z}}}\sinh\sqrt{\tilde{z}\tilde{y}}\ell & \cosh\sqrt{\tilde{z}\tilde{y}}\ell \end{bmatrix} \begin{bmatrix} \tilde{V}_r \\ \tilde{I}_r \end{bmatrix} \tag{1.6.7}$$

where $\tilde{V}_s$ and $\tilde{I}_s$ represent the voltage and current at the sending end of the line, and $\tilde{V}_r$ and $\tilde{I}_r$ represent the voltage and current at the receiving end of the line section of length ℓ.

Consider the three-phase line configuration in Figure 1.1.6. For a completely transposed line, with no ground wires, and for balanced three-phase operation, the effective single-phase characteristics for the line are:

$L = 1.4277$ [mH/km] [per Example 1.1.3 and Equation 1.1.63]

$P = 126.77$ [km/μF] [per Equations 1.2.42 and 1.2.45]

In order to accentuate the primary performance characteristics, we shall consider a line with no resistive losses so that:

$$\tilde{z} = j\omega L = j0.5382 \qquad [\Omega/km] @ 60\,Hz$$

$$\tilde{y} = j\omega C = j2.9739(10^{-6}) \qquad [S/km] @ 60\,Hz$$

$$\sqrt{\tilde{z}\tilde{y}} = j1.2651(10^{-3}) \qquad [rad/km] @ 60\,Hz$$

$$\sqrt{\frac{\tilde{z}}{\tilde{y}}} = 425.4 = Z_o \quad [\Omega] \qquad \sqrt{\frac{\tilde{y}}{\tilde{z}}} = 2.3507(10^{-3}) = Y_o \quad [S] \tag{1.6.8}$$

Because of the imaginary argument $\sqrt{\tilde{z}\tilde{y}}$, we rewrite Equation 1.6.7 in the form:

$$\begin{bmatrix} \tilde{V}_s \\ \tilde{I}_s \end{bmatrix} = \begin{bmatrix} \cos\dfrac{1.2651\ell}{1000} & j425.4\sin\dfrac{1.2651\ell}{1000} \\[2ex] j0.0023507\sin\dfrac{1.2651\ell}{1000} & \cos\dfrac{1.2651\ell}{1000} \end{bmatrix} \begin{bmatrix} \tilde{V}_r \\ \tilde{I}_r \end{bmatrix} \tag{1.6.9}$$

with ℓ in terms of [km].

Evaluating Equation 1.6.9 for various lengths of line we obtain:

$$\ell = 10\,km \quad \begin{bmatrix} \tilde{V}_s \\ \tilde{I}_s \end{bmatrix} = \begin{bmatrix} 0.9999 & j5.3816 \\ j2.974(10^{-5}) & 0.9999 \end{bmatrix} \begin{bmatrix} \tilde{V}_r \\ \tilde{I}_r \end{bmatrix}$$

$$\ell = 100\,km \quad \begin{bmatrix} \tilde{V}_s \\ \tilde{I}_s \end{bmatrix} = \begin{bmatrix} 0.9920 & j53.674 \\ j2.966(10^{-4}) & 0.9920 \end{bmatrix} \begin{bmatrix} \tilde{V}_r \\ \tilde{I}_r \end{bmatrix}$$

$$\ell = 1000\,km \quad \begin{bmatrix} \tilde{V}_s \\ \tilde{I}_s \end{bmatrix} = \begin{bmatrix} 0.3010 & j405.67 \\ j2.2417(10^{-3}) & 0.3010 \end{bmatrix} \begin{bmatrix} \tilde{V}_r \\ \tilde{I}_r \end{bmatrix}$$

$$\ell = 2000\,km \quad \begin{bmatrix} \tilde{V}_s \\ \tilde{I}_s \end{bmatrix} = \begin{bmatrix} -0.8189 & j244.18 \\ j1.3493(10^{-3}) & -0.8189 \end{bmatrix} \begin{bmatrix} \tilde{V}_r \\ \tilde{I}_r \end{bmatrix}$$

$$\ell = 2500\,km \quad \begin{bmatrix} \tilde{V}_s \\ \tilde{I}_s \end{bmatrix} = \begin{bmatrix} -0.9998 & -j8.9997 \\ -j0.0497(10^{-3}) & -0.9998 \end{bmatrix} \begin{bmatrix} \tilde{V}_r \\ \tilde{I}_r \end{bmatrix}$$

$$\tag{1.6.10}$$

If we write:

$$\begin{bmatrix} \tilde{V}_s \\ \tilde{I}_s \end{bmatrix} = \begin{bmatrix} \tilde{A} & \tilde{B} \\ \tilde{C} & \tilde{D} \end{bmatrix} \begin{bmatrix} \tilde{V}_r \\ \tilde{I}_r \end{bmatrix} \tag{1.6.11}$$

we observe in Equation 1.6.7 that:

$$\tilde{A} = \tilde{D}$$

$$\tilde{A}\tilde{D} - \tilde{B}\tilde{C} = 1.0 \tag{1.6.12}$$

We also observe that for line lengths up to about 100 km we have, for our example:

$$\tilde{A} = \tilde{D} \approx 1.0$$

$$\tilde{B} \approx \tilde{Z} = j\omega L \quad [\Omega]$$

$$\tilde{C} \approx \tilde{Y} = \frac{j\omega}{P} \quad [S] \tag{1.6.13}$$

We may write Equation 1.6.10 in terms of a power series (see Appendix H) as follows:

$$\begin{bmatrix} \tilde{V}_s \\ \tilde{I}_s \end{bmatrix} = \begin{bmatrix} \tilde{A} & \tilde{B} \\ \tilde{C} & \tilde{D} \end{bmatrix} \begin{bmatrix} \tilde{V}_r \\ \tilde{I}_r \end{bmatrix} \tag{1.6.14}$$

where:

$$\tilde{A} = 1 + \frac{\tilde{Z}\tilde{Y}}{2} + \frac{1}{6}\left(\frac{\tilde{Z}\tilde{Y}}{2}\right)^2 + \dots$$

$$\tilde{B} = \tilde{Z}\left[1 + \frac{\tilde{Z}\tilde{Y}}{6} + \frac{1}{30}\left(\frac{\tilde{Z}\tilde{Y}}{6}\right)^2 + \dots\right]$$

$$\tilde{C} = \tilde{Y}\left[1 + \frac{\tilde{Z}\tilde{Y}}{6} + \frac{1}{30}\left(\frac{\tilde{Z}\tilde{Y}}{6}\right)^2 + \dots\right]$$

$$\tilde{D} = 1 + \frac{\tilde{Z}\tilde{Y}}{2} + \frac{1}{6}\left(\frac{\tilde{Z}\tilde{Y}}{2}\right)^2 + \dots$$

As an approximate equivalent circuit representation for a transmission line, we might consider those shown in Figure 1.6.3. For the Pi network we have:

$$\begin{bmatrix} \tilde{V}_s \\ \tilde{I}_s \end{bmatrix} = \begin{bmatrix} 1 & 0 \\ \frac{\tilde{Y}}{2} & 1 \end{bmatrix} \begin{bmatrix} 1 & \tilde{Z} \\ 0 & 1 \end{bmatrix} \begin{bmatrix} 1 & 0 \\ \frac{\tilde{Y}}{2} & 1 \end{bmatrix} \begin{bmatrix} \tilde{V}_r \\ \tilde{I}_r \end{bmatrix}$$

$$= \begin{bmatrix} 1 + \frac{\tilde{Z}\tilde{Y}}{2} & \tilde{Z} \\ \tilde{Y}\left(1 + \frac{\tilde{Z}\tilde{Y}}{4}\right) & 1 + \frac{\tilde{Z}\tilde{Y}}{2} \end{bmatrix} \begin{bmatrix} \tilde{V}_r \\ \tilde{I}_r \end{bmatrix} \tag{1.6.15}$$

The reader may verify that the Pi network representation will give within 1 percent (error in Z) for our line up to about 200 km.

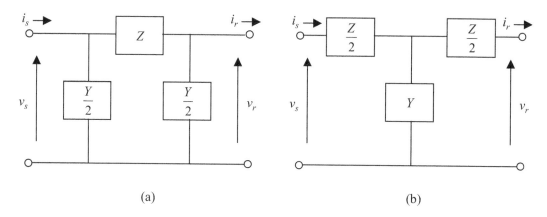

(a)   (b)

Figure 1.6.3: Short to Moderate Transmission Line Equivalent Circuit

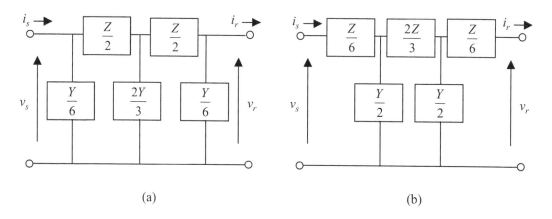

(a)   (b)

Figure 1.6.4: Moderate to Long Transmission Line Equivalent Network

For the Tee network we have:

$$
\begin{bmatrix} \tilde{V}_s \\ \tilde{I}_s \end{bmatrix} =
\begin{bmatrix} 1 & \dfrac{\tilde{Z}}{2} \\ 0 & 1 \end{bmatrix}
\begin{bmatrix} 1 & 0 \\ \tilde{Y} & 1 \end{bmatrix}
\begin{bmatrix} 1 & \dfrac{\tilde{Z}}{2} \\ 0 & 1 \end{bmatrix}
\begin{bmatrix} \tilde{V}_r \\ \tilde{I}_r \end{bmatrix}
$$

$$
=
\begin{bmatrix} 1 + \dfrac{\tilde{Z}\tilde{Y}}{2} & \tilde{Z}\left(1 + \dfrac{\tilde{Z}\tilde{Y}}{4}\right) \\[2ex] \tilde{Y} & 1 + \dfrac{\tilde{Z}\tilde{Y}}{2} \end{bmatrix}
\begin{bmatrix} \tilde{V}_r \\ \tilde{I}_r \end{bmatrix}
\tag{1.6.16}
$$

which is also within 1% (error in Y) for our line up to about 200 km.

For longer lines, a somewhat more accurate representation can be obtained with the equivalent network shown in Figure 1.6.4. For the Pi network we have:

$$
\begin{bmatrix} \tilde{V}_s \\ \tilde{I}_s \end{bmatrix} =
\begin{bmatrix} \tilde{A} & \tilde{B} \\ \tilde{C} & \tilde{D} \end{bmatrix}
\begin{bmatrix} \tilde{V}_r \\ \tilde{I}_r \end{bmatrix}
\tag{1.6.17}
$$

where:

$$\tilde{A} = 1 + \frac{\tilde{Z}\tilde{Y}}{2} + \frac{1}{9}\left(\frac{\tilde{Z}\tilde{Y}}{2}\right)^2$$

$$\tilde{B} = \tilde{Z}\left[1 + \frac{\tilde{Z}\tilde{Y}}{6}\right]$$

$$\tilde{C} = \tilde{Y}\left[1 + \frac{5}{18}\left(\frac{\tilde{Z}\tilde{Y}}{2}\right) + \frac{1}{54}\left(\frac{\tilde{Z}\tilde{Y}}{2}\right)^2\right]$$

$$\tilde{D} = 1 + \frac{\tilde{Z}\tilde{Y}}{2} + \frac{1}{9}\left(\frac{\tilde{Z}\tilde{Y}}{2}\right)^2$$

which is valid within 1% (error in Y) for our line up to about 475 km.

For the Tee network we have:

$$\begin{bmatrix} \tilde{V}_s \\ \tilde{I}_s \end{bmatrix} = \begin{bmatrix} \tilde{A} & \tilde{B} \\ \tilde{C} & \tilde{D} \end{bmatrix} \begin{bmatrix} \tilde{V}_r \\ \tilde{I}_r \end{bmatrix} \tag{1.6.18}$$

where:

$$\tilde{A} = 1 + \frac{\tilde{Z}\tilde{Y}}{2} + \frac{1}{9}\left(\frac{\tilde{Z}\tilde{Y}}{2}\right)^2$$

$$\tilde{B} = \tilde{Z}\left[1 + \frac{5}{18}\left(\frac{\tilde{Z}\tilde{Y}}{2}\right) + \frac{1}{54}\left(\frac{\tilde{Z}\tilde{Y}}{2}\right)^2\right]$$

$$\tilde{C} = \tilde{Y}\left[1 + \frac{\tilde{Z}\tilde{Y}}{6}\right]$$

$$\tilde{D} = 1 + \frac{\tilde{Z}\tilde{Y}}{2} + \frac{1}{9}\left(\frac{\tilde{Z}\tilde{Y}}{2}\right)^2$$

which is also valid within 1% (error in Z) for our line up to about 475 km.

It is of interest to compare, for 400 km of our line, the exact solution with the solution obtained with the equivalent network of Figure 1.6.4, and that obtained by two series equivalent networks of Figure 1.6.3 (each for 200 km of line).

For the exact solution we obtain:

$$\begin{bmatrix} \tilde{V}_s \\ \tilde{I}_s \end{bmatrix} = \begin{bmatrix} 0.8746 & j206.20 \\ j1.1394(10^{-3}) & 0.8746 \end{bmatrix} \begin{bmatrix} \tilde{V}_r \\ \tilde{I}_r \end{bmatrix} \tag{1.6.19}$$

For the Pi network (Figure 1.6.4, 400 km):

$$\begin{bmatrix} \tilde{V}_s \\ \tilde{I}_s \end{bmatrix} = \begin{bmatrix} 0.8738 & j206.09 \\ j1.1476(10^{-3}) & 0.8738 \end{bmatrix} \begin{bmatrix} \tilde{V}_r \\ \tilde{I}_r \end{bmatrix} \tag{1.6.20}$$

For the Pi networks in series (Figure 1.6.3, 200 km each):

$$\begin{bmatrix} \tilde{V}_s \\ \tilde{I}_s \end{bmatrix} = \begin{bmatrix} 0.9680 & j107.64 \\ j0.5853(10^{-3}) & 0.9680 \end{bmatrix}_2 \begin{bmatrix} \tilde{V}_r \\ \tilde{I}_r \end{bmatrix}$$

$$= \begin{bmatrix} 0.8740 & j208.39 \\ j1.1331(10^{-3}) & 0.8740 \end{bmatrix} \begin{bmatrix} \tilde{V}_r \\ \tilde{I}_r \end{bmatrix} \tag{1.6.21}$$

From a practical engineering point of view, recognizing that we cannot predict line parameters (i.e., line geometry) to less than 1% error, all three of the above results are essentially the same. The single equivalent network of Figure 1.6.3 is to be preferred because it is easy to remember. Furthermore, for all practical purposes at 60 Hz operation, for a 100 km line we may write approximately [per Equation 1.6.14]:

$$\begin{bmatrix} \tilde{V}_s \\ \tilde{I}_s \end{bmatrix} = \begin{bmatrix} 1.0 & \tilde{Z} \\ \tilde{Y} & 1.0 \end{bmatrix} \begin{bmatrix} \tilde{V}_r \\ \tilde{I}_r \end{bmatrix} \tag{1.6.22}$$

Longer lines can be represented by a series concatenation of 100 km line sections.

1.6.4   Short Line Equations

We can represent Equation 1.6.14 in a more enlightening manner by per-unitizing the quantities. Let us select:

$$MVA_{base} = 500\,MVA \quad KV_{base} = 345\,kV \quad Z_{base} = 238.05\,ohms \tag{1.6.23}$$

In terms of per unit quantities we have for Equation 1.6.14:

$$\ell = 10\,km \quad \begin{bmatrix} \tilde{V}_s \\ \tilde{I}_s \end{bmatrix} = \begin{bmatrix} 0.9999 & j0.0226 \\ j0.0070 & 0.9999 \end{bmatrix} \begin{bmatrix} \tilde{V}_r \\ \tilde{I}_r \end{bmatrix}$$

$$\ell = 100\,km \quad \begin{bmatrix} \tilde{V}_s \\ \tilde{I}_s \end{bmatrix} = \begin{bmatrix} 0.9920 & j0.2255 \\ j0.0706 & 0.9920 \end{bmatrix} \begin{bmatrix} \tilde{V}_r \\ \tilde{I}_r \end{bmatrix}$$

$$\ell = 1000\,km \quad \begin{bmatrix} \tilde{V}_s \\ \tilde{I}_s \end{bmatrix} = \begin{bmatrix} 0.3010 & j1.704 \\ j0.5366 & 0.3010 \end{bmatrix} \begin{bmatrix} \tilde{V}_r \\ \tilde{I}_r \end{bmatrix}$$

$$\ell = 2000\,km \quad \begin{bmatrix} \tilde{V}_s \\ \tilde{I}_s \end{bmatrix} = \begin{bmatrix} -0.8189 & j1.026 \\ j0.3212 & -0.8189 \end{bmatrix} \begin{bmatrix} \tilde{V}_r \\ \tilde{I}_r \end{bmatrix}$$

$$\ell = 2500\,km \quad \begin{bmatrix} \tilde{V}_s \\ \tilde{I}_s \end{bmatrix} = \begin{bmatrix} -0.9998 & -j0.0318 \\ -j0.0118 & -0.9998 \end{bmatrix} \begin{bmatrix} \tilde{V}_r \\ \tilde{I}_r \end{bmatrix} \tag{1.6.24}$$

For the 10 km line, for example, we observe that the line impedance drop is about 2.26%, and that the line-charging current is only 0.7% of the line current. Clearly we would simplify our equivalent network for the transmission line by ignoring the line capacitance in Figure 1.6.3. For the 100 km line, the line-charging current is about 7% of the line current. Perhaps we would want to employ the complete equivalent network in Figure 1.6.3.

For the 100 km line we note that the line impedance drop is 22.6%. We are alerted to performing an economic analysis to establish whether series capacitor compensation may be appropriate.

Finally, it is sometimes of interest to review the phasor power relationships at the sending and receiving ends. Using the form of Equation 1.6.11, we have:

$$(\tilde{V}_s) = \begin{bmatrix} \tilde{A} & \tilde{B} \end{bmatrix} \begin{bmatrix} \tilde{V}_r \\ \tilde{I}_r \end{bmatrix} = \begin{bmatrix} \tilde{V}_r & \tilde{I}_r \end{bmatrix} \begin{bmatrix} \tilde{A} \\ \tilde{B} \end{bmatrix} \quad (\tilde{I}_s) = \begin{bmatrix} \tilde{C} & \tilde{D} \end{bmatrix} \begin{bmatrix} \tilde{V}_r \\ \tilde{I}_r \end{bmatrix} \tag{1.6.25}$$

$$\tilde{S}_s = P_s + Q_s = \tilde{V}_s \tilde{I}_s^* \tag{1.6.26}$$

$$\tilde{S}_s = \begin{bmatrix} \tilde{V}_r & \tilde{I}_r \end{bmatrix} \begin{bmatrix} \tilde{A} \\ \tilde{B} \end{bmatrix} \begin{bmatrix} \tilde{C}^* & \tilde{D}^* \end{bmatrix} \begin{bmatrix} \tilde{V}_r^* \\ \tilde{I}_r^* \end{bmatrix}$$

$$= \begin{bmatrix} 1 & 1 \end{bmatrix} \begin{bmatrix} \tilde{A}\tilde{C}^* \mid \tilde{V}_r \mid^2 & \tilde{A}\tilde{D}^* \tilde{S}_r \\ \tilde{B}\tilde{C}^* \tilde{S}_r^* & \tilde{B}\tilde{D}^* \mid \tilde{I}_r \mid^2 \end{bmatrix} \begin{bmatrix} 1 \\ 1 \end{bmatrix}$$

$$= \tilde{A}\tilde{C}^* \mid \tilde{V}_r \mid^2 + \tilde{B}\tilde{D}^* \mid \tilde{I}_r \mid^2 + \tilde{A}\tilde{D}^* \tilde{S}_r + \tilde{B}\tilde{C}^* \tilde{S}_r^* \qquad (1.6.27)$$

$$\qquad (1.6.28)$$

For the 100 km line in Equation 1.6.24, we obtain:

$$\tilde{S}_s = \tilde{S}_r + j0.9920\{0.2255 \mid I_r \mid^2 - 0.0706 \mid V_r \mid^2 - 0.0321 Q_r\} \qquad (1.6.29)$$

1.7 References

[1] P. M. Anderson. *Analysis of Faulted Power Systems.* Iowa State University Press, Ames, Iowa, 1973.

[2] E. Clarke. *Circuit Analysis of AC Power Systems - Volume I.* John Wiley and Sons, Inc., New York, New York, 1943.

[3] E. Clarke. *Circuit Analysis of AC Power Systems - Volume II.* John Wiley and Sons, Inc., New York, New York, 1943.

[4] Westinghouse Electric Corporation. *Electrical Transmission and Distribution Reference Book.* Westinghouse Electric Corporation, East Pittsburgh, Pennsylvania.

[5] Electric Power Research Institute. *Transmission Line Reference Book.* Electric Power Research Institute, 3412 Millview Ave., Palo Alto, California.

[6] Electric Power Research Institute. *Transmission Line Reference Book - 345kV and Above.* Electric Power Research Institute, 3412 Millview Ave., Palo Alto, California.

[7] W. D. Stevenson. *Elements of Power System Analysis, 4th Ed.* McGraw Hill, New York, New York, 1982.

[8] J. Zaborsky and J. Rittenhouse. *Electric Power Transmission - Volume I.* The Rensselaer Bookstore, Troy, New York, 1969.

[9] J. Zaborsky and J. Rittenhouse. *Electric Power Transmission - Volume II.* The Rensselaer Bookstore, Troy, New York, 1969.

1.8 Exercises

1.1 Consider the four parallel conductors in free space, illustrated in Figure 1.3.2 (pg. 35), sufficiently long so that receiving end and sending end effects may be neglected, and double all spacings between conductors and maintain the same GMR's.

(a) For $I1 = 1000$ amperes, and $I_2 = I_3 = I_4 = 0$, and beginning with Ampere's circuital law, calculate $|H|$, $|B|$, and the flux linking conductors 1, 2, 3, and 4 for 1 km line length.

(b) For $I_2 = -1000$ amperes, and $I_1 = I_3 = I_4 = 0$, calculate the flux linking conductors 1, 2, 3, and 4 for 1 km line length.

(c) Repeat (b) with $I_1 = -I_2 = 1000$ amperes, $I_3 = I_4 = 0$. Observe that this is the first physically sensible solution.

(d) Let conductors 3 and 4 be connected together at both receiving and sending ends of the lines. What is the magnitude of the flux linking the loop formed by conductors 3 and 4 when $I_1 = -I_2 = 1000$ amperes direct current?

(e) What is the flux linking a loop formed by conductors 1 and 2 for $I_1 = -I_2 = 1000$ amperes direct current?

1.2 Consider the same overhead conductor configuration described in Exercise 1.3.1. Let the remote end of conductors 1 and 2 be connected, and let a voltage be applied between these conductors at the near end: $e_{12} = 120\sqrt{2}sin\omega t$ @ 60 Hz Let all conductors be 10 km in length, and ignore conductor resistance.

(a) With only loop 1-2 energized, and conductors 3 and 4 open, calculate I_1 and I_2. Also calculate ΔV_3 and ΔV_4.

(b) Repeat (a) except that conductors 3 and 4 are connected at the near end and remote end to form a closed loop. Calculate I_1, I_2, I_3, and I_4.

(c) As far as the source e_{12} is concerned, what can you say about the effective input reactance with loop 3-4 open and loop 3-4 closed?

(d) In retrospect, from the basic 4x4 system of equations: $[\Lambda] = [L][I]$ how can you determine the effective 2x2 system of equations:

$$\begin{bmatrix} \Lambda_I \\ \Lambda_{II} \end{bmatrix} = \begin{bmatrix} L_{I-I} & L_{I-II} \\ L_{II-I} & L_{II-II} \end{bmatrix} \begin{bmatrix} I_1 \\ I_3 \end{bmatrix}$$

where I represents loop 1-2 and II represents loop 3-4?

1.3 Consider 1.5 km of the single-circuit three-phase line configuration shown in Figure 1.1.6. Let there be a fence parallel to the line, 5 m to the right of phase "b" conductor, with fence posts (insulated from ground) with a wire of GMR = 0.01 m stretched along the fence on top of the fence posts at 1.0 m above ground.

(a) Expand Equation 1.1.60 to include an equation for the fence wire. Calculate the flux linking the fence wire for a balanced three-phase current of 100 amperes, rms, and 60 Hz (positive sequence).

(b) Repeat (a) but for $I_a = I_b = I_c = 100$ amperes, all currents being in phase (zero sequence).

(c) Evaluate the current in the fence if it is solidly grounded at both ends (ignoring all resistive losses, let $\Lambda_f = 0$) for the system conditions defined in (a).

(d) Repeat (c) but for the system conditions in (b).

(e) Let the fence be solidly grounded at one end and grounded through a 100 ohm resistance at the other end. Using the results obtained in (b) and (d), determine the Thevenin-Helmholtz equivalent network and calculate the watts consumed in the resistor.

1.4 Consider the arrangement shown in Figure 1.1.8 on page 18. Replace $d_{ca} = 10$ m and $d_{ab} = 10$ m to $d_{ca} = 8$ m and $d_{ab} = 12$ m.

(a) Evaluate the inductance matrix (ignoring ground wires) for the untransposed line in terms of mH/km.

(b) Evaluate the inductance matrix (ignoring ground wires) for the line transposed according to Figure 1.1.7(a) in terms of mH/km.

(c) Evaluate the inductance matrix (ignoring ground wires) for the line transposed according to Figure 1.1.7(b) in terms of mH/km.

1.5 Consider the arrangement shown in Figure 1.1.8 on page 18. Replace $d_{ca} = 10$ m and $d_{ab} = 10$ m to $d_{ca} = 8$ m and $d_{ab} = 12$ m.

(a) Evaluate the 5x5 inductance matrix (including ground wires) for the untransposed line in terms of mH/km.

(b) Evaluate the 5x5 inductance matrix (including ground wires) for the line transposed according to Figure 1.1.7(a) in terms of mH/km.

(c) Evaluate the 5x5 inductance matrix (including ground wires) for the line transposed according to Figure 1.1.7(b) in terms of mH/km.

(d) Reduce the 5x5 matrices in (a), (b), and (c) to equivalent 3x3 matrices.

1.6 For a balanced set of three-phase currents, compare the influence of the ground wires in 1.5(a), 1.5(b), and 1.5(c).

1.7 Invert the 3x3 inductance matrices [L] obtained in Exercise 1.5(d):

(a) By the cofactor method.

(b) By the row and column elimination method employed in Exercise 1.5(d), using the augmented matrix, written in partitioned form as:

$$\begin{bmatrix} 0 & U \\ U & L \end{bmatrix}$$

1.8 Reduce the inductance matrix in Equation 1.1.74 to a 3x3 matrix:

(a) By the subtraction of rows only, per Equation 1.1.77, and a subsequent reduction as in Exercise 1.5(d).

(b) By the subtraction of rows and columns, per Equation 1.1.78, and a subsequent reduction as in Exercise 1.5(d).

1.9 A single conductor above earth is energized at 250 kV (dc) with respect to earth. The transmission line conductor has a radius of 2 cm and is at a height of 20 m above earth. Another conductor with radius 0.3 cm is located directly above the transmission line conductor at a height of 22 m above earth.

(a) With the topmost conductor totally insulated from earth, and with no net electrical charge residing on it, determine the voltage between the topmost conductor and earth caused by electrostatic induction. Determine also the voltage difference between the two overhead conductors.

(b) If the topmost conductor is continuously grounded (as would be the case for a ground wire), determine the charge density on the ground wire.

(c) Assume that 3000 kV/m is sufficiently electric field gradient at the surface of the conductor to cause initiation of corona. With the topmost conductor continuously grounded, and employing the charge density already determined in part (b), determine whether the transmission line conductor or the ground wire will go into corona.

(d) Are you surprised with the result in (c)?

1.10 Consider the horizontal line configuration shown in Figure 1.1.6 (where we now let $r_a = r_b = r_c = 0.010 m$). Beginning with Equation 1.2.35:

(a) Evaluate the equivalent line capacitances.

(b) For 50 km of line operating at a balanced 345 kV, determine the line-charging MVA.

1.11 Consider the horizontal line configuration shown in Figure 1.1.8 (where we now let $r_a = r_b = r_c = 0.010m$). Beginning with Equation 1.2.47:

   (a) Evaluate the equivalent line capacitances.

   (b) For 50 km of line operating at a balanced 345 kV, determine the line-charging MVA.

Chapter 2

Single-Phase Transformers

The voltage at the terminals of a synchronous generator is of the order of 22 kV. A three-phase generator rated 500 MVA, 22kV, 60 Hz, has a rated terminal current of 13,122 amperes per phase. It is not practical to carry currents of such magnitudes over considerable distances. Therefore, generator step-up transformers are employed to transform from 22 kV to a much higher voltage, say 345 kV or 500 kV, so that the current to be transmitted is considerably smaller. Therefore, the transmission line $I^2 \times R$ losses will be smaller and the drop in voltage along the line will be smaller.

Transmission voltages up to 765 kV are employed for transmitting large blocks of power over very long distances. For transmitting smaller blocks of power over short distances, such high voltage lines are too expensive and lower voltages are employed. Therefore, between the generating station and the ultimate user, there are many power transformers and various levels of voltage.

It is therefore important to understand the performance characteristics of transformers in order that they may be properly selected and in order that they may be properly operated. One important aspect of transformer performance concerns the insulation of windings, the overvoltage stresses placed upon this insulation due to lightning or switching in particular, and how this insulation may be protected from such overvoltages. We shall not concern ourselves with these aspects in this book.

The other major aspect of transformer performance deals with the performance characteristics of the copper and iron in the transformer. This requires a background in magnetic circuit analysis as presented in textbooks on electric machinery to which the reader is referred. We shall, however, briefly review some salient aspects of single-phase two-winding transformers in order to establish an approach that is particularly useful in analyzing some of the more complex transformer connections.

2.1 Ideal Single-Phase Two-Winding Transformer

Consider the closed iron-core magnetic circuit shown in Figure 2.1.1. When a voltage $v_1(t)$ is applied to the first winding, a magnetic flux is established in the iron-core such that the flux linking the first winding is equal to:

$$\lambda_1(t) = \int v_1(t)dt \qquad [V \cdot s] \qquad (2.1.1)$$

according to Faraday's law if we ignore the resistance of the winding at this point. In order to establish the proper magnetic flux in the core we require some exciting current according to Ampere's law.

In place of the classical form of Ampere's circuital law that we employed in Chapter 1, we may now employ the simpler form:

$$MMF = N \cdot i = \Re \cdot \Phi \qquad [A] \qquad (2.1.2)$$

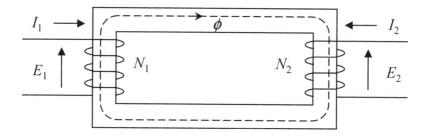

Figure 2.1.1: Ideal Transformer

which is sometimes referred to as the *magnetic circuit Ohm's law*. Equation 2.1.2 is correct only when all of the flux ϕ is confined within the iron-core, and if the flux density is uniform across the cross section of the core. That closed iron-core magnetic circuits satisfy these constraints with only a small error is established in textbooks on electric machinery. The term $\Re$ is called the (magnetic) *reluctance* of the core:

$$\Re = \frac{\ell}{\mu \cdot A} \qquad [H]^{-1} \tag{2.1.3}$$

and its inverse is called the (magnetic) *permeance* of the core:

$$\wp = \frac{\mu \cdot A}{\ell} \qquad [H] \tag{2.1.4}$$

with:

$$\mu = \mu_o \cdot \mu_r \qquad [H/m] \tag{2.1.5}$$
$$\mu_o = 4\pi(10^{-7}) \qquad [H/m] \tag{2.1.6}$$

μ_r = relative magnetic permeability
A = cross-sectional area of the core [m^2]
ℓ = mean length of closed flux line [m]

Example 2.1.1
Given a closed iron-core circuit with:

$$A = 0.10[m^2] \qquad \ell = 4[m] \qquad \mu_r = 5000$$

The magnetic permeability of this core is:

$$\wp = \frac{5000(4\pi)(10^{-7})(0.10)}{4} = 157(10^{-6}) \qquad [H]$$

Let the voltage applied to the first winding be 120 V rms @ 60 Hz so that:

$$v_1(t) = 120\sqrt{2}\cos\omega t \qquad [V]$$

so that:

$$\lambda_1(t) = \frac{\sqrt{2}}{\pi} \sin \omega t \qquad [V.s]$$

If the winding has 1000 turns, then:

$$\Phi_1(t) = \frac{\sqrt{2}}{1000\pi} \sin \omega t \qquad [Wb]$$

Then:

$$N_1 i_1(t) = \Re \Phi_1(t) = \frac{\Phi_1(t)}{\wp} = \frac{\sqrt{2}}{1000\pi(157)(10^{-6})} \sin \omega t$$

and:

$$i_1(t) = 2.03(10^{-3})\sqrt{2} \sin \omega t \qquad [A]$$

Example 2.1.2

From another point of view we could have written:

$$L = N^2 \wp = 157 \qquad [H]$$
$$X = \wp L = 59.19(10^3) \qquad [\Omega]$$
$$I = \frac{V}{X} = 2.03(10^{-3}) \qquad [A]$$

Example 2.1.1 illustrates how Faraday's law and Ampere's law are employed to determine the exciting current in the winding from a magnetic circuit point of view. It should be carefully noted that the required exciting current is extremely small.

Example 2.1.2 illustrates how we may determine the exciting current from an an electric circuit point of view. Here we see that the inductive reactance of the winding is so high that the current is very small. The relationship between inductive reactance and permeance of the magnetic circuit should be carefully noted.

Let the second winding in the magnetic circuit shown in Figure 2.1.1 also carry a current. We consider first the idealized situation where all of the magnetic flux ϕ_m is confined within the core and the flux density within the core is everywhere uniform. Since we have two sources of MMF acting simultaneously upon a single core, we have:

$$\overline{\phi}_m = \frac{N_1 \cdot \tilde{I}_1 + N_2 \cdot \tilde{I}_2}{\Re_m} \qquad [Wb] \tag{2.1.7}$$

and according to Faraday's law:

$$\tilde{E}_1 = \frac{d\tilde{\lambda}_1}{dt} = N_1 \cdot \frac{d\tilde{\Phi}_m}{dt} \qquad [V] \tag{2.1.8}$$

$$\tilde{E}_2 = \frac{d\tilde{\lambda}_2}{dt} = N_2 \cdot \frac{d\tilde{\Phi}_m}{dt} \qquad [V] \tag{2.1.9}$$

For such an *ideal* transformer (neglecting iron and winding losses and neglecting leakage flux) we have:

$$\frac{\tilde{E}_1}{\tilde{E}_2} = \frac{N_1}{N_2} \qquad (2.1.10)$$

Thus, the voltage induced in the windings of an ideal transformer are directly proportional to the turns ratios. This is perhaps best remembered if we rewrite Equations 2.1.8 and 2.1.9 in the form:

$$\frac{\tilde{E}_1}{N_1} = \frac{\tilde{E}_2}{N_2} = \frac{d\tilde{\Phi}_m}{dt} \qquad [V/turn] \qquad (2.1.11)$$

If the reluctance of the magnetic core may be assumed to be vanishingly small, the MMF required to sustain the necessary flux (as required by Faraday's law) will also become vanishingly small, according to the results in Example 2.1.1, and in the limit:

$$N_1 \cdot \tilde{I}_1 + N_2 \cdot \tilde{I}_2 = 0 \qquad (2.1.12)$$

or:

$$\frac{\tilde{I}_1}{\tilde{I}_2} = -\frac{N_2}{N_1} \qquad (2.1.13)$$

Comparing the phasor power input to each of these windings, we find:

$$\tilde{S}_1 = \tilde{E}_1 \cdot \tilde{I}_1^* = P_1 + jQ_1 \qquad [VA] \qquad (2.1.14)$$

$$\tilde{S}_2 = \tilde{E}_2 \cdot \tilde{I}_2^* = (\frac{N_2}{N_1} \cdot \tilde{E}_1) \cdot (-\frac{N_1}{N_2} \cdot \tilde{I}_1^*)$$

$$= -\tilde{E}_1 \cdot \tilde{I}_1^* = -(P_1 + jQ_1) \qquad [VA] \qquad (2.1.15)$$

Such an idealized device, the ideal transformer, transmits a given amount of active and reactive power without loss while transforming from one voltage level to the other exactly according to the turns ratio of the transformer, and provides electrical isolation between the circuits connected to the two windings. While it is only an idealized or hypothetical concept, it is a very useful network element in circuit analysis. It is also, to some extent, an approximation to an actual power transformer.

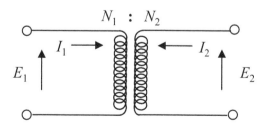

Figure 2.1.2: Ideal Transformer Equivalent Network

Figure 2.1.2 shows the network representation of an ideal transformer whose characteristics are:

In a transformer with no copper losses and no iron losses, and with no leakage flux, the primary and secondary voltages are exactly proportional to the ratio of the turns of the respective windings.

In a transformer with no copper losses and no iron losses, and with no leakage flux, and with an infinitely permeable core, the summation of ampere turns (magnetizing currents) is zero and power is transferred with no loss.

Consider the network shown in Figure 2.1.3, where an ideal transformer is employed to supply load at a voltage different than that available at the source. On the basis of Equations 2.1.11 and 2.1.13, we may write:

$$\tilde{I}_1 = -\left(\frac{N_2}{N_1}\right)\tilde{I}_2 = \left(\frac{N_2}{N_1}\right)\cdot\left(\frac{\tilde{E}_2}{R}\right) = \left(\frac{N_2}{N_1}\right)^2\cdot\left(\frac{\tilde{E}_1}{R}\right) \qquad (2.1.16)$$

and:

$$\frac{\tilde{E}_1}{\tilde{I}_1} = \left(\frac{N_1}{N_2}\right)^2\cdot R \qquad (2.1.17)$$

Equation 2.1.17 states how an ideal transformer may be employed for impedance matching purposes, and Figure 2.1.4 illustrates how an ideal transformer may be transferred in an equivalent circuit and/or how a resistance (impedance in general) may be reflected from one side of the transformer to the other and how the ideal transformer may be eliminated from the network entirely if and when it is not necessary to identify values for $\tilde{E}_1$ and $\tilde{E}_2$ for example and when isolation is not necessary.

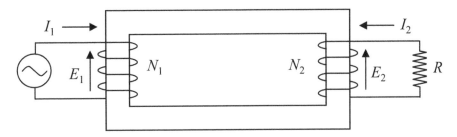

Figure 2.1.3: Network Employing Ideal Transformer

2.2 Practical Two-Winding Transformer

Practical ferromagnetic cores have finite permeabilities and require some small but finite net magnetizing current to establish a desired magnetic flux. For a pure sinusoidal excitation we have:

$$\tilde{E}_1 = \frac{d\tilde{\lambda}_1}{dt} = j\omega N_1\tilde{\Phi}_m \qquad [V] \qquad (2.2.1)$$

$$\tilde{E}_2 = \frac{d\tilde{\lambda}_2}{dt} = j\omega N_2\tilde{\Phi}_m \qquad [V] \qquad (2.2.2)$$

$$\tilde{\Phi}_m = (N_1\tilde{I}_1 + N_2\tilde{I}_2)\cdot\wp_m \qquad [Wb] \qquad (2.2.3)$$

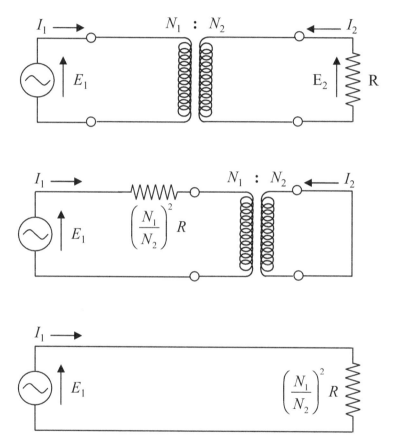

Figure 2.1.4: Transferring Ideal Transformers

On the basis of these three equations we may therefore write:

$$\frac{\tilde{E}_1}{N_1} = \frac{\tilde{E}_2}{N_2} = j\omega\wp_m \cdot (N_1\tilde{I}_1 + N_2\tilde{I}_2) \qquad [V/turn] \qquad (2.2.4)$$

The equivalent circuit for this transformer, as presented by Equation 2.2.4 is shown in Figure 2.2.1. This form of the equivalent circuit (employing ideal $N_1 : 1$ and $1 : N_2$ transformers) stresses the relationship between volts per turn $\tilde{E}_1/N_1$ and $\tilde{E}_2/N_2$ on a common magnetic circuit, and the summation or net ampere turns acting upon the permeance $\wp_m$ of the magnetic circuit to produce the necessary flux.

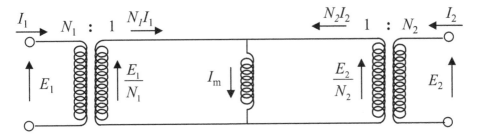

Figure 2.2.1: Transformer Equivalent Circuit: Magnetizing Current Only

Since eddy currents and hysteresis losses are also directly related to the magnitude of the flux density, these may be included in an equivalent network representation as shown in Figure 2.2.2.

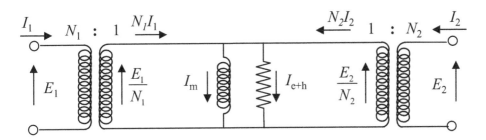

Figure 2.2.2: Transformer Equivalent Circuit: Exciting Current Only

Finally, we recall the leakage flux concept. Flux may leave the core and return over an air path even if the core is infinitely permeable. Figure 2.2.3 illustrates the various manners in which the flux components may be visualized in an transformer. From a physical point of view relative to Faraday's law, it is the total flux-linkages associated with the winding that is responsible for the voltage appearing between the terminals of the winding (ignoring losses). Any partitioning of these total flux-linkages into mutual and leakage flux-linkages is purely arbitrary and is done only for the convenience of permitting superposition in the analysis.

Observe that the relationships in Equations 2.1.10 and 2.1.11 are valid even if the reluctance of the core is not zero. In practical cores of this type, the leakage flux will become more pronounced as the reluctance of the iron becomes greater, but we have neglected leakage flux in this magnetic circuit.

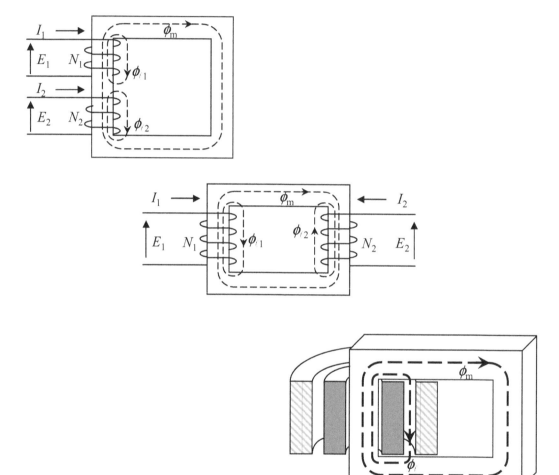

Figure 2.2.3: Mutual and Leakage Flux Component Representation

Considering the partitioning of the total flux into mutual and leakage flux we have:

$$\tilde{\Phi}_{\ell 1} = N_1 \tilde{I}_1 \wp_{\ell 1} \qquad [Wb] \tag{2.2.5}$$

$$\tilde{\Phi}_{\ell 2} = N_2 \tilde{I}_2 \wp_{\ell 2} \qquad [Wb] \tag{2.2.6}$$

$$\tilde{\Phi}_m = (N_1 \tilde{I}_1 + N_2 \tilde{I}_2) \cdot \wp_m \qquad [Wb] \tag{2.2.7}$$

so that:

$$\begin{bmatrix} \frac{\tilde{\lambda}_1}{N_1} \\ \frac{\tilde{\lambda}_2}{N_2} \end{bmatrix} = \begin{bmatrix} (\wp_{\ell 1} + \wp_m) & \wp_m \\ \wp_m & (\wp_{\ell 2} + \wp_m) \end{bmatrix} \cdot \begin{bmatrix} N_1 \tilde{I}_1 \\ N_2 \tilde{I}_2 \end{bmatrix} \tag{2.2.8}$$

and:

$$\begin{bmatrix} \frac{\tilde{E}_1}{N_1} \\ \frac{\tilde{E}_2}{N_2} \end{bmatrix} = (j\omega) \cdot \begin{bmatrix} (\wp_{\ell 1} + \wp_m) & \wp_m \\ \wp_m & (\wp_{\ell 2} + \wp_m) \end{bmatrix} \cdot \begin{bmatrix} N_1 \tilde{I}_1 \\ N_2 \tilde{I}_2 \end{bmatrix} \tag{2.2.9}$$

whereby the eddy current and hysteresis losses have been ignored but can easily be introduced directly into the equivalent network shown in Figure 2.2.4.

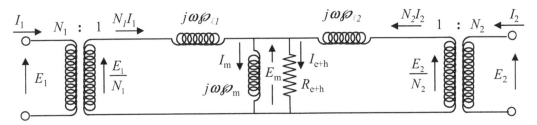

Figure 2.2.4: Transformer Equivalent Circuit including Leakage Flux Effects

And finally, recognizing that the primary and secondary windings have some electrical resistance, we add this to obtain the equivalent circuit shown in Figure 2.2.5.

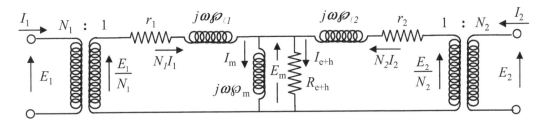

Figure 2.2.5: Complete Transformer Equivalent Circuit

It is interesting to observe that the relatively few principles developed in Chapter 1 have enabled us to analyze the performance of a transformer directly and to present its electrical characteristics in the form of the equivalent network shown in Figure 2.2.5. However, we are not at all in a position at this point to design such a transformer. We leave this for a later section.

The complete transformer equivalent equivalent circuit shown in Figure 2.2.5 is the link between the internal performance characteristics of a transformer and its external characteristics. The value of understanding these relationships cannot be overemphasized. In review, $\tilde{E}_m$ represents the voltage induced in both windings due to $\tilde{\Phi}_m$, and $X_m = \omega \wp_m$ and $R_{e\&h}$ represent the magnetic core characteristics. The terms $\omega \wp_{\ell 1}$ and $\omega \wp_{\ell 2}$ represent the leakage reactances per turn associated with each winding and r_1 and r_2 represent resistances per turn of the windings.

The somewhat unusual employment of the $N_1 : 1$ and $1 : N_2$ ideal transformers proves to be very valuable from a conceptual and analytical point of view. Thus, all of the parameters indicated in Figure 2.2.5 are in terms of units expressed on a *per turn* basis. From an external point of view, if the transformer is viewed from the primary winding terminals, all of the impedance units appear to be multiplied by N_1^2, the voltages by N_1, and the currents by $1/N_1$ in keeping with what was said in connection with Figure 2.1.4.

Consider a two-winding transformer supplying power to a load. Figure 2.2.6(a) illustrates the complete or exact equivalent network, and Figure 2.2.6(b) illustrates the phasor relationships between the various voltages and currents. Since the ideal transformers transfer voltages and currents without phase shift, we begin the phasor diagram with $\tilde{E}_2/N_2$ as reference and locate $N_2\tilde{I}_2$ according to the magnitude and power factor of the load. The voltage induced by the mutual flux $\tilde{\Phi}_m$ is represented by $\tilde{E}_m$, and the leakage flux $\tilde{\Phi}_{\ell 2}$ induces a voltage which, together with $N_2\tilde{I}_2r_2$, reduces $\tilde{E}_m$ to $\tilde{E}_2/N_2$.

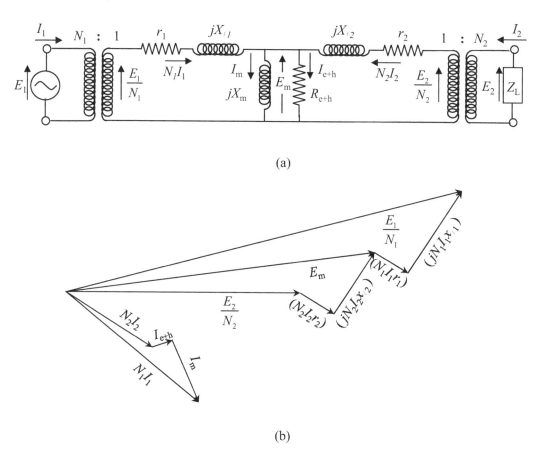

(a)

(b)

Figure 2.2.6: Two-Winding Transformer under Load

The primary $N_1\tilde{I}_1$ must be slightly larger than $N_2\tilde{I}_2$ to account for the exciting current components $\tilde{I}_m$ and $\tilde{I}_{e\&h}$. The $N_1\tilde{I}_1r_1$ drop and induced voltage in the primary winding due to the leakage flux $\tilde{\Phi}_{\ell 1}$ is in a direction to increase the voltage above $\tilde{E}_m$ to $\tilde{E}_1/N_1$. Thus, there appears to be a voltage drop from primary to secondary side:

$$\Delta\tilde{V} = N_1\tilde{I}_1(r_1 + jx_{\ell 1}) + N_2\tilde{I}_2(r_2 + jx_{\ell 2}) \qquad [V] \qquad (2.2.10)$$

proportional in magnitude to the load current and dependent upon the leakage fluxes of the transformer, the resistance of the windings, and to the exciting current required by the transformer.

The transformer designer is certainly able to calculate, with reasonable accuracy, all of the elements shown in Figure 2.2.5. On the other hand, a transformer may already be in existence for which the design data has been lost. It therefore becomes desirable to establish some practical test procedures for determining the values of the elements in the equivalent circuit.

2.2.1   Open-Circuit Test

An open-circuit test consists in energizing the one winding while allowing the other winding to remain open-circuited. For example, if we energize the primary winding and let the secondary winding to be open-circuited, then $\tilde{I}_2 = 0$ and because of the ideal $1 : N_2$ transformer, $N_2\tilde{I}_2 = 0$. Thus, according to Figure 2.2.5, the only excitation of the magnetic circuit is that due to the $N_1\tilde{I}_1$, which produces $\tilde{\Phi}_m$ in the iron-core, and $\tilde{\Phi}_{\ell 1}$ in the leakage flux path. For such a singly-excited magnetic circuit, the flux in the core is several hundred times that of the leakage flux so that the induced voltage in the primary winding is almost entirely due to the flux in the core, and the $N_1\tilde{I}_1 r_1$ and $j\omega\wp_{\ell 1}N_1\tilde{I}_1$ voltages can be assumed to be negligible.

With these reasonable assumptions, the effective equivalent network for a transformer under open-circuit test is shown in Figure 2.2.7. The open-circuit test permits a practical determination of the exciting current required, usually *at rated voltage*. An appropriate power factor measurement permits identification of the hysteresis loss and eddy current loss components and the magnetizing current component. These are conventionally expressed in terms of $R_{e\&h}$ and the magnetizing reactance $X_m = \omega\wp_m$.

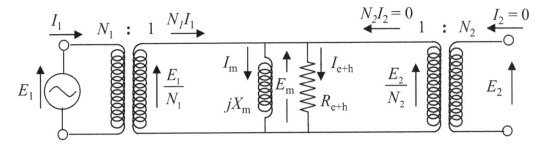

Figure 2.2.7: Practical Open-Circuit Equivalent Network

> The open-circuit test permits a practical evaluation of the exciting current of the transformer. Separation into core loss and magnetizing current components can also be obtained.

It is furthermore apparent from Figure 2.2.7 that:

> The open-circuit test permits the practical evaluation of the turns ratio of the transformer.

By transferring the exciting branch impedance to the primary side of the $N_1 : 1$ ideal transformer we obtain the circuit shown in Figure 2.2.8(a). If the open-circuit test had been performed by energizing the secondary winding and opening the primary winding, a similar transfer of exciting impedance to the secondary side results in the circuit shown in Figure 2.2.8(b).

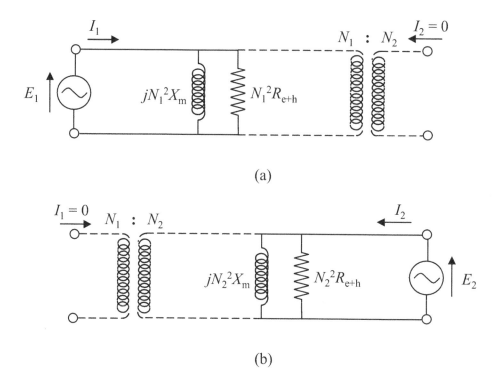

(a)

(b)

Figure 2.2.8: Exciting Branch Impedance: (a) As Seen from the Primary, (b) As Seen from the Secondary

If we let:

$$\frac{\tilde{E}_1}{\tilde{I}_1} = {}_1\tilde{Z}_e \qquad [\Omega] \tag{2.2.11}$$

$$\frac{\tilde{E}_2}{\tilde{I}_2} = {}_2\tilde{Z}_e \qquad [\Omega] \tag{2.2.12}$$

represent the exciting branch impedance measured at the primary and secondary windings respectively, we observe that:

$$\frac{\tilde{Z}_{e_1}}{\tilde{Z}_{e_2}} = \left(\frac{N_1}{N_2}\right)^2 \tag{2.2.13}$$

Therefore, it becomes necessary to specify, in an open-circuit test, which winding is energized when the exciting impedance is *expressed in ohms*. We say that ${}_1\tilde{Z}_e$ is the exciting impedance referred to (or seen from) the primary winding, and ${}_2\tilde{Z}_e$ as referred to (or seen from) the secondary winding.

We shall shortly introduce the concept of per unit quantities whereby the ideal $N_1 : 1$ and $1 : N_2$ transformers can be eliminated (generally) so that the per unit (dimensionless) exciting impedance is numerically the same when viewed from either winding.

If the open-circuit test is performed for a range of applied voltages up to, and perhaps a bit beyond rated voltage, the transformer saturation curve shown in Figure 2.2.9 is obtained. Such test data provides valuable information to the designer. A significant deviation of test results from predictions might indicate changes in quality of core material, changes in processing core material, deviations from expected space factor, excessive air gaps at the joints, etc.

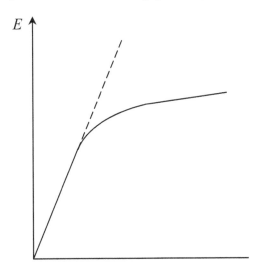

Figure 2.2.9: Transformer Open-Circuit Saturation Curve

In order to optimize the use of the core material, the designer will attempt to employ as high a flux density B_{max} as possible. On the saturation curve shown in Figure 2.2.9, rated voltage will ordinarily be near the knee of the saturation curve, just somewhat above where the curve deviates from the straight line. Operation much above this point causes an excessive increase in required excitation current and core losses. Since the transformer ordinarily remains continuously excited, whether it carries a significant load or not, these no-load losses represent a fixed cost and must be kept reasonably small. Furthermore, mechanical magnetostrictive forces in steel laminations give rise to audible noise and mechanical vibrations, and these become more severe with increased flux density.

2.2.2 Short-Circuit Test

A short-circuit test consists in energizing one winding (at much less than rated voltage, but at rated current) while allowing the other winding to be short-circuited. For reasons already cited in Section 2.2.1, both $R_{e\&h}$ and $j\omega\wp_m$ are very much greater than $j\omega\wp_{\ell1}$ and $j\omega\wp_{\ell2}$ so that the effective equivalent circuit reduces to that shown in Figure 2.2.10 for the secondary winding short-circuited. This is equivalent to recognizing that the exciting current is such a negligible portion of the transformer (short-circuit) current that the magnetizing impedance branch may be taken to be infinitely large (open-circuited). This is no more than recognizing that the permeance of the iron-core is very large.

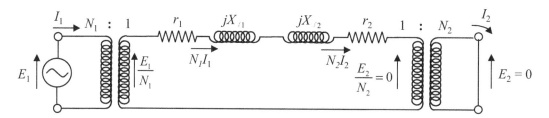

Figure 2.2.10: Transformer Short-Circuit Equivalent Network

With the *secondary* winding short-circuited we have:

$$\frac{\tilde{E}_1}{\tilde{I}_1} = N_1^2[(r_1 + r_2) + j\omega(\wp_{\ell1} + \wp_{\ell2})] = {}_1\tilde{Z}_{s-c} \tag{2.2.14}$$

or with the *primary* winding short-circuited

$$\frac{\tilde{E}_2}{\tilde{I}_2} = N_2^2[(r_1 + r_2) + j\omega(\wp_{\ell1} + \wp_{\ell2})] = {}_2\tilde{Z}_{s-c} \tag{2.2.15}$$

so that:

$$\frac{{}_1\tilde{Z}_{s-c}}{{}_2\tilde{Z}_{s-c}} = \left(\frac{N_1}{N_2}\right)^2 \tag{2.2.16}$$

Since the power transformer designer attempts to minimize the winding losses, r_1 and r_2 are generally negligible compared to the leakage reactances.

2.2.3 Approximate Equivalent Network for Two-Winding Transformers

We have demonstrated that the performance of a practical two-winding power transformer, operating at conventional power system frequencies, can be completely represented by the exact equivalent circuit shown in Figure 2.2.5. We have also established how an open-circuit test and short-circuit test enable us to determine the numerical values for all of the branch elements in the network with one exception, namely that only the sum $(\wp_{\ell1} + \wp_{\ell2})$ is determinable, not the individual elements. This presents no practical difficulties, however.

We refer again to Figure 2.2.6 representing a two-winding transformer under load. In order to better evaluate the influence of various loading conditions upon the transformer, some typical numerical values (relative) for a practical power transformer are illustrated in Figure 2.2.11, where the load impedance has been reflected through $1 : N_2$ ideal transformer for the sake of convenience. The load impedance, taken as $Z_L = 1.0$, represents in relative magnitude a load equal to rated kVA of the transformer.

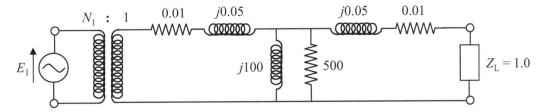

Figure 2.2.11: Two-Winding Transformer under Load

Now it is very apparent that for a transformer loadings from full load down to even 1/5 load ($Z_L = 5.0$), the current required for exciting the transformer is such a small portion of the load current that it may be neglected in the analysis. Therefore, it is not surprising that the commonly accepted approximate equivalent network is simply a series reactance directly related to the leakage flux of the transformer. The resistance losses in the winding are usually quite minimal.

Since this is essentially the same sort of conclusion that we arrived at in the discussion of the short-circuit test, the element $X_\ell = X_{\ell 1} + X_{\ell 2}$ is sometimes called the *leakage reactance* of the transformer, and it is also called the *short-circuit reactance* of the transformer. We see quite clearly now why the short-circuit test is perfectly satisfactory for determining the values of the practical series reactance equivalent network and that it is not necessary in this case to know the $X_{\ell 1}$ and $X_{\ell 2}$ components explicitly. Figure 2.2.12 illustrates the approximate equivalent circuit of the transformer.

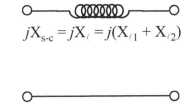

$$jX_{s-c} = jX_\ell = j(X_{\ell 1} + X_{\ell 2})$$

Figure 2.2.12: Approximate Equivalent Circuit

2.3 Per Unit Quantities

In power system analysis involving various voltage levels, it has been found more convenient to employ relative numerical values rather than actual volts and amperes. Thus, a statement that a given transformer is carrying 80% or .80 per unit of its rating is often more enlightening than the actual kVA being carried. More important, when transformer characteristics are expressed in terms of per unit of the transformer rated voltage, current, or kVA, the numerical per unit values for, say, short-circuit reactance, do not vary considerably over a large range of transformer ratings and sizes. In any numerical problem, worked out in terms of per unit values, we can ascertain at a glance whether an error has been committed in the analysis simply by knowing the typical per unit values of the network parameters.

The method is really quite simple. We select a system of base or reference values for kVA, voltage, amperes, ohms, etc., and then take the actual kVA, volts, amperes, ohms, etc., given or appearing in a given situation and divide them by the appropriate base quantities to obtain dimensionless per unit quantities. In principle it makes little difference what numerical values are selected for the reference or base system of units. In practice, however, a commonly accepted procedure has evolved over the years and we shall align ourselves with this practice.

In selecting a reference or base system, we wish to be able to employ the same laws of circuit analysis, without exception, in dealing with per unit values as we are accustomed to use in dealing with actual physical numbers. We first select a base kVA. This is a simple and straightforward matter when we are dealing with only a single piece of equipment, such as a transformer in our case. We logically select the rated kVA of our transformer as kVA base. For our introductory purposes here we shall assume that the transformer has only a single rated kVA. We postpone until a later point how to deal with transformers that may have two or three rated kVAs depending upon several stages of forced cooling that may be available.

The reader may perhaps wonder what to do when the study concerns an integrated power system involving electrical equipment spanning a large range of kVA ratings. The important point is that only a single base kVA may be selected, which will then apply to every portion of the entire power system. Exactly what numerical value is selected is, in that instance, quite arbitrary. Invariably a nice round number like 500 MVA or 1000 MVA will be selected depending upon the largest generating unit in the system and/or depending upon whether the study pertains to nominal operating conditions or whether the study involves short circuits.

We next select a voltage base. In the strictest sense we select a voltage base that corresponds to the rated voltage of one of the windings (any winding) of a power transformer. Subsequently, and in single-phase systems, we will select all other voltage bases according to the turns ratios of the transformer windings. We shall see shortly why this is advantageous.

Thereafter, all other base quantities for current, impedance, etc., are no longer free for selection but must be calculated according to the conventional laws of electric circuit analysis. The following example will help to clarify the procedure.

Example 2.3.1

Given a single-phase transformer with the following rating:

$$150 \ kVA, \qquad 2400V/240V, \qquad 60 \ Hz$$

and with short-circuit reactances:

$$_H X_{s-c} = 1.017 \ [\Omega] \qquad _L X_{s-c} = 0.01017 \ [\Omega]$$

We first *select*:

$$kVA_{base} = kVA_{rated} = 150 kVA$$

and we observe that this base is common to both windings. Next, we *select* base voltages equal to the rated voltages of the windings, recognizing that in the case of conventional two-winding power transformers, the rated winding voltages are related to the winding turns ratios as defined by Equation 2.2.4. Thus:

$$kV_{H-base} = 2.4kV \qquad kV_{L-base} = 0.24kV$$

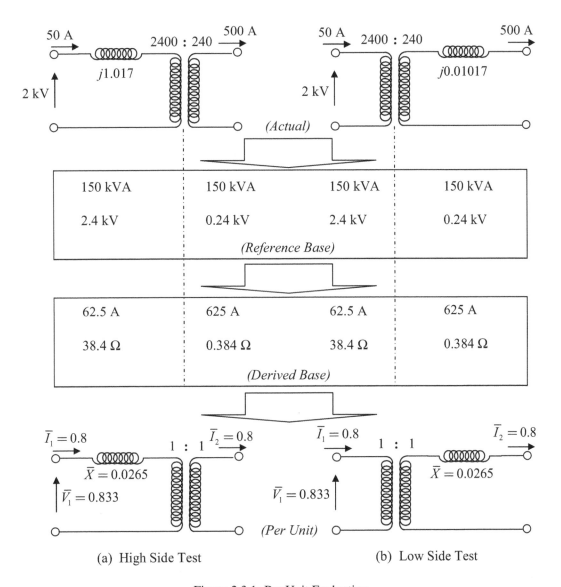

Figure 2.3.1: Per Unit Evaluation

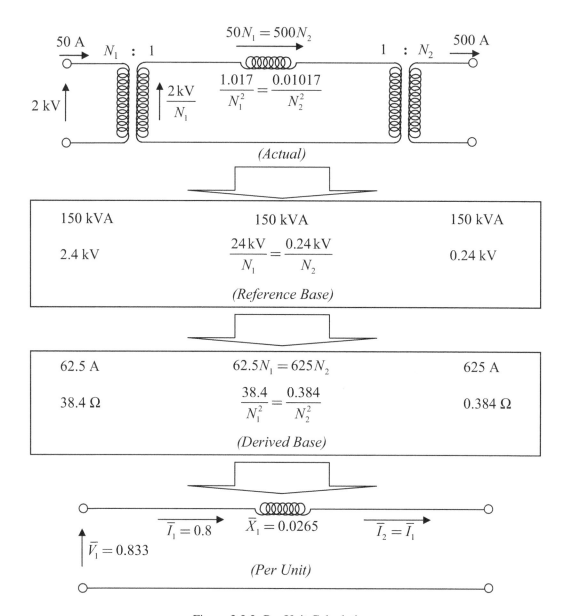

Figure 2.3.2: Per Unit Calculations

We then *calculate* the other base quantities:

$$I_{H-base} = \frac{kVA_{base}}{kV_{H-base}} = \frac{150}{2.4} \qquad I_{L-base} = \frac{kVA_{base}}{kV_{L-base}} = \frac{150}{0.24}$$
$$= 62.5\ [A] \qquad\qquad = 625\ [A]$$
$$Z_{H-base} = \frac{(kV_{H-base})^2}{MVA_{base}} \qquad Z_{L-base} = \frac{(kV_{L-base})^2}{MVA_{base}}$$
$$= 38.4\ [\Omega] \qquad\qquad = 0.384\ [\Omega]$$

Now let the transformer be energized at 2 kV on the high-voltage winding and carry a current of 50 amperes in the high-voltage winding. It is suggested that a set of three diagrams be drawn as illustrated in Figure 2.3.1. Figure 2.3.1(a) should be a complete network diagram of the actual system showing the location of all impedance elements and the ideal transformers. Figure 2.3.1(c) should be exactly the same except that actual physical quantities for volts, amps, and impedance should be entered in Figure 2.3.1(a), while corresponding per unit values will be entered in Figure 2.3.2.

Figure 2.3.1(b) shows no impedance elements, but merely lists the base quantities for volts, amperes, etc., appropriate to an isolated portion of the system.

The per unit values are then obtained by dividing actual system quantities by the appropriate base quantities and entered into the Figure 2.3.2. The actual turns ratio of the transformer is divided by the turns ratio of the base system to obtain the turns ratio of the ideal transformer in the per unit diagram. It should be apparent that selecting voltage bases according to the turns ratio of the actual transformer results in a 1:1 ideal transformer in the per unit diagram which may, of course, be replaced by a direct connection unless isolation must be retained for other reasons.

From the per unit diagram we readily see that the transformer is energized at only 83.3% of its rated voltage and the primary is carrying only 80% of its rated current.

It is left for the reader to verify that the actual impedances may be given with reference to (as seen from) either winding without affecting the resultant per unit values.

2.4   Transformer Polarity Designation

In Figure 2.4.1(a), it is apparent that both windings are wound in the same direction on a common core. The voltages induced in the windings due to the mutual flux Φ_m will produce voltages of like polarity at the end of the windings indicated by the dots or the polarity marks. In Figure 2.4.1(b) it is not quite as apparent at first glance, but a moment's reflection identifies the ends of the two windings that have a common polarity of voltage induced by the mutual flux, and these are designated with the polarity marks. Figure 2.4.1(c) is the same as Figure 2.4.1(b) as far as the polarity marks are concerned.

With the aid of polarity marks, a phasor diagram relating all the voltages and currents can be constructed, much in the same manner as the developed in Figure 2.2.6, in a rather straightforward manner for the arrangements shown in Figures 2.4.1(a) and (b), particularly since the sense of V_2 and I_2 are correctly indicated for output quantities.

If the output voltage and current happen to be defined as indicated in Figure 2.4.1(c), and if this tends to cause confusion, a simple approach is to sketch the phasor diagram for the conventional designations [as shown in Figure 2.4.1(b)], and then to superimpose phasors $V_s = -V_2$ and $I_s = -I_2$ on the phasor diagram.

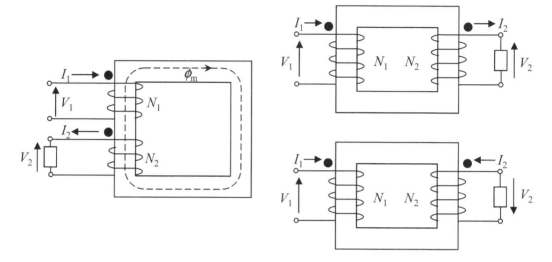

Figure 2.4.1: Transformer Polarity Designations

2.5   Transformers with Taps

It is often desirable to provide taps in the primary and/or secondary windings of a transformer in order to obtain variable output voltages. Figure 2.5.1(a) shows a transformer with taps in the secondary winding. Depending upon the construction of the transformer windings, tapping out sections of windings may influence the net leakage flux by introducing cross-flux leakage. For most practical calculations, this effect is minor and can usually be neglected. Thus, when we assume $\wp_{\ell 1}$ and $\wp_{\ell 2}$ to remain constant for all tap positions, then X_ℓ ohms referred to the primary winding (in this case) remains constant. Since the base quantities on the primary winding side are not influenced by the changes in tap position on the secondary side, it is clear that $\bar{X}_{s-c}$ remains constant as calculated on the primary side. The reader is encouraged to verify that the change in base quantities on the secondary side as a result of tap changes will be such (provided that the base impedance is readjusted for each tap position change) that, together with the corresponding changes in $_2X_\ell$ ohms reflected to the secondary winding, per unit $\bar{X}_{s-c}$ also remains the same when calculated on the secondary side.

> The per unit values in the per unit equivalent circuit of a transformer remain unchanged, regardless of changes in tap position, provided that corresponding changes in base impedance are introduced.

2.5.1   Tap-Changing Transformer Represented in Terms of Constant Base Impedance

In power system studies, where the transformer is only one of many components, it is generally undesirable to vary system base impedance with changes in tap position. To do so would necessitate varying all per unit system line impedance values with changes in tap position of a transformer.

Figure 2.5.2 illustrates an alternate, but preferable, method in which the base impedance is held constant (for a nominal tap position). Consequently an ideal transformer is required in the per unit equivalent network. This has the advantage that all network (fixed) impedances $\bar{Z}_s$ remain constant in per unit. In addition, if the transformer short-circuit reactance is referred to the nontap side of the

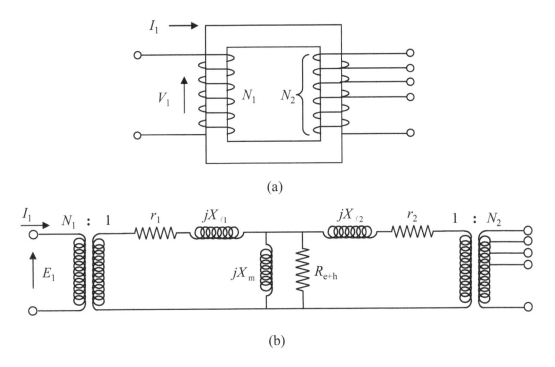

Figure 2.5.1: Transformer with Provisions for Taps

transformer, its value in per unit will also remain constant despite variations in the transformer tap position.

Occasionally (but very rarely) it may be necessary to deviate from the most convenient per unit representation (Figure 2.5.2 of a tap-changing transformer. If necessary, however, Figure 2.5.3 illustrates some alternative per unit representations with the following characteristics:

(a) Raise and lower is similar, $\bar{X}_{s-c}$ remains constant

(b) Raise and lower is similar, $\bar{X}_{s-c}$ is variable

(c) Raise and lower is inverse, $\bar{X}_{s-c}$ remains constant

(d) Raise and lower is inverse, $\bar{X}_{s-c}$ is variable

2.5.2   Closed-Loop Circuits with Remanent Turns Ratio

According to Section 2.3, voltage bases should be established on the basis of transformer-winding turns ratio. As a consequence, the equivalent transformer in the per unit equivalent network will have a 1:1 turns ratio.

Upon occasion this rule must be violated. Consider the double-circuit line (loop circuit) illustrated in Figure 2.5.4. Perhaps one line has different conductors than the other, and/or it may be of greater overall length. This may require, as shown, one transformer, with a 13.2/110 turns ratio while the other transformer has a turns ratio of 13.2/100. If the turns ratio rule is followed, beginning at the generator, for example, a clockwise path would require that the kV base for "b" would be 110 kV, while a counterclockwise path would require a kV base for "a" to be 100 kV, clearly an incompatibility. It becomes apparent that an ideal transformer will be required between a-b in the per unit diagram in order to resolve this discrepancy. Whether this should be a step-up or step-down

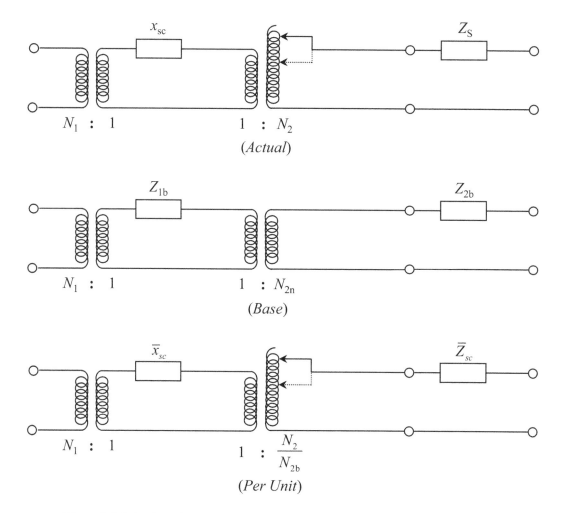

Figure 2.5.2: Preferred per Unit Equivalent Network for Tap-Changing Transformer

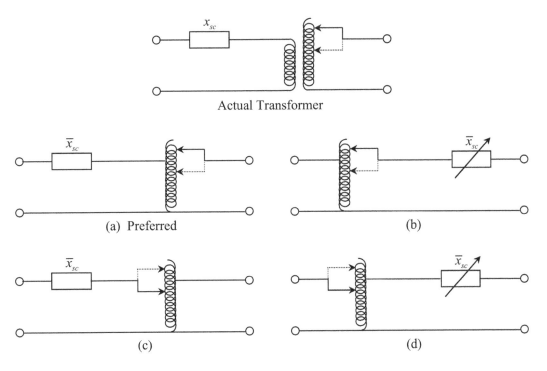

Actual Transformer

(a) Preferred

(b)

(c)

(d)

Figure 2.5.3: Alternatives to Figure 2.5.2

ideal transformer in the per unit network often causes considerable confusion...but this need not be so.

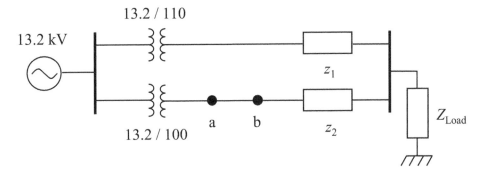

Figure 2.5.4: Loop Circuit with Remanent Turns Ratio

The matter is most easily resolved by inserting an ideal 1:1 (or 110/100) transformer in the actual circuit between a-b. Then, by properly selecting base quantities as shown in Figure 2.5.5, the appropriate transformer for the per unit network is determined directly. The transformer may be shifted around the loop to any most favorable location without affecting the results. It will become apparent in the next section that the 1.1/1 ideal transformer required in the per unit diagram may equally well be represented by an ideal 1.1/1 autotransformer as shown in Figure 2.5.5.

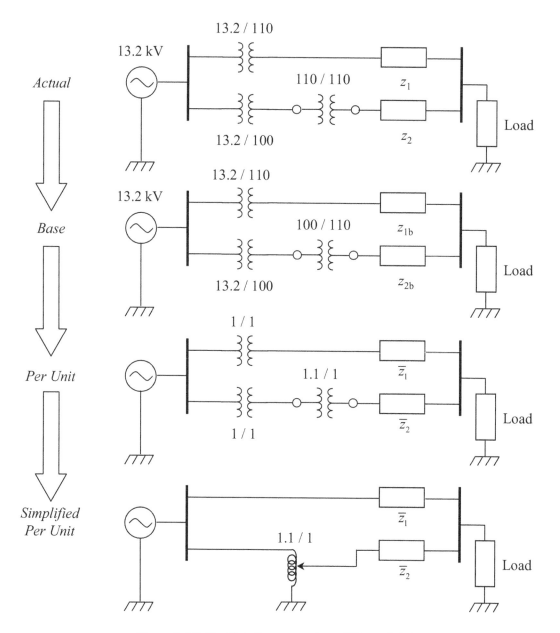

Figure 2.5.5: Per Unit Representation of Figure 2.5.4

2.6   Autotransformers

An ordinary two-winding transformer may be reconnected as shown in Figure 2.6.1(a). Such an arrangement is called an *autotransformer*. Observe that the primary and secondary circuits are no longer conductively isolated. The autotransformer has some properties that make it desirable in certain applications. For example, let V_1, V_2, I_1, and I_2 be rated values of voltage and current for the individual windings so that the volt-ampere rating of the original transformer is $V_1 I_1 = V_2 I_2$.

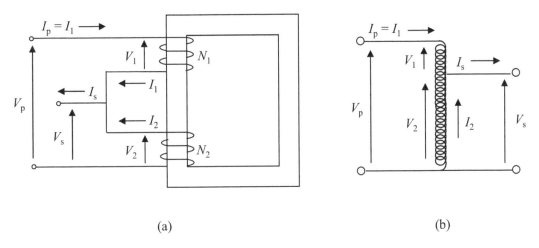

(a)   (b)

Figure 2.6.1: The Autotransformer

With the connection shown in Figure 2.6.1(a), the *input* to the transformer is now:

$$V_P I_P = (V_1 + V_2) I_1 = V_1 I_1 \left(1 + \frac{N_2}{N_1} \right) \tag{2.6.1}$$

and the *output* is:

$$V_S I_S = V_2 (I_1 + I_2) = V_1 I_1 \left(1 + \frac{N_2}{N_1} \right) \tag{2.6.2}$$

Thus, when an ordinary two-winding transformer is reconnected as an autotransformer, the kVA transfer (output) capability is increased by the ratio $(1 + N_2/N_1)$ without exceeding the rating of either winding. This is explained by the fact that only $V_1 I_1 = V_2 I_2$ is transferred from primary to secondary electromagnetically by transformer action, and the remaining $V_2 I_1$ is due to primary current proceeding conductively into the load directly.

Figure 2.6.2(a) illustrates the approximate equivalent network for the autotransformer, and in Figure 2.6.2(b) the leakage reactance has been referred to the high voltage side. It is apparent that, as seen from the high voltage side, the leakage reactance in ohms of the autotransformer is the same as that for the original two-winding transformer. However, if $\bar{X}_{s-c}$ represents the per unit short-circuit reactance of the two-winding transformer, then on the same $V_1 I_1 = V_2 I_2$ base, the per unit reactance of the autotransformer is now:

$$\bar{X} = \bar{X}_{s-c} \left(\frac{V_1}{V_P} \right)^2 = \bar{X}_{s-c} \left(\frac{V_1}{V_1 + V_2} \right)^2 = \frac{\bar{X}_{s-c}}{\left(1 + \frac{N_2}{N_1} \right)^2} \tag{2.6.3}$$

and furthermore, recognizing the greater kVA output capability of the autotransformer and employing this output kVA as the new kVA base, we have:

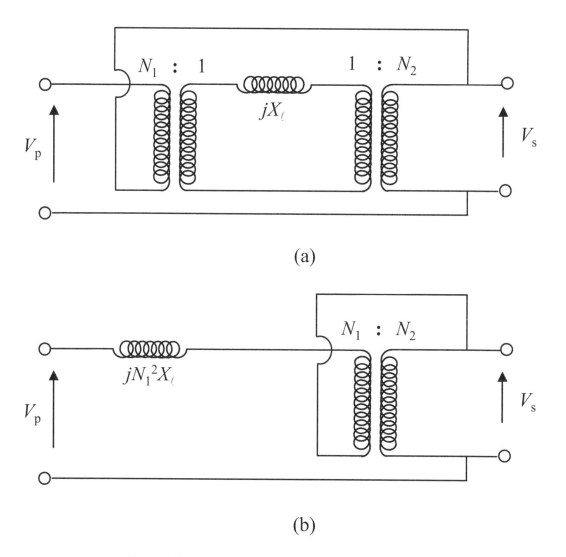

(a)

(b)

Figure 2.6.2: Equivalent Network for the Autotransformer

$$\bar{X} = \frac{\bar{X}_{s-c}}{\left(1 + \frac{N_2}{N_1}\right)} \tag{2.6.4}$$

The *turns ratio* of the autotransformer is :

$$N = \frac{N_s}{N_P} = \frac{N_2}{N_1 + N_2} \tag{2.6.5}$$

and we define the *coratio* of the autotransformer as:

$$C.R. = 1 - N = \frac{N_1}{N_1 + N_2} \tag{2.6.6}$$

In Figure 2.6.1(b), winding N_1 is referred to as the *series* winding, and winding N_2 is called the *common* winding, and the two taken together as the *series-common* winding.

When the turns ratio of the autotransformer is near unity, the coratio is very small. Consequently the per unit short-circuit reactance of the autotransformer is very small compared to that of a two-winding transformer. Therefore, in particular applications where it is desired to transform to a moderately higher or lower voltage level, the autotransformer is employed because of its relatively low short-circuit reactance. On the other hand, care must be taken to ensure that excessive short-circuit currents do not appear in the windings under fault conditions.

2.7 Multiwinding Transformers

In a two-winding transformer it is conceptually convenient to define the total flux linking a winding in terms of mutual and leakage flux components, as reflected in Equation 2.2.8, whereby the leakage flux component is defined as that part of the total flux, which links only one of the windings and not the other. The permeance of the leakage flux path lies between the pair of windings.

In a multiwinding transformer the concept of leakage flux components associated with a *single* winding is no longer applicable. For example, in a concentric three-winding construction illustrated in Figure 2.7.1, the leakage flux component between the inside winding (closest to the core) and the outside winding must necessarily link the intermediate winding and therefore no longer conforms, rigorously, to our definition of a leakage flux component.

In order to develop an equivalent circuit for a multiwinding transformer, we are forced to make some concessions. Attempting to construct a complete equivalent circuit with an allocation of magnetizing impedances amongst the different windings valid for all operating conditions is essentially impossible. We therefore ignore all magnetizing impedances and develop an equivalent circuit, consisting of a number of lumped impedance elements, valid for normal load and short-circuit conditions (see Section 2.2.3).

The equivalent network must have as many terminal pairs as there are pairs of windings. In most cases this requires n independent terminals plus one reference terminal. Also, in order to avoid the necessity of isolation transformers at each terminal pair reflecting the actual turns ratios amongst the windings, we shall require that all short-circuit reactances between *pairs* of transformer windings be expressed in terms of *per unit*.

For an n-winding transformer, n(n-1)/2 significantly different short-circuit tests between pairs of windings are possible. As many equations, and equivalent network elements, are required (in general) for a satisfactory equivalent network. Evaluation of the elements in an equivalent network is a rather tedious process, especially when the transformer has more than four windings. The procedure will be presented in detail for a three-winding transformer.

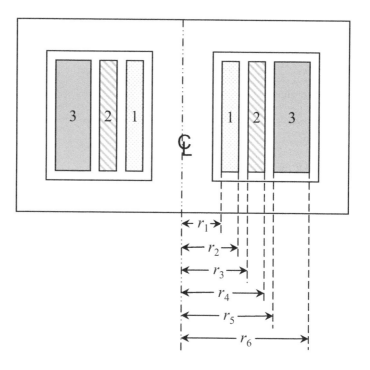

Figure 2.7.1: Three-Winding Transformer with Concentric Windings

2.7.1   Three-Winding Transformer

For a three-winding transformer we require an equivalent network that has three (numbered) terminals and a reference terminal.

We are able to perform three significantly different short-circuit tests, between pairs of windings, on a three-winding transformer, to obtain $\bar{X}_{1-2}$, $\bar{X}_{2-3}$, and $\bar{X}_{3-1}$. Alternatively, if we know the geometry of the winding arrangement as given in Figure 2.7.1, we may evaluate the associated permeances as follows:

$$\wp_{1-2} = k_{12}[\frac{(r_2 - r_1)}{3} + (r_3 - r_2) + \frac{(r_4 - r_3)}{3}]$$

$$\wp_{2-3} = k_{23}[\frac{(r_4 - r_3)}{3} + (r_5 - r_4) + \frac{(r_6 - r_5)}{3}]$$

$$\wp_{3-1} = k_{31}[\frac{(r_2 - r_1)}{3} + (r_5 - r_2) + \frac{(r_6 - r_5)}{3}] \qquad (2.7.1)$$

The most common equivalent network representation for a three-winding transformer is the wye connection of network elements shown in Figure 2.7.2, although an equivalent delta connection could equally well be selected.

The network elements must satisfy the equations corresponding to the three short-circuit tests as follows:

$$\begin{bmatrix} \bar{X}_{1-2} \\ \bar{X}_{2-3} \\ \bar{X}_{3-1} \end{bmatrix} = \begin{bmatrix} 1 & 1 & 0 \\ 0 & 1 & 1 \\ 1 & 0 & 1 \end{bmatrix} \begin{bmatrix} X_1 \\ X_2 \\ X_3 \end{bmatrix} \qquad (2.7.2)$$

with the result:

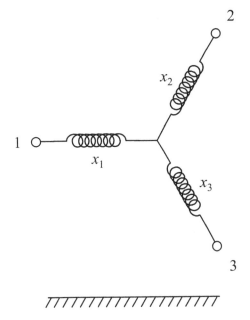

Figure 2.7.2: Equivalent Network for a Three-Winding Transformer

$$\begin{bmatrix} X_1 \\ X_2 \\ X_3 \end{bmatrix} = \left(\frac{1}{2}\right) \begin{bmatrix} 1 & -1 & 1 \\ 1 & 1 & -1 \\ -1 & 1 & 1 \end{bmatrix} \begin{bmatrix} \bar{X}_{1-2} \\ \bar{X}_{2-3} \\ \bar{X}_{3-1} \end{bmatrix} \qquad (2.7.3)$$

The alert reader will observe that:

$$X_u = \frac{1}{2}[\bar{X}_{u-v} + \bar{X}_{u-w} - \bar{X}_{v-w}] \qquad (2.7.4)$$

which is a sequence easily remembered.

For a three-winding transformer with concentric windings, as shown in Figure 2.7.1, we might typically have:

$$\bar{X}_{1-2} = 0.067 \qquad \bar{X}_{2-3} = 0.075 \qquad (2.7.5)$$

On the basis of Equation 2.7.1, we have:

$$\wp_{3-1} = \wp_{1-2} + \wp_{2-3} + k_{31}[\frac{(r_4 - r_3)}{3}] \qquad (2.7.6)$$

which shows that $\wp_{3-1}$ is always somewhat bigger than ($\wp_{1-2} + \wp_{2-3}$). Thus, we may have:

$$\wp_{3-1} = 0.155 \qquad (2.7.7)$$

Then, on the basis of Equation 2.7.3:

$$X_1 = 0.0735 \qquad X_2 = -0.0065 \qquad X_3 = 0.0815 \qquad (2.7.8)$$

That X_2 is a negative number emphasizes the fact that the elements in the equivalent network in Figure 2.7.2 are *not leakage reactance* components associated with any given winding (or any other physical reality), but only numerically reflect the total performance of the three-winding transformer under near-rated load conditions.

2.7.2   Four-Winding Transformer

An equivalent network for a single-phase four-winding transformer, with magnetizing current neglected, we must have four terminals with respect to a reference, and at least (in general) six significantly different circuit elements. The equivalent network shown in Figure 2.7.3, proposed by F.M. Starr, has eight elements although only six are significantly different. For analog representation purposes, this form of network has the advantage of being free from negative elements.

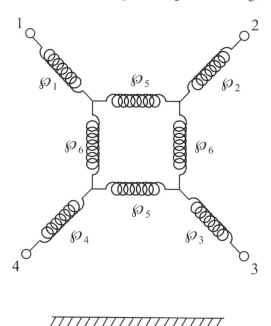

Figure 2.7.3: Equivalent Network for a Four-Winding Transformer

Relating equivalent network performance to short-circuit tests we have:

$$\bar{P}_{1-2} = \bar{P}_1 + \bar{P}_2 + p \qquad \bar{P}_{3-4} = \bar{P}_3 + \bar{P}_4 + p$$

$$\bar{P}_{2-3} = \bar{P}_2 + \bar{P}_3 + q \qquad \bar{P}_{4-1} = \bar{P}_4 + \bar{P}_1 + q$$

$$\bar{P}_{3-1} = \bar{P}_3 + \bar{P}_1 + \frac{1}{2}(\bar{P}_5 + \bar{P}_6)$$

$$\bar{P}_{4-2} = \bar{P}_4 + \bar{P}_2 + \frac{1}{2}(\bar{P}_5 + \bar{P}_6) \tag{2.7.9}$$

with:

$$p = \frac{1}{2}\left(\frac{\bar{P}_5^2 + 2\bar{P}_5\bar{P}_6}{\bar{P}_5 + \bar{P}_6}\right) \qquad q = \frac{1}{2}\left(\frac{\bar{P}_6^2 + 2\bar{P}_5\bar{P}_6}{\bar{P}_5 + \bar{P}_6}\right) \tag{2.7.10}$$

Now:

$$\bar{P}_{1-3} + \bar{P}_{2-4} - \bar{P}_{2-3} - \bar{P}_{4-1} = \bar{P}_5 + \bar{P}_6 - 2q = \frac{\bar{P}_5^2}{\bar{P}_5 + \bar{P}_6} = K_1 \tag{2.7.11}$$

$$\bar{P}_{1-3} + \bar{P}_{2-4} - \bar{P}_{1-2} - \bar{P}_{3-4} = \bar{P}_5 + \bar{P}_6 - 2p = \frac{\bar{P}_6^2}{\bar{P}_5 + \bar{P}_6} = K_2 \qquad (2.7.12)$$

so that:

$$\bar{P}_5 = \sqrt{K_1 K_2} + K_1 \qquad \bar{P}_6 = \sqrt{K_1 K_2} + K_2 \qquad (2.7.13)$$

The remaining four elements are found to be:

$$\bar{P}_1 = \frac{1}{2}(\bar{P}_{1-2} + \bar{P}_{1-4} - \bar{P}_{2-4} - \sqrt{K_1 K_2})$$

$$\bar{P}_2 = \frac{1}{2}(\bar{P}_{1-2} + \bar{P}_{2-3} - \bar{P}_{1-3} - \sqrt{K_1 K_2})$$

$$\bar{P}_3 = \frac{1}{2}(\bar{P}_{2-3} + \bar{P}_{3-4} - \bar{P}_{2-4} - \sqrt{K_1 K_2})$$

$$\bar{P}_4 = \frac{1}{2}(\bar{P}_{1-4} + \bar{P}_{3-4} - \bar{P}_{1-3} - \sqrt{K_1 K_2}) \qquad (2.7.14)$$

2.7.3 Five-Winding Transformer

In a five-winding transformer there are ten possible short-circuit tests between windings, taken two at a time with the other windings open. An equivalent network for a five-winding transformer, with magnetizing current neglected, must therefore have five terminals (with respect to a reference) and (in general) ten independent network elements.

The equivalent circuit illustrated in Figure 2.7.4, developed by L.C. Aicher, Jr., satisfies these conditions. The results are:

$$
\begin{aligned}
&\underline{with} &&\underline{then}\\
L &= \bar{P}_{1-3} + \bar{P}_{2-4} - \bar{P}_{1-4} - \bar{P}_{2-3} & \bar{P}_6 &= \frac{L}{2}\left(\frac{MQ}{PS} + \frac{Q}{S} + 1\right) + \frac{M}{2}\left(\frac{Q}{P} + 1\right)\\
M &= \bar{P}_{1-4} + \bar{P}_{2-5} - \bar{P}_{1-5} - \bar{P}_{2-4} & \bar{P}_7 &= \frac{P}{M}\bar{P}_6\\
P &= \bar{P}_{2-4} + \bar{P}_{3-5} - \bar{P}_{3-4} - \bar{P}_{2-5} & \bar{P}_8 &= \frac{S}{Q}\bar{P}_7\\
Q &= \bar{P}_{1-3} + \bar{P}_{2-5} - \bar{P}_{1-2} - \bar{P}_{3-5} & \bar{P}_9 &= \frac{M}{L}\bar{P}_8\\
S &= \bar{P}_{1-4} + \bar{P}_{3-5} - \bar{P}_{1-3} - \bar{P}_{4-5} & \bar{P}_{10} &= \frac{S}{L}\bar{P}_6
\end{aligned}
\qquad (2.7.15)
$$

$$
\begin{aligned}
&\underline{with} &&\underline{then}\\
\bar{P}_t &= \bar{P}_6 + \bar{P}_7 + \bar{P}_8 + \bar{P}_9 + \bar{P}_{10} & \bar{P}_1 &= \frac{N}{2} - \frac{\bar{P}_6 \bar{P}_{10}}{\bar{P}_t}\\
K &= \bar{P}_{1-2} + \bar{P}_{2-3} - \bar{P}_{1-3} & \bar{P}_2 &= \frac{K}{2} - \frac{\bar{P}_6 \bar{P}_7}{\bar{P}_t}\\
N &= \bar{P}_{1-2} + \bar{P}_{1-5} - \bar{P}_{2-5} & \bar{P}_3 &= \frac{O}{2} - \frac{\bar{P}_7 \bar{P}_8}{\bar{P}_t}\\
O &= \bar{P}_{2-3} + \bar{P}_{3-4} - \bar{P}_{2-4} & \bar{P}_4 &= \frac{R}{2} - \frac{\bar{P}_8 \bar{P}_9}{\bar{P}_t}\\
R &= \bar{P}_{3-4} + \bar{P}_{4-5} - \bar{P}_{3-5} & \bar{P}_5 &= \frac{T}{2} - \frac{\bar{P}_9 \bar{P}_{10}}{\bar{P}_t}\\
T &= \bar{P}_{1-5} + \bar{P}_{4-5} - \bar{P}_{1-4}
\end{aligned}
\qquad (2.7.16)
$$

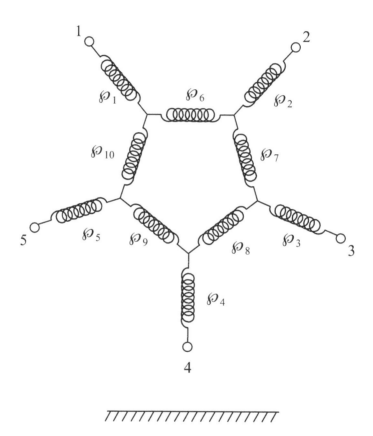

Figure 2.7.4: Equivalent Network for a Five-Winding Transformer

2.8   Magnetic Energy in Transformers

As with any electromagnetic device, there exists stored magnetic energy. For sinusoidal currents and voltages, this energy oscillates between the source and the device(vars). It is of interest to ascertain where, in the transformer, this energy resides.

For a *two-winding* transformer we have:

$$
\begin{bmatrix} \dfrac{\tilde{\Lambda}_1}{N_1} \\[2mm] \dfrac{\Lambda_2}{N_2} \end{bmatrix} = \begin{bmatrix} (\wp_{\ell 1} + \wp_m) & \wp_m \\ \wp_m & (\wp_{\ell 2} + \wp_m) \end{bmatrix} \begin{bmatrix} N_1 \tilde{I}_1 \\ N_2 \tilde{I}_2 \end{bmatrix}
\tag{2.8.1}
$$

Then, on the basis of Equation F.3.5 in Appendix F, we may write for phasor *energy:*

$$
\begin{aligned}
W_{mag} &= \frac{1}{2} \left[N\tilde{I}^* \right]^T \left[\frac{\tilde{\Lambda}}{N} \right] \\[2mm]
&= \left(\frac{1}{2} \right) \begin{bmatrix} 1 & 1 \end{bmatrix} \begin{bmatrix} N_1^2 \tilde{I}_1^* \tilde{I}_1 (\wp_{\ell 1} + \wp_m) & N_1 N_2 \tilde{I}_1^* \tilde{I}_2 (\wp_m) \\ N_2 N_1 \tilde{I}_2^* \tilde{I}_1 (\wp_m) & N_2^2 \tilde{I}_2^* \tilde{I}_2 (\wp_{\ell 2} + \wp_m) \end{bmatrix} \begin{bmatrix} 1 \\ 1 \end{bmatrix} \\[2mm]
&= \left(\frac{1}{2} \right) \{ N_1^2 \mid I_1 \mid^2 \wp_{\ell 1} + N_2^2 \mid I_2 \mid^2 \wp_{\ell 2} + (N_1 \tilde{I}_1^* \\
&\qquad + N_2 \tilde{I}_2^*)(N_1 \tilde{I}_1 + N_2 \tilde{I}_2) \wp_m \} \qquad [J]
\end{aligned}
\tag{2.8.2}
$$

Let:

$$
(N_1 \tilde{I}_1 + N_2 \tilde{I}_2) = \tilde{\epsilon}
\tag{2.8.3}
$$

$$
(N_1 \tilde{I}_1 + N_2 \tilde{I}_2) \wp_m = \tilde{\epsilon} \wp_m = \tilde{\phi}_m
\tag{2.8.4}
$$

where $\tilde{\phi}_m$ is a finite number determined by the applied voltage and Faraday's law. Then:

$$
\left(\frac{1}{2} \right) (N_1 \tilde{I}_1^* + N_2 \tilde{I}_2^*)(N_1 \tilde{I}_1 + N_2 \tilde{I}_2) \wp_m = \frac{\tilde{\epsilon}^* \tilde{\epsilon}}{2} \wp_m = \frac{\tilde{\epsilon}^* \tilde{\Phi}_m}{2} \to 0
\tag{2.8.5}
$$

Thus:

$$
\begin{aligned}
W_{mag} &= \left(\frac{1}{2} \right) \{ N_1^2 \left| I_1 \right|^2 \wp_{\ell 1} + N_2^2 \left| I_2 \right|^2 \wp_{\ell 2} \} \\[2mm]
&= -\left(\frac{1}{2} \right) (N_1 \left| I_1 \right|)(N_2 \left| I_2 \right|)(\wp_{\ell 1} + \wp_{\ell 2}) \\[2mm]
&= -\left(\frac{1}{2} \right) (N_1 \left| I_1 \right|)(N_2 \left| I_2 \right|) \wp_{1-2} \quad [J]
\end{aligned}
\tag{2.8.6}
$$

> With negligible error, the magnetic energy in a practical power transformer resides entirely in the leakage flux paths of the transformer.

In Appendix F we show that for multiwinding transformers we have:

$$
W_{mag} = -\frac{1}{4} \sum_{j=1}^{n} \sum_{k=j+1}^{n} N_j N_k \left(\tilde{I}_j^* \tilde{I}_k + \tilde{I}_j \tilde{I}_k^* \right) \wp_{j-k} \quad [J]
\tag{2.8.7}
$$

2.9 Magnetic Energy Method for Reconnected Windings

Occasionally an n-winding transformer is externally connected to form an *effective* m-winding transformer. Given the short-circuit reactances between all pairs of actual windings, it is desired to find the *effective* short-circuit reactance of the reconnected transformer. In 1936, A.N. Garin and K.K. Paluev presented a so-called "phasor power method," which is an extremely powerful method for accomplishing this goal.

The same benefits accrue by employing the "magnetic energy method" described in Section 2.8. The basis of this latter method lies in the fact that the magnetic energy residing in a transformer is the same whether viewed as the *actual* n-winding transformer or as an *effective* m-winding transformer.

We shall consider several examples of *effective two-winding* transformers. This is also the first step in dealing with effective multiwinding transformers according to the methods described in Section 2.7. Parameters for the actual n-winding transformer will be identified with *numerical* subscripts; parameters for the effective two-winding transformer will be identified by alphabetical subscripts (a,b). According to Section 2.8 we have:

$$W_{mag} = -\frac{1}{4}\sum_{j=1}^{n}\sum_{k=j+1}^{n}N_jN_k\left(\tilde{I}_j^*\tilde{I}_k + \tilde{I}_j\tilde{I}_k^*\right)\wp_{j-k} \tag{2.9.1}$$

$$= -\frac{1}{4}N_aN_b\left(\tilde{I}_a^*\tilde{I}_b + \tilde{I}_a\tilde{I}_b^*\right)\wp_{a-b}\quad [J]$$

On the basis that $N_a\tilde{I}_a + N_b\tilde{I}_b = 0$, we find that I_b is in phase with I_a. Since our reconnected transformer will always be a two-winding transformer, we shall find that all winding currents are also related to I_a according to winding turns ratios and will also therefore be in phase with I_a. Thus, we write:

$$\wp_{a-b} = \frac{\sum_{j=1}^{n}\sum_{k=j+1}^{n}(N_jI_j)(N_kI_k)\wp_{j-k}}{(N_aI_a)(N_bI_b)} \tag{2.9.2}$$

We shall demonstrate the method with a few selected examples.

2.9.1 Autotransformer

Consider an ordinary two-winding transformer reconnected as an autotransformer. Thus:

$$\wp_{a-b} = \frac{(N_1I_1)(N_2I_2)\wp_{1-2}}{(N_aI_a)(N_bI_b)} \tag{2.9.3}$$

First we observe that:

$$N_aI_a + N_bI_b = 0 \tag{2.9.4}$$

$$N_1I_1 + N_2I_2 = 0 \tag{2.9.5}$$

Then we relate terminal currents to winding currents per Figure 2.9.1:

$$I_a = I_1 \tag{2.9.6}$$

Therefore:

$$\wp_{a-b} = \left(\frac{N_1}{N_2}\right)^2\wp_{1-2} \tag{2.9.7}$$

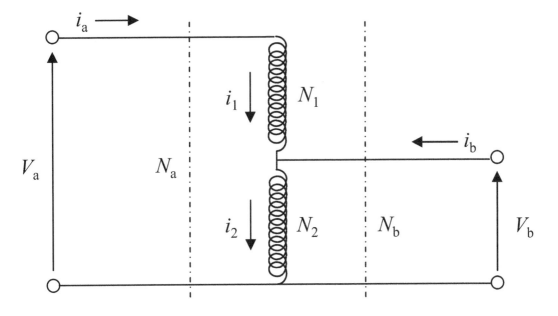

Figure 2.9.1: Autotransformer

Finally we relate terminal voltages to winding voltages:

$$\frac{V_a}{N_a} = \frac{V_b}{N_b} = \frac{V_2}{N_2} = \frac{V_1 + V_2}{N_a} = \frac{N_1 + N_2}{N_a}\frac{V_2}{N_2} \tag{2.9.8}$$

so that:

$$N_a = N_1 + N_2 \tag{2.9.9}$$

and:

$$\wp_{a-b} = \left(\frac{N_1}{N_1 + N_2}\right)^2 \wp_{1-2} \tag{2.9.10}$$

In the terms of per unit, on the original MVA and rated winding voltages bases, we may write directly (per Section F.2 in Appendix F):

$$\bar{X}_{a-b} = \left(\frac{N_1}{N_1 + N2}\right)^2 \bar{X}_{1-2} = (C.R)^2 \bar{X}_{1-2} \tag{2.9.11}$$

2.9.2   Three-Winding to Two-Winding

Consider the arrangement shown in Figure 2.9.2:

$$\wp_{a-b} = \frac{(N_1 I_1)(N_2 I_2)\wp_{1-2} + (N_2 I_2)(N_3 I_3)\wp_{2-3} + (N_3 I_3)(N_1 I_1)\wp_{3-1}}{(N_a I_a)(N_b I_b)} \tag{2.9.12}$$

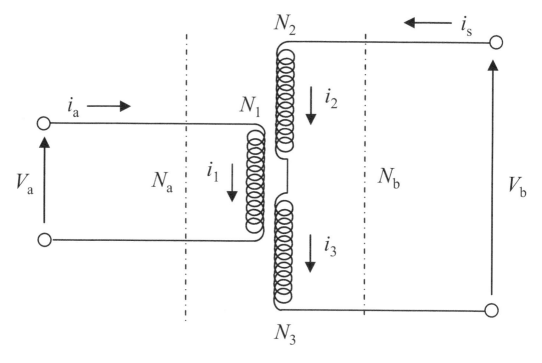

Figure 2.9.2: Three-Winding to Two-Winding Transformer

$$N_a I_a + N_b I_b = 0 \qquad \frac{V_a}{N_a} = \frac{V_b}{N_b}$$

$$N_1 I_1 + N_2 I_2 + N_3 I_3 = 0 \qquad \frac{V_1}{N_1} = \frac{V_2}{N_2} = \frac{V_3}{N_3} \qquad (2.9.13)$$

$$I_1 = I_a$$
$$I_2 = I_b = -\left(\frac{N_a}{N_b}\right) I_a$$
$$I_3 = I_b = -\left(\frac{N_a}{N_b}\right) I_a \qquad (2.9.14)$$

$$\wp_{a-b} =$$

$$\frac{1}{-(N_a)^2}\left\{-\left(\frac{N_1 N_2 N_a}{N_b}\right)\wp_{1-2} + \left(\frac{N_2 N_3 N_a^2}{N_b^2}\right)\wp_{2-3} - \left(\frac{N_3 N_1 N_a}{N_b}\right)\wp_{3-1}\right\} \qquad (2.9.15)$$

Then:

$$N_a = N_1$$

$$\frac{N_a}{N_b} = \frac{V_a}{V_b} = \frac{V_1}{V_2 + V_3} = \frac{N_1}{N_2 + N_3} \tag{2.9.16}$$

so that:

$$\wp_{a-b} = \frac{N_2}{N_2 + N_3}\wp_{1-2} - \frac{N_2 N_3}{(N_2 + N_3)^2}\wp_{2-3} + \frac{N_3}{N_2 + N_3}\wp_{3-1} \tag{2.9.17}$$

and in terms of per unit, on the original MVA and rated winding voltage bases, we may write directly (per Section F.2):

$$\bar{X}_{a-b} = \frac{N_2}{N_2 + N_3}\bar{X}_{1-2} - \frac{N_2 N_3}{(N_2 + N_3)^2}\bar{X}_{2-3} + \frac{N_3}{N_2 + N_3}\bar{X}_{3-1} \tag{2.9.18}$$

2.9.3 Transformer with Load Ratio Control

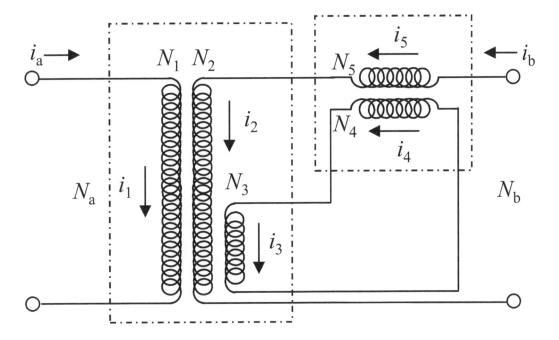

Figure 2.9.3: Transformer with Load Ratio Control

Consider the arrangement shown in Figure 2.9.3. Here we have a main power transformer with a tertiary winding which, via a second series transformer (with automatic tap-changing capability not shown), compensates for line voltage drop as the load changes. Now we have:

$$wp_{a-b} =$$

$$\frac{(N_1 I_1)(N_2 I_2)\wp_{1-2} + (N_2 I_2)(N_3 I_3)\wp_{2-3}}{(N_a I_a)(N_b I_b)}$$
$$+ \quad \frac{(N_3 I_3)(N_1 I_1)\wp_{3-1} + (N_4 I_4)(N_5 I_5)\wp_{4-5}}{(N_a I_a)(N_b I_b)} \tag{2.9.19}$$

Now:

$$N_a I_a + N_b I_b = 0 \qquad \frac{V_a}{N_a} = \frac{V_b}{N_b}$$

$$N_1 I_1 + N_2 I_2 + N_3 I_3 = 0 \qquad \frac{V_1}{N_1} = \frac{V_2}{N_2} = \frac{V_3}{N_3}$$

$$N_4 I_4 + N_5 I_5 = 0 \qquad \frac{V_4}{N_4} = \frac{V_5}{N_5} \tag{2.9.20}$$

Therefore:

$$\wp_{a-b} =$$

$$\frac{(N_1 I_1)(N_2 I_2)\wp_{1-2} + (N_2 I_2)(N_3 I_3)\wp_{2-3}}{-(N_a I_a)^2}$$
$$+ \quad \frac{(N_3 I_3)(N_1 I_1)\wp_{3-1} - (N_4 I_4)^2 \wp_{4-5}}{-(N_a I_a)^2} \tag{2.9.21}$$

Then, relating terminal and winding currents, we have:

$$I_1 = I_a$$
$$I_2 = I_5 = I_b = -\left(\frac{N_a}{N_b}\right) I_a$$
$$I_3 = I_4 = -\left(\frac{N_5}{N_4}\right) I_5 = \frac{N_5 N_a}{N_4 N_b} I_a \tag{2.9.22}$$

Therefore:

$$\wp_{a-b} = \frac{-N_1 N_2 N_4 N_a N_b \wp_{1-2} - N_2 N_3 N_5 N_a^2 \wp_{2-3}}{-N_4 (N_a N_b)^2}$$
$$+ \frac{+N_1 N_3 N_5 N_a N_b \wp_{3-1} - N_4 (N_5 N_a)^2 \wp_{4-5}}{-N_4 (N_a N_b)^2} \tag{2.9.23}$$

Then, relating terminal and winding voltages, we have:

$$V_a = V_1 \qquad N_s = N_1$$

$$V_b = \frac{N_b}{N_a} \qquad V_a = \frac{N_b}{N_1} V_1$$

$$= V_2 + V_5 = V_2 + \frac{N_5}{N_4} V_4$$

$$= V_2 - \frac{N_5}{N_4} V_3 = \frac{(N_2 - \frac{N_3 N_5}{N_4}) V_1}{N_1} \tag{2.9.24}$$

so that:

$$N_b = N_2 - \frac{N_3 N_5}{N_4} = \left(\frac{N_2 N_4 - N_3 N_5}{N_4} \right) \tag{2.9.25}$$

Let:

$$q = \frac{N_2 N_4}{N_2 N_4 - N_3 N_5} = \frac{1}{1 - \frac{N_3 N_5}{N_2 N_4}} \tag{2.9.26}$$

$$r = \frac{N_3 N_5}{N_2 N_4 - N_3 N_5} = \frac{1}{\frac{N_2 N_4}{N_3 N_5} - 1} \tag{2.9.27}$$

Hence:

$$P_{a-b} = q P_{1-2} - r P_{3-1} + q r P_{2-3} + \left(\frac{N_4}{N_3} \right)^2 r^2 P_{4-5} \tag{2.9.28}$$

This may also be expressed in terms of reactances as:

$$_a X_{a-b} = q \, _1 X_{1-2} - r \, _1 X_{3-1} + q r \, _1 X_{2-3} + \left(\frac{N_1}{N_3} \right)^2 r^2 \, _4 X_{4-5} \tag{2.9.29}$$

and in terms of per unit:

$$\bar{X}_{a-b} = q \bar{X}_{1-2} - r \bar{X}_{3-1} + q r \bar{X}_{2-3} + r^2 \bar{X}_{4-5} \left(\frac{_4 Z_{base}}{_3 Z_{base}} \right) \tag{2.9.30}$$

where $\bar{X}_{4-5}$ is presumed to have been expressed on its own $_4 Z_{base}$

2.10 Exercises

2.1 A 25 kVA transformer supplies a load of 12 kW at a power factor of 0.6 lagging. How many kVA can be added before the transformer is at rated load?

(a) When the added load is at 1.0 P.F.?

(b) When the added load is at 0.866 P.F. leading?

2.2 A 500 kVA transformer is at full rated load with an overall load P.F. of 0.6 lagging. A purely capacitive load is added to bring the net load P.F. to 0.9 lagging.

(a) How many kVAR of capacitive load are required?

(b) What percent of rated load is the transformer carrying?

2.3 Prove that the impedance of a transformer may be shown on either the primary or secondary side in the equivalent circuit if the impedance is expressed in per unit.

2.4 The following data is given for a 1kVA, 208/120 volt, single-phase transformer.

Short-Circuit Test High Voltage Side	N-Load Test Low Voltage Side
V = 10.4 [V]	V = 100 %
I = 4.81 [A]	I = 3 %
P = 20 [W]	P = 1 %

(a) For the short-circuit test, determine the voltage, current, and power in per unit.

(b) Convert the no-load test data to volts, amps, and watts on the high voltage side.

(c) Determine the per unit values of $(r_1 + r_2)$, $(X_1 + X_2)$, $(Z_1 + Z_2)$, $R_{e\&h}$, X_m, and Z_{exc}.

(d) In ohms, determine $R_{e\&h}$, X_m, and Z_{exc} on HV and LV sides.

(e) In ohms, determine $(r_1 + r_2)$, $(X_1 + X_2)$, and $(Z_1 + Z_2)$ on HV and LV sides.

2.5 In an open-circuit test of a 10 kVA, 2400/240 volt distribution transformer, the low side volts, amperes, and watts are found to be 240 [V], 0.75[A], and 72[W]. The low side is short-circuited. When 67[V] are applied to the high side, rated current of 4.17 [A] flows and the power is 146 [W]. Determine $(\bar{R}_1 + \bar{R}_2)$, $(\bar{X}_1 + \bar{X}_2)$, $R_{e\&h}$, and X_m.

2.6 A transformer has a secondary resistance of 0.25%, a secondary leakage reactance of 3.0 %, and exciting current of 3.1%, and a power factor at no load of 12%. Compute the angle between the primary and secondary currents when this transformer delivers its rated kVA at 0.8 P.F.:

(a) to an inductive load.

(b) to a capacitive load.

2.7 At no load, with 110[V], 30 Hz impressed on a transformer, the current is 1.2 [A] and the power is 53 [W], of which 14[W] is dissipated in eddy currents. What will be the current and power at no load if 220 [V], 60 Hz is impressed on the transformer?

2.8 An autotransformer, 115 kV/230 kV, and 0.08 per unit impedance on its through (output) base has an equivalent (through, output) rating of 200 MVA. What is the per unit impedance, on the rated winding base, between the two separate windings that form the autotransformer?

2.9 An autotransformer has the ratio 2 to 1. That is, the terminal B is half-way between the terminals A and C. The resistance of the entire winding is 0.10 ohms, and the resistance of the common part BC is 0.04 ohms. If the exciting current is neglected, what is the copper loss at an output of 10 [A]?

2.10 Specify the tests required for the equivalent circuit of a three-winding transformer. Using the results of the tests, derive the equations for the three impedances of the equivalent circuit referred to a common base S.

2.11 Given the single-phase, three-winding transformer shown Figure 2.10.1 with a rating of:

$$4.42MVA, \qquad V_P/V_F/V_C = 3006\ V/1169\ V/167\ V$$

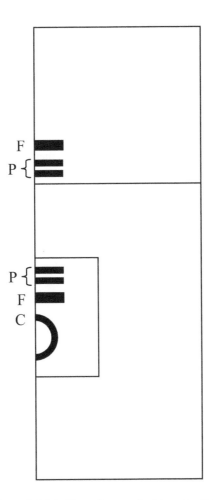

Figure 2.10.1: Transformer for Exercise 2.11

and the short-circuit reactances:

$$_P X_{C-P} = 0.250 \ [\Omega] \qquad _F X_{C-F} = 0.029 \ [\Omega] \qquad _F X_{P-F} = 0.066 \ [\Omega]$$

eqns

(a) Establish the three sets of base quantities for MVA, kV, and Z.

(b) Express the short-circuit reactances in per unit.

(c) Sketch the equivalent circuit for a three-winding transformer and evaluate $\bar{X}_1$, $\bar{X}_2$, and $\bar{X}_3$ for the equivalent circuit.

2.12 Given a single-phase, three-winding transformer with the following ratings:

Primary	14.4 kVA	480 V
Secondary	9.6 kVA	240 V
Tertiary	4.8 kVA	120 V

and with:

$$\bar{X}_{p-s} = 0.12 \qquad \bar{X}_{s-t} = 0.12 \qquad \bar{X}_{p-t} = 0.28$$

all on a common 14.4 kVA base and rated voltage bases. The windings are reconnected such that the input remains at 480 V, with the other two windings connected in a series for an output at 360 V, and an output at 120 V.

(a) Sketch the equivalent circuit and evaluate the elements for the equivalent circuit employing the magnetic energy method.

(b) What is the maximum total load that can be carried if the 360 V load and the 120 V load are of equal kVA and the same power factor? Assume all short-circuit reactances to be equal to zero.

2.13 Given the same conditions as in Exercise 2.12. Now, however, we wish to have outputs at 360 V, 120 V, and 240 V. Sketch the equivalent circuit and evaluate the elements for the equivalent circuit employing the magnetic energy method.

Chapter 3

Balanced Three-Phase Systems

In Chapter 1, the impedance matrix for a section of the three-phase transmission line was developed. For the general *unsymmetrical* and *untransposed* line, this representation is very unwieldy, particularly when it represents only a portion of a much larger system. The analysis would become much more manageable if the three phases could be treated separately with no mutual coupling between them. The problem would then be one of resolving a single-phase system, a relatively simple task. Can a three-phase system be analyzed in terms of an equivalent single-phase system? The answer is yes when:

    (a) the three-phase electrical *system* is *symmetrical* (when mutual coupling between the phases is nonexistent or, at least, when the mutual coupling between pairs of phases is identical),

    (b) and when a balanced or *symmetrical* system of source *voltages* is applied

This will form the subject of this chapter.

In other cases, where one or relatively few dissymmetries exist in the system, it is often possible to resolve the mutually coupled three-phase system into three uncoupled single-phase systems, the latter being simply interconnected at the points of dissymmetry. This can sometimes be accomplished by a change of variables. The symmetrical component transformation, and the Clarke α, β, 0 component transformation are two ways in which this can be done. These topics will be taken up in subsequent chapters.

3.1 Wye-Connected Loads

By common acceptance, three-phase loads are ordinarily represented by three single-phase impedances with no mutual coupling between phases. Figure 3.1.1 shows the representation for a wye-connected load, with a neutral (ground) return through Z_{nL}. The associated voltage equations are:

$$\begin{bmatrix} \tilde{E}_a - \tilde{V}_{nS} - \tilde{V}_{nL} \\ \tilde{E}_b - \tilde{V}_{nS} - \tilde{V}_{nL} \\ \tilde{E}_c - \tilde{V}_{nS} - \tilde{V}_{nL} \end{bmatrix} = \begin{bmatrix} \tilde{Z}_a & 0 & 0 \\ 0 & \tilde{Z}_b & 0 \\ 0 & 0 & \tilde{Z}_c \end{bmatrix} \begin{bmatrix} \tilde{I}_a \\ \tilde{I}_b \\ \tilde{I}_c \end{bmatrix} \qquad (3.1.1)$$

so that:

$$\begin{bmatrix} \tilde{I}_a \\ \tilde{I}_b \\ \tilde{I}_c \end{bmatrix} = \begin{bmatrix} \tilde{Z}_a^{-1} & 0 & 0 \\ 0 & \tilde{Z}_b^{-1} & 0 \\ 0 & 0 & \tilde{Z}_c^{-1} \end{bmatrix} \begin{bmatrix} \tilde{E}_a - \tilde{V}_{nS} - \tilde{V}_{nL} \\ \tilde{E}_b - \tilde{V}_{nS} - \tilde{V}_{nL} \\ \tilde{E}_c - \tilde{V}_{nS} - \tilde{V}_{nL} \end{bmatrix} \qquad (3.1.2)$$

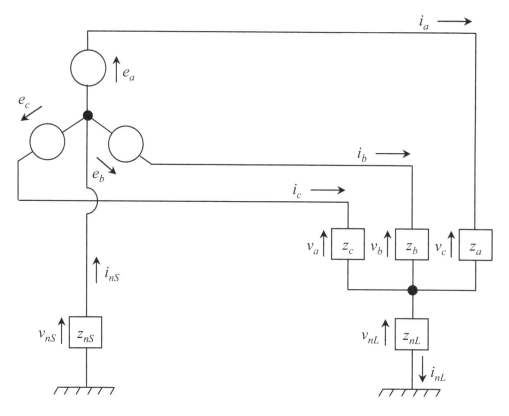

Figure 3.1.1: Wye-Connected Loads

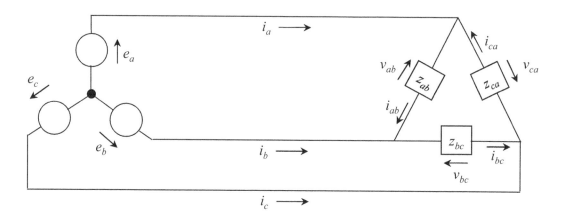

Figure 3.1.2: Delta-Connected Loads

with the auxiliary equation:

$$(\tilde{V}_{nS} + \tilde{V}_{nL}) = (\tilde{I}_a + \tilde{I}_b + \tilde{I}_c)(\tilde{Z}_{nS} + \tilde{Z}_{nL}) \tag{3.1.3}$$

According to Equation 3.1.2 we have:

$$(\tilde{I}_a + \tilde{I}_b + \tilde{I}_c) = \left(\frac{\tilde{E}_a}{\tilde{Z}_a} + \frac{\tilde{E}_b}{\tilde{Z}_b} + \frac{\tilde{E}_c}{\tilde{Z}_c}\right) - (\tilde{V}_{nS} + \tilde{V}_{nL})\left(\frac{1}{\tilde{Z}_a} + \frac{1}{\tilde{Z}_b} + \frac{1}{\tilde{Z}_c}\right) \tag{3.1.4}$$

For the special case of a *balanced* load ($\tilde{Z}_a = \tilde{Z}_b = \tilde{Z}_c$), and *balanced source voltages* ($\tilde{E}_b = a^2\tilde{E}_a, \tilde{E}_c = a\tilde{E}_a$), we obtain:

$$(\tilde{I}_a + \tilde{I}_b + \tilde{I}_c) = (1 + a^2 + a)\frac{\tilde{E}_a}{\tilde{Z}_a} - 3\left(\frac{\tilde{V}_{nS} + \tilde{V}_{nL}}{\tilde{Z}_a}\right) \tag{3.1.5}$$

and Equation 3.1.3 becomes:

$$(\tilde{V}_{nS} + \tilde{V}_{nL}) = -3\left(\frac{\tilde{Z}_{nS} + \tilde{Z}_{nL}}{\tilde{Z}_a}\right)(\tilde{V}_{nS} + \tilde{V}_{nL}) \tag{3.1.6}$$

and:

$$(\tilde{V}_{nS} + \tilde{V}_{nL}) = 0 \tag{3.1.7}$$

Therefore:

$$
\begin{aligned}
\tilde{I}_a &= \frac{\tilde{E}_a}{\tilde{Z}_a} \\
\tilde{I}_b &= a^2\tilde{I}_a = a^2\frac{\tilde{E}_a}{\tilde{Z}_a} \\
\tilde{I}_c &= a\tilde{I}_a = a\frac{\tilde{E}_a}{\tilde{Z}_a}
\end{aligned} \tag{3.1.8}
$$

A symmetrical three-phase wye-connected load, connected to a balanced set of three-phase voltages. behaves like three *independent* single-phase loads with resultant three-phase currents that are also balanced. Therefore, the analysis under such conditions can be carried out on a single-phase basis, "a"-phase generally employed as a reference, and the other phase quantities related by the 1, a, a^2 operators.

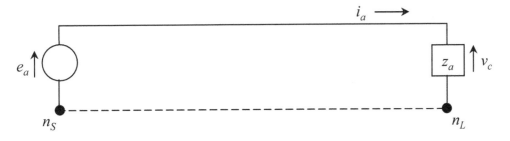

Figure 3.1.3: Wye-Connected Loads

Figure 3.1.3 illustrates the appropriate single-phase network representation. Under the assumed conditions of symmetry, it is observed that $\tilde{V}_{nS} = \tilde{V}_{nL} = 0$, $(\tilde{I}_a + \tilde{I}_b + \tilde{I}_c) = 0$. Consequently, the performance of the three-phase system is not altered by the introduction or omission of a neutral-return conductor between the source neutral nS and the load neutral nL. This hypothetical neutral-return path is used as the actual return path (shown dotted) in Figure 3.1.3.

3.2   Delta-Connected Loads

Given a delta-connected load as shown in Figure 3.1.2, and via the straightforward expedient of a delta-wye conversion, all of the conclusions in the previous section are directly applicable. When the delta load is *perfectly balanced*, the following concepts should be helpful in translating between wye and delta quantities. Based upon the voltage triangle, there should be no difficulty in remembering that, for a balanced three-phase voltage source:

$$V_{LL} = \sqrt{3}V_{LN} \qquad [V] \tag{3.2.1}$$

Since the volt-amperes per leg must be the same for the wye and delta (being one-third of the total three-phase volt-amperes) we have:

$$I_{LL} = \frac{V_{LN}I_{LN}}{V_{LL}} = \frac{I_{LN}}{\sqrt{3}} \tag{3.2.2}$$

Then:

$$Z_{LL} = \frac{V_{LL}}{I_{LL}} = \frac{\sqrt{3}V_{LN}}{\frac{I_{LN}}{\sqrt{3}}} = 3Z_{LN} \tag{3.2.3}$$

3.3   Three-Phase Per Unit System

The mechanics of per unitizing in three-phase systems are identical to those established for single-phase systems. There are at best, some additional options. Clearly, for balanced three-phase systems, totally in *wye* format, the single-phase equivalent representation reflects one phase of the actual system. Using for base quantities:

$$MVA_{1\Phi} \;=\; \frac{MVA_{3\Phi}}{3} \tag{3.3.1}$$

$$kV_{1\Phi} \;=\; kV_{LN} \tag{3.3.2}$$

$$kI_{1\Phi} \;=\; kI_L = \frac{MVA_{1\Phi}}{kV_{LN}} = kI_{LN} \tag{3.3.3}$$

$$Z_{1\Phi} \;=\; \frac{(kV_{LN})^2}{MVA_{1\Phi}} = \frac{(kV_{LL})^2}{MVA_{3\Phi}} = Z_{LN} \tag{3.3.4}$$

Thus, nothing has changed from our previous single-phase per unit system. However, we do indeed have two additional options open to us, namely in the selections for MVA and kV bases. Three-phase machinery is generally rated on a three-phase MVA and a line-to-line voltage. Therefore, if we choose, we might select *base* quantities as follows:

$$MVA_{base} = MVA_{3\Phi} \tag{3.3.5}$$

$$kV_{base} = kV_{LL} \tag{3.3.6}$$

so that:

$$kI_{1\Phi} = \frac{MVA_{3\Phi}}{\sqrt{3}kV_{LL}} = kI_{LN} \tag{3.3.7}$$

$$Z_{1\Phi} = \frac{(kV_{LL})^2}{MVA_{3\Phi}} = Z_{LN} \tag{3.3.8}$$

On the surface, Equations 3.3.5 and 3.3.6 may appear inconsistent with the single-phase representation of a balanced three-phase system. For the electric power engineer, who prefers this system, this presents no problems since he implicitly recognizes that 1.0 p.u. $MVA_{3\Phi}$ is equivalent to 1.0 p.u. $MVA_{1\Phi}$, and 1.0 p.u. kV_{LL} is equivalent to 1.0 p.u. kV_{LN}.

Where extensive delta-connected loads exist, the electric power engineer will define delta base quantities according to Equations 3.2.1 to 3.2.3. By expressing actual delta load quantities in per unit, based on delta quantities, he recognizes implicitly, for example, that 1.0 p.u. Z_{LL} is equivalent to 1.0 p.u. Z_{LN}, the latter of course expressed on line-neutral base quantities.

> The magnitude of all electrical quantities in a delta configuration, when expressed in per unit on delta base quantities, are numerically equal to the corresponding per unit values of the equivalent single-phase representation, expressed in per unit on wye base quantities. Phase angle relationships also remain the same.

3.4 Transmission Lines

The voltage drop across a fully *transposed* transmission line section (without ground wires), shown in Figure 3.4.1, is given by:

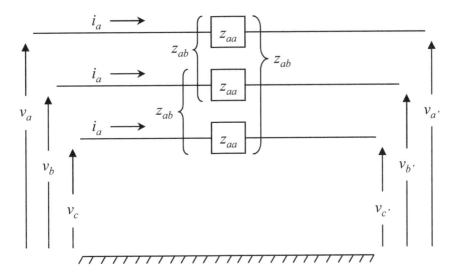

Figure 3.4.1: Fully Transposed Transmission Line Segment

$$\begin{bmatrix} \Delta \tilde{V}_a \\ \Delta \tilde{V}_b \\ \Delta \tilde{V}_c \end{bmatrix} = \begin{bmatrix} \tilde{V}_a - \tilde{V}'_a \\ \tilde{V}_b - \tilde{V}'_b \\ \tilde{V}_c - \tilde{V}'_c \end{bmatrix} = \begin{bmatrix} \tilde{Z}_{aa} & \tilde{Z}_{ab} & \tilde{Z}_{ab} \\ \tilde{Z}_{ab} & \tilde{Z}_{aa} & \tilde{Z}_{ab} \\ \tilde{Z}_{ab} & \tilde{Z}_{ab} & \tilde{Z}_{aa} \end{bmatrix} \begin{bmatrix} \tilde{I}_a \\ \tilde{I}_b \\ \tilde{I}_c \end{bmatrix} \tag{3.4.1}$$

$$\begin{bmatrix} \Delta \tilde{V}_a \\ \Delta \tilde{V}_b \\ \Delta \tilde{V}_c \end{bmatrix} = (\tilde{Z}_{aa} - \tilde{Z}_{ab}) \begin{bmatrix} 1 & 0 & 0 \\ 0 & 1 & 0 \\ 0 & 0 & 1 \end{bmatrix} \begin{bmatrix} \tilde{I}_a \\ \tilde{I}_b \\ \tilde{I}_c \end{bmatrix} + (\tilde{Z}_{ab}) \begin{bmatrix} 1 & 1 & 1 \\ 1 & 1 & 1 \\ 1 & 1 & 1 \end{bmatrix} \begin{bmatrix} \tilde{I}_a \\ \tilde{I}_b \\ \tilde{I}_c \end{bmatrix} \qquad (3.4.2)$$

and for balanced three-phase currents:

$$\begin{bmatrix} \Delta \tilde{V}_a \\ \Delta \tilde{V}_b \\ \Delta \tilde{V}_c \end{bmatrix} = (\tilde{Z}_{aa} - \tilde{Z}_{ab})(\tilde{I}_a) \begin{bmatrix} 1 \\ a^2 \\ a \end{bmatrix} \qquad (3.4.3)$$

Under these conditions, a section of transmission line behaves like a series impedance, per phase, of $(\tilde{Z}_{aa} - \tilde{Z}_{ab})$ with no mutual coupling to the other phases. Figure 3.4.2 shows the single-phase representation of such a three-phase transmission line section.

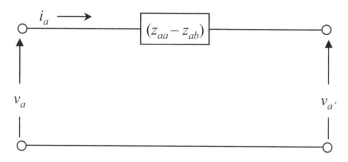

Figure 3.4.2: Single-Phase Representation for Figure 3.4.1

A few words are in order regarding the influence of earth upon $(\tilde{Z}_{aa} - \tilde{Z}_{ab})$. The influence of earth is accounted for by the addition of a term $\frac{\tilde{Z}_e}{3}$ to $\tilde{Z}_{aa}$ and $\tilde{Z}_{ab}$. It is apparent that earth has no effect upon the effective impedance $(\tilde{Z}_{aa} - \tilde{Z}_{ab})$ of a *symmetrical* transmission line when the line is operating under *balanced* voltage and current conditions.

The influence of ground wires is somewhat different. Since the ground wire correction is not necessarily the same for *all* $\tilde{Z}_{aa}$ and $\tilde{Z}_{ab}$, a simple single-phase representation is not rigorously possible. However, since the influence correction terms are *nearly* of equal magnitude, as a first order approximation the influence of ground wires may be neglected when a *symmetrical* transmission line section is operating under *balanced current* conditions.

Example 3.4.1

An 80 km, 115 kV three-phase transmission line has a line impedance of 0.2 + j 0.5 ohms per km. If the receiving end voltage is maintained at 115 kV and the line is supplying a 50 MVA balanced load at 1.0 P.F., calculate the voltage at the sending end of the line, neglecting the line's capacitive susceptance. Select:

$$MVA_{base} = 50MVA \qquad kV_{base} = 115kV$$

Then:

$$kI_{base} = \frac{50}{\sqrt{3}(115)} = 0.251 \,[kA]$$

$$Z_{base} = (kV)^2/MVA = (115)^2/50 = 265 \,[\Omega]$$

In per unit:

$$\bar{Z}_{line} = (0.2 + j0.5)(\frac{80}{265}) = 0.06 + j0.15$$

Since $\bar{P}$ and $\bar{V}$ are both equal to unity at the receiving end, the per unit current at the load must also be equal to unity; i.e.:

$$\tilde{I}_a = 1.0 + j0$$

The sending end voltage $\tilde{V}_a$ is then:

$$\tilde{V}_a = 1.0 + (1.0 + j0)(0.06 + 0.15) = 1.06 + j0.15 = 1.07\angle 8.05^o$$

Now:

$$\tilde{S}_a = \tilde{V}_a\tilde{I}_a{}^* = (1.06 + j0.15)(1.0 - j0)$$

and in actual (physical) quantities:

$$\tilde{S}_a = (1.06 + j0.15)(50) = 53 + j7.5 \qquad [MVA]$$

3.5 Equivalent Circuits for Y-Y and Δ-Δ Transformers

For the purposes of this chapter we shall assume that a three-phase transformer is geometrically symmetrical with respect to the three phases; the implications of this statement will become clearer as we proceed.

Consider first a three-phase bank composed of three identical single-phase transformers, connected either wye-wye or delta-delta as shown in Figure 3.5.1. On the basis of the per unit equivalent networks for single-phase transformers, we find that Figure 3.5.2 represents the per unit equivalent networks for the configurations shown in Figure 3.5.1. Ideal 1:1 transformers are shown merely for the purposes of reflecting the isolation between the primary and secondary windings.

According to Figure 3.5.2(a), a wye-wye bank of three single-phase transformers is equivalent to a balanced series line impedance equal to the short-circuit impedance of each transformer. Under balanced three-phase operation, the neutral connection carries no current and may therefore be omitted without affecting the performance.

The delta-delta connection, shown in Figure 3.5.2(b), ultimately can be translated into the same equivalent network as the wye-wye connection. For example, consider the line and load impedances to the right of terminals A, B, and C, to be represented in terms of an equivalent delta impedance. After reflecting this network through the ideal transformers, it appears as an impedance in series with $\bar{Z}_{S.C.}$ in the form of a delta. This may be translated into an equivalent wye, thus being of the same form as Figure 3.5.2(a). As a matter of fact, since $\bar{Z}_{S.C.}$ is given on the single-phase line-to-line voltage base of the system, it has the same numerical value when expressed on the system line-to-neutral voltage base in the equivalent wye representation.

Since both the wye-wye and delta-delta transformer banks can be represented by balanced three-phase series impedances, an analysis can be performed on a single-phase basis as described in the previous sections. Furthermore, all of the principles developed for single-phase transformers can be applied directly in studies of balanced and symmetrical three-phase systems; i.e., except for magnetizing current characteristics, which will be discussed shortly.

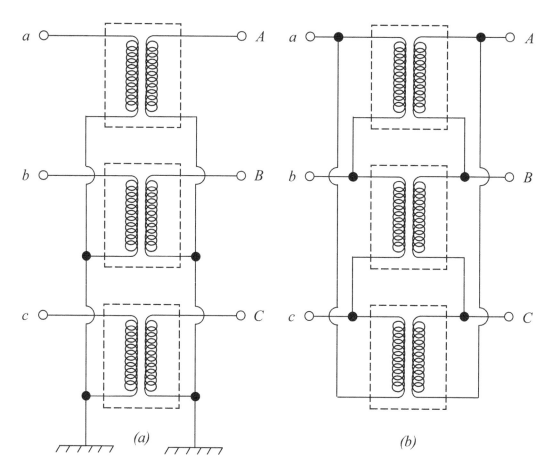

Figure 3.5.1: Three-Phase Transformer Bank Connections

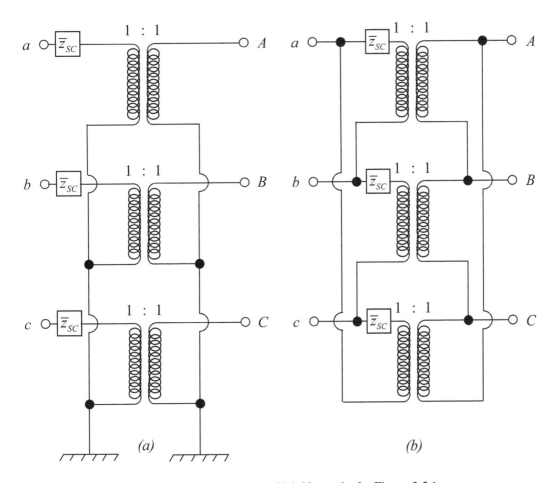

Figure 3.5.2: Equivalent per Unit Networks for Figure 3.5.1

3.6   Wye-Delta Connected Transformers

Figure 3.6.1 shows two possible arrangements for wye-delta connected transformer banks. If the wye side is energized by a balanced set of voltages, and if the delta side is short-circuited, the impedance as seen from the wye side is that of a balanced set of wye-connected impedances equal to $\bar{Z}_{S.C.}$ of the transformer in each phase.

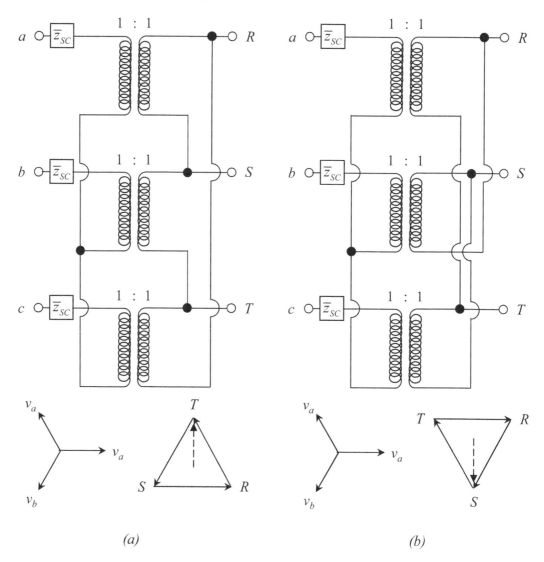

Figure 3.6.1: Forms of Wye-Delta Transformer Connections

Conversely, if the wye side is short-circuited, a delta-connected set of impedances equal to $\bar{Z}_{S.C.}$ is seen from the delta side of the transformer. This may be replaced by an equivalent wye-connected set of impedances. Consequently:

> For wye-delta or delta-wye connected transformers, a single-phase equivalent network for the transformer may be used. The magnitudes of the resultant currents in per unit will be correct for both the primary and secondary sides.

The following example should help to clarify some of the above conclusions.

Example 3.6.1

A three-phase 120 MVA, 230 kV/69 kV, wye-delta transformer has a short-circuit reactance $\bar{X}_{s.c.} = 0.12$ per unit on its rated MVA and voltage base. This transformer is to be represented in a single-phase (line-to-neutral) equivalent circuit. For the equivalent system, let us select:

$$MVA_{base} = 120\,MVA \quad k_H V_{base} = 230\,kV \quad k_L V_{base} = 66\,kV$$

whereby we have arbitrarily (just to be difficult) selected 66 kV as the low voltage base. The per unit equivalent circuit is shown in Figure 3.6.2. This equivalent network may be used directly in a system analysis and will correctly reflect the correct magnitudes of voltages and currents. To this extent, the wye-delta representation is no different than that for the wye-wye and delta-delta configurations. For many types of system studies, this representation is completely satisfactory.

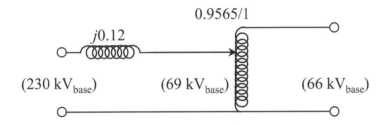

Figure 3.6.2: Per Unit Equivalent Circuit for Example 3.6.1

3.6.1 Phase-Shift in Wye-Delta Connections

For some purposes, however, it may also be necessary to recognize phase shifts when they occur. In protective relaying, where input and output quantities of a transformer are compared (differential relaying), the phase relationships of these quantities cannot be ignored.

In Figure 3.6.1(a), we observe that there is a shift in phase of the line-to-neutral voltage from the wye to the delta side with R, S, and T to emphasize the fact that, since there is isolation from the wye side, we are free to choose any designations that we wish for the delta side. Thus, A may be assigned to R, S, or T.

According to most standards for terminal designations, and adopted by protective relay engineers, the line-to-neutral voltage V_R in Figure 3.6.1(a) would be designated as V_A, since this voltage is closest in phase angle to V_a on the wye side. Thus, V_A lags V_a by 30 degrees.

For hand calculations, the entire network analysis is carried out, ignoring phase shifts. After the per unit voltages and currents have been evaluated and entered in their appropriate places in the network, the phase-shift adjustment is made. This is accomplished by identifying some convenient point in the network, as a reference, and then proceeding left and right through the network, assigning appropriate $+j$ operators where required.

Example 3.6.2

Consider the network shown in Figure 3.6.3(a) with a line impedance of 66 ohms reactance, a wye-delta, $\bar{X}_{s.c.} = 0.12$, transformer, 120 MVA, 230 kV/69 kV, supplying a 60 MVA load at 1.0 P.F.

$$\text{Bases:} \qquad 120\ \text{MVA}$$

$$230\ \text{kV} \qquad\qquad 69\ \text{kV}$$
$$440.83\ \text{ohms} \qquad 39.675\ \text{ohms}$$

$$\text{Per Unit:}$$
$$\bar{X}_{line} = 0.15 \qquad \bar{R}_{load} = 2.00$$

According to Figure 3.6.3(b):

$$\bar{I} = \frac{1.0}{2 + j0.27} = 0.491 - j0.066 \qquad \bar{V} = 0.982 - j0.132$$

If we select $\bar{E}_a = 1.0$ as reference, the appropriate phase-shifted values are shown in Figure 3.6.3(c). There is no inconsistency between Figures 3.6.4(a) and 3.6.3(c); in the latter we have entered $\bar{V}_a$ (not -j $\bar{V}_A$). Also observe that even after the phase shift:

$$\frac{\bar{V}_A}{\bar{I}_A} = (\frac{0.132 + j0.982}{0.066 + j0.491} = 2.0 + j0$$

which is as it must be.

For analytical purposes, it is considerably more convenient to assign A to T. In this event, $\bar{V}_A$ leads $\bar{V}_a$ by 90 degrees. In calculations this is accomplished by indicating a (+j) operator to the voltage, which is considerably simpler than employing ($\frac{\sqrt{3}-j1}{2}$) as an operator for the 30 degree phase shift.

As stated earlier, for three-phase *balanced* operation it may be *desirable* to properly account for this phase shift. Under an *unbalanced* operation (as we shall see later), it is *absolutely essential* that we account for this phase shift.

Figure 3.6.4(a) shows how the voltage phase shift, for the connection in Figure 3.6.1(a), may be properly indicated. For example, let $\bar{V}_a = 0.93$. Then $-j\bar{V}_A = 0.93$, or $\bar{V}_a = j0.93$ resulting in the proper phase shift.

In Figure 3.6.4(a), the impedance to the right of the network is given by:

$$\bar{Z}_{eff} = \frac{\bar{V}_a}{\bar{I}_a} \tag{3.6.1}$$

which must be equal to:

$$\bar{Z}_{eff} = \frac{-j\bar{V}_A}{-j\bar{I}_A} = \frac{\bar{V}_A}{\bar{I}_A} \tag{3.6.2}$$

From another point of view:

$$\tilde{\bar{S}}_a = \tilde{\bar{V}}_a \tilde{\bar{I}}_a^* \tag{3.6.3}$$
$$\tilde{\bar{S}}_A = (-j\tilde{\bar{V}}_A)(-j\tilde{\bar{I}}_A)^* = \tilde{\bar{S}}_a \tag{3.6.4}$$

It is left as an exercise for the reader to demonstrate, on the basis of Figure 3.6.1(a), that if $\bar{V}_b$ and $\bar{V}_c$ are interchanged (reverse phase rotation), that the line-to-neutral voltages on the delta side will lag by 90 degrees, rather than lead the line-to-neutral voltages on the wye side by 90 degrees. The phase rotation will also be reversed on the delta side (as expected).

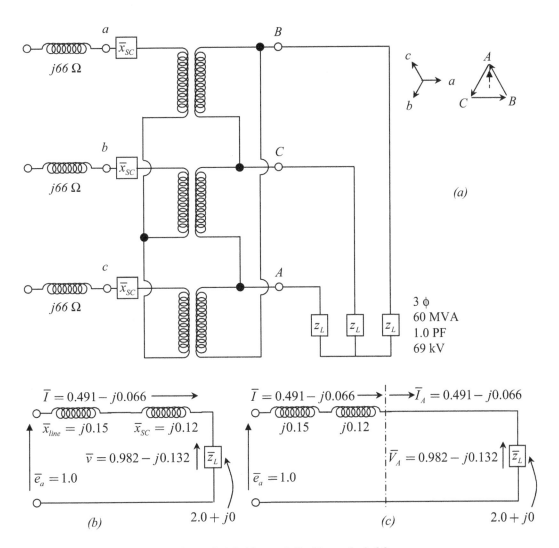

(a)

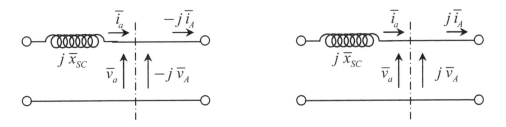

Figure 3.6.3: Network for Example 3.6.2

Figure 3.6.4: Per Unit Diagram Showing Phase Shifts Corresponding to Figures 3.6.1 (a) and (b), Respectively

3.7   Magnetizing Currents in Three-Phase Transformers

Under ordinary load conditions, the exciting current required by transformers is only a fraction of a percent of rated current. Therefore, the exciting current is considered negligible, and the magnetizing impedance branch is not included in the equivalent network for single-phase transformers. However, when single-phase transformers are connected to form a bank of three-phase transformers, problems may arise if the triplen harmonics in the magnetizing current are not permitted to develop properly.

Figure 3.7.1 illustrates the distorted magnetizing current required in a single-phase transformer when the applied voltage is a pure sinusoid. In terms of a Fourier series we have:

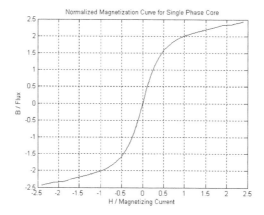

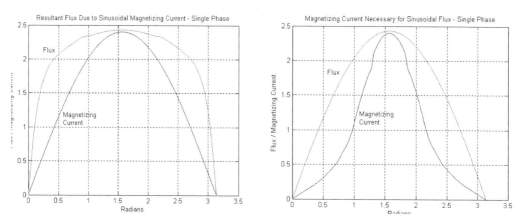

Figure 3.7.1: Distorted Magnetizing Current Required for Sinusoidal Flux and Distorted Flux Resulting from Sinusoidal Magnetizing Current in Single-Phase Transformer

$$i_m(t) = i_1 sin\omega t + i_3 sin3\omega t + i_5 sin5\omega t \ldots \tag{3.7.1}$$

For practical transformer flux densities, and with some allowance for air gaps at the butt joints of the laminations, the third harmonic component is about 16% to 17% of the fundamental. Consider now the connection of three single-phase transformers as shown in Figure 3.7.2. Since third harmonic currents are in phase, in a three-phase system, they require a neutral return in order to be able to flow.

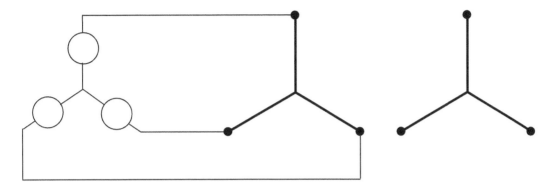

Figure 3.7.2: Ungrounded Wye-Wye Connected Transformer Bank

When the triplen harmonic magnetizing current components are not permitted to develop, the flux in the core will no longer be sinusoidal. Consequently, the voltages induced in the windings will not be purely sinusoidal but will contain triplen harmonics. A 16% to 17% third harmonic flux (due to the missing 16% to 17% third harmonic magnetizing current) will cause a 50% third harmonic voltage. The waveshape of the flux, due to the distortion, will appear flat-topped because the triplen harmonics will tend to subtract from the peak of the wave, whereas the voltage waveshape will appear to be peaked since the triplen harmonics will tend to add to the peak of the wave. The line-to-line voltages remain sinusoidal, but the line-to-neutral (winding) voltages will be distorted. Since the distorted peak line-to-neutral voltage may be 150% of the rated fundamental frequency peak voltage, the voltage stress on the insulation is increased by 50% above its normal value, thereby contributing to loss of life of the insulation.

3.7.1   Wye-Wye Connection

For the unloaded transformer arrangement shown in Figure 3.7.2, the line currents will be distorted, consisting of all of the required harmonics (for sinusoidal flux) *except* for the third, ninth, etc. (all triplen) harmonics. Consequently the flux will be sinusoidal except for triplen harmonics. All windings linked by this flux will be sinusoidal *except* for the addition of triplen harmonics.

If the voltage source is assumed to be an infinite bus, the line-to-line voltages at the transformer must clearly also be sinusoidal. Following Kirchhoff's voltage law, if we trace the voltages from the source neutral to the transformer winding and to the transformer neutral, we find that the voltage between the source neutral and the transformer neutral contains only triplen harmonics.

The combination of fundamental frequency and triplen-harmonic flux induces in each phase (winding) a distorted voltage, so that a voltmeter connected between a line terminal and neutral terminal of each transformer indicates an rms value of about $\sqrt{(1.0)^2 + (0.5)^2} = 1.12$ of the normal rms value.

The same conditions as above exist for the open-delta secondary connection illustrated in Figure 3.7.3. If a voltmeter is connected across the open in the delta, no fundamental frequency voltage will be measured (since they must add to zero), but three times the third harmonic voltages will be measured. Thus, the voltmeter should indicate about 1.5 times normal phase voltage.

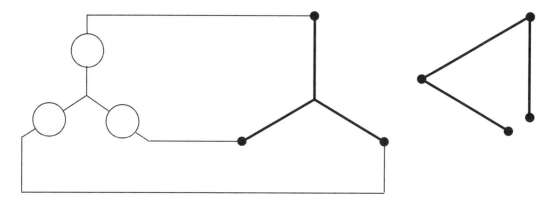

Figure 3.7.3: Ungrounded Wye, Open-Delta Connected Transformer Bank

Transformer Neutral Grounding

If the primary neutral of the transformer is grounded (assuming that the infinite bus source neutral is also grounded), the total required magnetizing current can flow, and the triplen voltage problems are eliminated. When the primary neutral is ungrounded, however, various problems can occur depending upon the transformer secondary connections.

Secondary Neutral Ungrounded

When the secondary neutral is also ungrounded, we might have the situation shown in Figure 3.7.4, where the secondary is connected to a section of an unloaded (open-circuited) overhead transmission line and the line-to-ground voltage is equal to the fundamental frequency plus the third harmonic voltage. The third harmonic line-to-ground voltage on the transmission line depends upon the ratio of the transmission line-to-ground capacitance to the transformer winding-to-ground capacitance (ignoring $3X_m$ since $3X_m << X_{w-g}$):

$$V_{3\ell-g} = \frac{V_{3t}}{1 + \frac{C_{\ell-g}}{C_{w-g}}} \tag{3.7.2}$$

The longer the transmission line, the smaller $V_{3\ell-g}$ will be.

Secondary Neutral Grounded

If the secondary neutral is grounded, we have the situation depicted in Figure 3.7.5. The transformer circuit now accounts for three times the magnetizing reactance through which the third harmonic must flow.

    With the system neutral ungrounded, the third harmonic line to ground voltage depends upon the relative magnitudes of $3X_m$ and $X_{c\ell-g}$:

$$V_{3\ell-g} = \frac{X_{c\ell-g}}{X_{c\ell-g} - 3X_m} V_{3t} \tag{3.7.3}$$

For short lines where $X_{c\ell-g} > 3X_m$, we have $V_{3\ell-g} > V_{3t}$. When $X_{c\ell-g} = 3X_m$, a resonance condition occurs and the third harmonic voltage may rise to very dangerous values. As much as three times normal induced phase voltage has been observed under these conditions. For very long lines where $X_{c\ell-g} << 3X_m$, we have a low impedance between line and ground so that the third

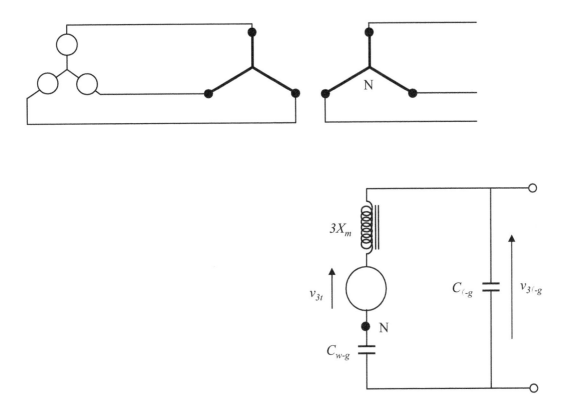

Figure 3.7.4: Third Harmonic Voltage with Secondary Neutral Ungrounded

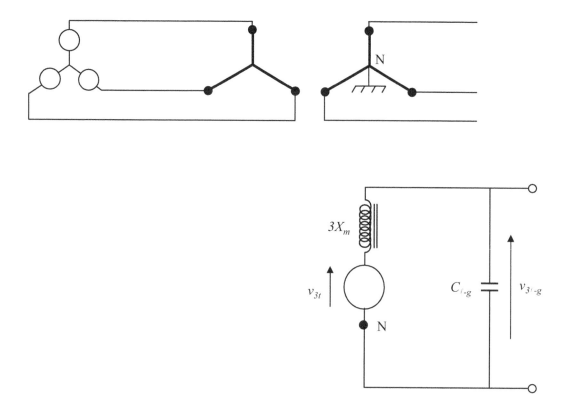

Figure 3.7.5: Third Harmonic Voltage with Secondary Neutral Grounded

harmonic currents may now flow freely. The cumulative effect of reduction in V_{3t} as third harmonic currents are permitted to flow causes a nearly fundamental frequency line-to-ground voltage.

Three-Phase Core-Type Transformers

The disadvantages cited for the wye-wye transformer connections can be significantly reduced if, instead of three single-phase transformers, a single three-phase, three-legged, core-type transformer is employed. The suppressed third harmonic ampere turns are all in phase, and therefore attempt to cause third harmonic flux to flow as shown in Figure 3.7.6. Since this flux must leave the top of the core and return, through air, through the bottom of the core, the magnetic reluctance to third harmonic flux is considerably greater than that to the balanced, three-phase, fundamental frequency flux. In addition, the third harmonic flux links the outer steel tank. The induced third harmonic voltage causes a third harmonic current to circulate in the tank wall. The transformer tank, under these conditions, behaves like a high reactance third winding and accounts, to some extent, for the missing third harmonic current. The third harmonic voltage content is reduced from 50% for single-phase units to about 3.5%.

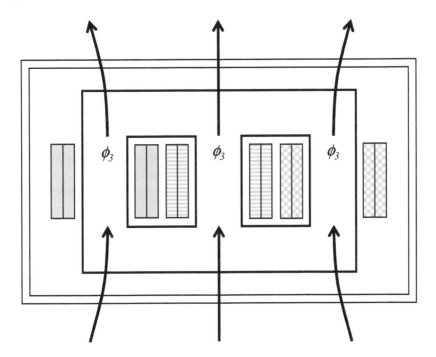

Figure 3.7.6: Triplen Harmonic Flux Distribution in Three-Legged Core-Type Transformer

3.7.2   Wye-Delta Connection

For the wye-delta connection shown in Figure 3.7.7, the line currents on the wye side are exactly the same as those for Figure 3.7.4. The tendency for a third harmonic flux and voltage component in the delta winding causes a third harmonic component of current to flow around the delta-connected windings. Since the mutual flux linking all windings is determined by the sum of all MMFs acting upon the flux path, and since the delta windings supply the otherwise missing third harmonic current, the mutual flux will be sinusoidal.

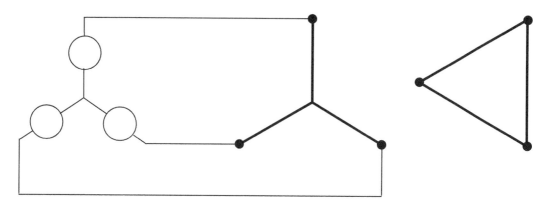

Figure 3.7.7: Wye-Delta Connection Providing Triplen Harmonic Currents

If a corner of the delta is opened, the voltage appearing across the opening will be three times the third harmonic component of voltage in each winding. No fundamental frequency component of voltage will be observed since they add to zero if the windings are symmetrical and the currents are balanced.

If the neutral of the wye is grounded, the required third harmonic currents will divide between the wye side and the delta side according to the leakage reactances of the windings. It is interesting to note that a grounded-wye/delta connection (unloaded) may be employed in the system described in Section 3.7.1.1 (System Neutral Ungrounded) to supply the missing third harmonic current. In this situation, the third harmonic in the system line-to-ground voltage would appear across the wye and delta. The circulating third harmonic currents in the delta would remove the third harmonic voltage in the wye, and hence in the system. The kVA rating of such a grounded wye-delta transformer would be very small, provided that this transformer bank was not required to carry any load.

3.7.3   Delta-Delta Connection

The line currents to a delta-connected winding contain all the required harmonics except the triplen. As in the case for the wye-delta connection, the required triplen currents will flow around the delta windings. Therefore, the flux is sinusoidal with no distortion. In the delta-delta connection, the triplen currents are shared between the delta windings. In delta-wye (ungrounded) connections, all of the triplen flow in the delta. In delta-wye (grounded) connections, the triplen currents are again shared by the windings.

3.7.4   Wye-Wye with Delta Tertiary

When two windings are connected wye-wye (both ungrounded) a small third (tertiary) winding may be added to supply the missing triplen harmonics. No voltage distortion is then present.

Occasionally, when two windings are connected wye-wye (with one or both neutrals ungrounded), the triplen harmonic currents flowing along the lines and returning through the earth may cause telephone interference. A properly designed tertiary winding, connected in delta, will significantly reduce, if not eliminate, the triplen currents returning through the earth. A schematic of such is shown in Figure 3.7.8.

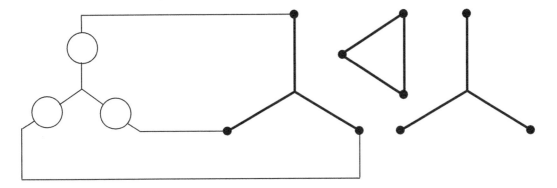

Figure 3.7.8: Wye-Wye with Delta Tertiary Providing Triplen Harmonic Currents

3.8   Steady-State Power Transfer

Under balanced three-phase operation of symmetrical (fully transposed) three-phase networks, all components of the network can be represented on a phase-to-neutral basis. The performance predictions therefore require only a single-phase analysis.

After having determined the electromagnetic characteristics of various network components, we now proceed to study how these characteristics, together with the practical constraints of system voltage control, limit the actual real power that can be transferred from one point of the network to another.

3.8.1   Phasor Equations

To begin, we shall consider the power transfer capabilities of the simple series impedance element of the network illustrated in Figure 3.8.1(a). All parameters are to be taken as phasor quantities such that:

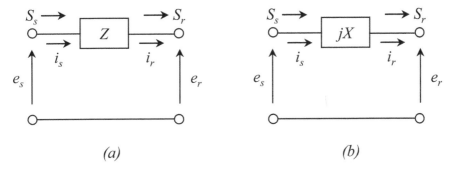

(a)   *(b)*

Figure 3.8.1: Network Power Transfer Elements

$$\tilde{e}_s = |e_s|\,\varepsilon^{j\delta_s} \tag{3.8.1}$$

$$\tilde{e}_r = |e_r|\,\varepsilon^{j\delta_r} \tag{3.8.2}$$

$$\tilde{Z} = |Z|\,\varepsilon^{j\Phi} \tag{3.8.3}$$

Now:

$$\tilde{S}_s = \tilde{e}_s \tilde{I}_s^* = \frac{\tilde{e}_s(\tilde{e}_s^* - \tilde{e}_r^*)}{\tilde{Z}^*} \tag{3.8.4}$$

$$\tilde{S}_r = \tilde{e}_r \tilde{I}_r^* = \frac{\tilde{e}_r(\tilde{e}_s^* - \tilde{e}_r^*)}{\tilde{Z}^*} \tag{3.8.5}$$

and in expanded form:

$$\tilde{S}_s = \frac{\varepsilon^{j\Phi}\left[|\,e_s\,|^2 - |\,e_s\,|\,|\,e_r\,|\,\varepsilon^{j\delta_{sr}}\right]}{|\,Z\,|} \tag{3.8.6}$$

$$\tilde{S}_r = \frac{\varepsilon^{j\Phi}\left[-|\,e_r\,|^2 + |\,e_s\,|\,|\,e_r\,|\,\varepsilon^{-j\delta_{sr}}\right]}{|\,Z\,|} \tag{3.8.7}$$

where:

$$\delta_{sr} = \delta_s - \delta_r \tag{3.8.8}$$

In order to more easily bring to light some salient aspects, let $\tilde{Z}$ be purely imaginary (i.e., $\tilde{Z} = jX$) as shown in Figure 3.8.1(b). Then we have:

$$\tilde{S}_s = \frac{j\left[|\,e_s\,|^2 - |\,e_s\,|\,|\,e_r\,|\,\varepsilon^{j\delta_{sr}}\right]}{|\,X\,|} \tag{3.8.9}$$

$$\tilde{S}_r = \frac{j\left[-|\,e_r\,|^2 + |\,e_s\,|\,|\,e_r\,|\,\varepsilon^{-j\delta_{sr}}\right]}{|\,X\,|} \tag{3.8.10}$$

and:

$$\tilde{S}_s = \frac{|\,e_s\,|\,|\,e_r\,|}{X}\sin\delta_{sr} + j\frac{|\,e_s\,|^2 - |\,e_s\,|\,|\,e_r\,|\cos\delta_{sr}}{X} \tag{3.8.11}$$

$$\tilde{S}_r = \frac{|\,e_s\,|\,|\,e_r\,|}{X}\sin\delta_{sr} - j\frac{|\,e_r\,|^2 - |\,e_s\,|\,|\,e_r\,|\cos\delta_{sr}}{X} \tag{3.8.12}$$

Observe that:

$$P_s = P_r = \frac{|\,e_s\,|\,|\,e_r\,|}{X}\sin\delta_{sr} \tag{3.8.13}$$

Also, if we take:

$$
\begin{aligned}
|\,i_r\,|^2 X &= \tilde{i}_r \tilde{i}_r^* X = \left(\frac{\tilde{e}_s - \tilde{e}_r}{jX}\right)\left(\frac{\tilde{e}_s^* - \tilde{e}_r^*}{-jX}\right)X \\
&= \frac{|\,e_s\,|^2 + |\,e_r\,|^2 - |\,e_s\,|\,|\,e_r\,|\,\varepsilon^{j\delta_{sr}} - |\,e_r\,|\,|\,e_s\,|\,\varepsilon^{-j\delta_{sr}}}{X} \\
&= \frac{|\,e_s\,|^2 + |\,e_r\,|^2 - 2\,|\,e_s\,|\,|\,e_r\,|\cos\delta_{sr}}{X} \tag{3.8.14}
\end{aligned}
$$

we observe that:

$$Q_s - Q_r = |\,i_r\,|^2 X = |\,i_s\,|^2 X \tag{3.8.15}$$

3.8.2   Steady-State Power Transfer Limit Power Angle Curve

Equation 3.8.13 can be expressed graphically as shown in Figure 3.8.2. We observe that for a fixed $\mid e_s \mid, \mid e_r \mid$, and X, that the real power which can be transferred across a pure reactance element is determined by the power angle δ_{sr} (phase difference between $\tilde{e}_s$ and $\tilde{e}_r$) and has a maximum when $\delta_{sr} = \pi/2$. We define $\mid e_s \mid\mid e_r / X$ as the *steady-state power transfer limit* of this pure reactance element. Due to practical voltage constraints, the only practical way of increasing the power transfer capability of a transmission line section is to reduce its effective series reactance. We shall show later that the internal impedance of a synchronous generator can also be represented by a series inductive reactance element (synchronous reactance) under steady-state operation. For a synchronous machine we may write:

$$P_{max} = \frac{\mid E_i \mid\mid e_t \mid}{X_d} \tag{3.8.16}$$

where $\mid e_t \mid$ represents the magnitude of the generator terminal voltage, and $\mid E_i \mid$ represents the generator open-circuit voltage for a given field winding excitation. We shall show later how appropriate excitation systems may be employed to increase the steady-state stability limit of a synchronous generator.

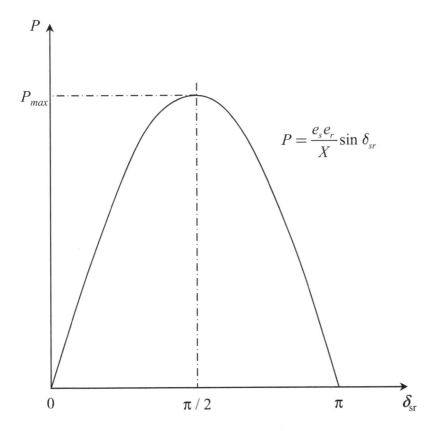

Figure 3.8.2: Power Angle Curve

3.8.3   Circle Diagrams

The total phasor power relationships in Equations 3.8.9 and 3.8.10 may also be expressed graphically, in terms of circle diagrams, as shown in Figure 3.8.3. The center of the $\tilde{S}_s$ circle lies at $j\frac{|e_s|^2}{X}$, and the center of the $\tilde{S}_r$ circle lies at $-j\frac{|e_r|^2}{X}$. The radii for both circles are $\frac{|e_s||e_r|}{X}$. In view of this latter fact, it is apparent from the circle diagrams that for all δ_{sr}, $P_s = P_r$.

The circle diagrams also permit an enlightened observation of the increased Q requirement of the line as we attempt to increase P. For $P = 0$, $\delta_{sr} = 0$, we have:

$$Q_{s0} = \frac{|e_s|^2 - |e_s||e_r|}{X} \tag{3.8.17}$$

$$Q_{r0} = -\frac{|e_r|^2 - |e_s||e_r|}{X} \tag{3.8.18}$$

so that:

$$Q_{s0} - Q_{r0} = \frac{(|e_s| - |e_r|)^2}{X} \tag{3.8.19}$$

At the power limit we have:

$$Q_{s-lim} = \frac{|e_s|^2}{X} \tag{3.8.20}$$

$$Q_{r-lim} = -\frac{|e_r|^2}{X} \tag{3.8.21}$$

and:

$$Q_{s-lim} - Q_{r-lim} = \frac{|e_s|^2 + |e_r|^2}{X} \tag{3.8.22}$$

Furthermore, on the basis of Equation 3.8.19, we may determine directly:

$$|i_{r0}|^2 = |i_{s0}|^2 = \frac{(|e_s| - |e_r|)^2}{X^2} \tag{3.8.23}$$

and at the power transfer limit per Equation 3.8.22:

$$|i_{r-lim}|^2 = |i_{s-lim}|^2 = \frac{|e_s|^2 + |e_r|^2}{X^2} \tag{3.8.24}$$

Figure 3.8.4 illustrates the appropriate power circle diagrams for the more general Z series impedance element shown in Figure 3.8.1(a).

For the even more general transmission line representation described in Chapter 1 (Section 1.6.4) we have, on the basis of Equations 1.6.11 and 1.6.12:

$$\tilde{I}_s = \frac{\tilde{A}\tilde{V}_s - \tilde{V}_r}{\tilde{B}} \qquad \tilde{I}_r = \frac{\tilde{V}_s - \tilde{A}\tilde{V}_r}{\tilde{B}} \tag{3.8.25}$$

$$\tilde{S}_s = \frac{\left[\tilde{A}^*|e_s|^2 - |e_s||e_r|\,\varepsilon^{j\delta_{sr}}\right]}{\tilde{B}^*} \tag{3.8.26}$$

$$\tilde{S}_r = \frac{\left[\tilde{A}^*|e_r|^2 - |e_r||e_s|\,\varepsilon^{-j\delta_{sr}}\right]}{\tilde{B}^*} \tag{3.8.27}$$

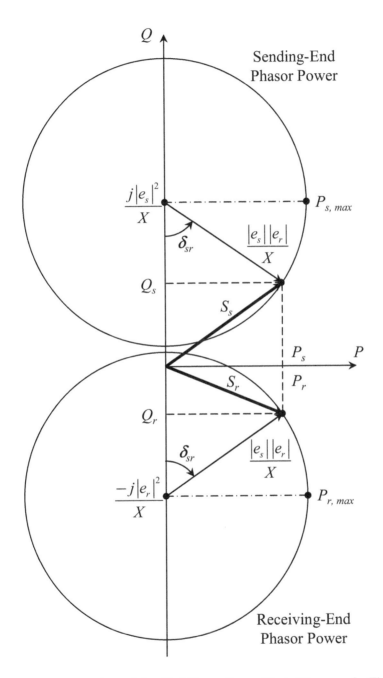

Figure 3.8.3: Sending-End and Receiving-End Phasor Power Circle Diagrams for Figure 3.8.1(b)

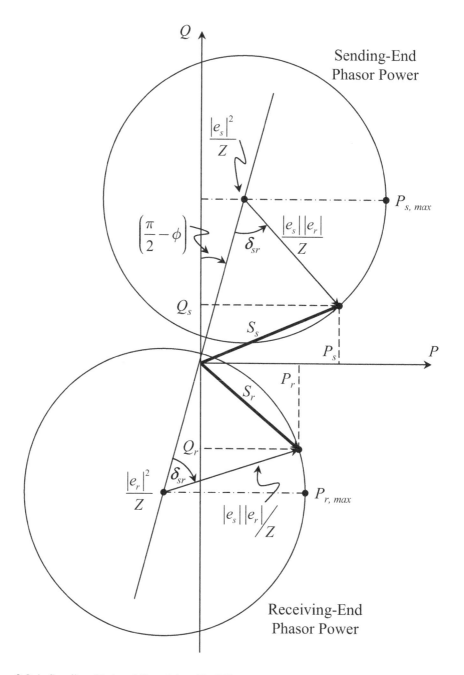

Figure 3.8.4: Sending-End and Receiving-End Phasor Power Circle Diagrams for Figure 3.8.1(a)

3.8.4 Steady-State Power Transfer Limit with Intermediate Voltage Regulation

We shall have occasion, later in our studies, to consider the situation depicted in Figure 3.8.5. Specifically, we wish to know the power transfer limit when the voltage e_t is regulated and maintained at a fixed value. We consider the phasor power $\tilde{S}_t = \tilde{e}_t \tilde{I}_t^*$, and since it may be expressed in three different ways we have:

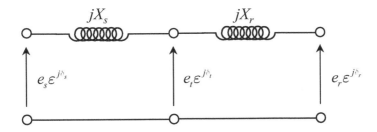

Figure 3.8.5: Circuit with Intermediate Voltage Regulation

$$\tilde{S}_t = \frac{\tilde{e}_t(\tilde{e}_s^* - \tilde{e}_t^*)}{-jX_s} = j\frac{|e_t|}{X_s}\left[|e_s|\,\varepsilon^{-j\delta_{st}} - |e_t|\right] \tag{3.8.28}$$

$$\tilde{S}_t = \frac{\tilde{e}_t(\tilde{e}_t^* - \tilde{e}_r^*)}{-jX_r} = j\frac{|e_t|}{X_r}\left[-|e_r|\,\varepsilon^{j\delta_{tr}} + |e_t|\right] \tag{3.8.29}$$

$$\tilde{S}_t = \frac{\tilde{e}_t(\tilde{e}_s^* - \tilde{e}_r^*)}{-j(X_s + X_r)} = j\frac{|e_t|}{X_s + X_r}\left[|e_s|\,\varepsilon^{-j\delta_{st}} - |e_r|\,\varepsilon^{j\delta_{tr}}\right] \tag{3.8.30}$$

Now:

$$P_{max} = \frac{|e_s||e_r|}{(X_s + X_r)} \tag{3.8.31}$$

for:

$$\delta_{srm} = \frac{\pi}{2} = \delta_{stm} + \delta_{trm} \tag{3.8.32}$$

Then on the basis of Equation 3.8.28, we have:

$$P_{max} = \frac{|e_t||e_s|}{X_s}\sin\delta_{stm} \tag{3.8.33}$$

which, with Equation 3.8.31, leads us to:

$$
\begin{aligned}
|e_r| &= \frac{|e_t|(X_s + X_r)}{X_s}\sin\delta_{stm} \\
&= \frac{|e_t|(X_s + X_r)}{2jX_s}\left(\varepsilon^{j\delta_{stm}} - \varepsilon^{-j\delta_{stm}}\right)
\end{aligned}
\tag{3.8.34}
$$

Substituting Equation 3.8.34 into Equation 3.8.29, we obtain, after simplifying:

$$\tilde{S}_t = j\frac{|e_t|^2}{2}\left[\left(\frac{1}{X_r} - \frac{1}{X_s}\right) + \left(\frac{1}{X_r} + \frac{1}{X_s}\right)\varepsilon^{-j2\delta_{stm}}\right] \tag{3.8.35}$$

Thus, the locus for $P_{\max}$, which depends upon the relationship between $\tilde{e}_s$ and $\tilde{e}_r$ for fixed $\tilde{e}_t$, is a circle with the center at:

$$C = j\frac{|e_t|^2}{2}\left[\frac{1}{X_r} + \frac{1}{X_s}\right] \tag{3.8.36}$$

and radius:

$$R = \frac{|e_t|^2}{2}\left[\frac{1}{X_r} + \frac{1}{X_s}\right] \tag{3.8.37}$$

Figure 3.8.6 shows the real power transfer limit circle diagram for the arrangement in Figure 3.8.5.

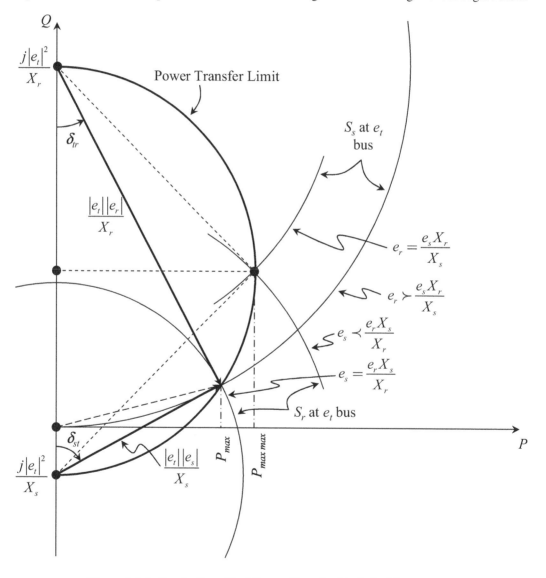

Figure 3.8.6: Steady-State Real Power Transfer Limit for Figure 3.8.5

It is left for the reader to observe that the above analysis could have been carried out graphically in the following manner. With the center at $j\frac{|e_t|^2}{X_r}$, plot the family of concentric sending end power circles with radii $\frac{|e_t||e_r|}{X_r}$. With the center at $-j\frac{|e_t|^2}{X_s}$, plot the family of concentric receiving end power circles with radii $\frac{|e_t||e_s|}{X_s}$. A pair of radii must meet in order to have a common P_{max}. Since $\delta_{stm} + \delta_{trm} = \pi/2$, the resultant right triangle must have its vertex on a semicircle.

Observe that the maximum P_{max}, which can be transferred occurs when $\delta_{stm} = \delta_{trm} = \pi/4$, which occurs when $\frac{|e_r|}{X_r} = \frac{|e_s|}{X_s}$. It is common practice to plot this real power transfer limit in normalized form (i.e., $\frac{P}{|e_t|^2}$ vs. $\frac{Q}{|e_t|^2}$) as shown in Figure 3.8.6.

When e_s and X_s represent synchronous generator quantities, we shall refer to this real power transfer limit circle diagram as the *"steady-state stability limit"* of a generator, since this is the limit beyond which the machine falls out of synchronism with respect to a power system.

3.8.5   Surge Impedance Loading

We have seen in the previous sections that practical electric power networks have limits to their capability for transferring real power. Having arrived at the ability to predict these limits, we might conclude that we should attempt to load all network elements up to and as close as possible to these limits. This is neither practical nor possible. A synchronous generator with a per unit synchronous reactance of 1.6 and a rated power factor of 0.90 would operate at a power angle of about 40 degrees, and the phase angle along a transmission line should not normally exceed about 30 degrees. While we shall ultimately be in a position to quite accurately determine the maximum practical loading limit of any specific transmission line section, it is of interest at this point to review past history and determine to what extent existing transmission lines are loaded. The surge impedance loading (SIL) of a line has been adopted as a practical measure for comparisons. SIL is the load that a three-phase line would carry if each phase were terminated in a resistance equal to the surge impedance of the line ($Z_o = \sqrt{\frac{L}{C}}$). Thus:

$$SIL = \frac{(kV_{LL})^2}{Z_o}$$

Under this operating condition, the line charging MVA is about equal to the series inductive MVA of the line. Table 3.8.1 is a representative practice in the United States for *uncompensated* lines.

	Representative Line Loading as a Function of Line Length and SIL	
	Line Length (km)	Line Loading (SIL)
	80	3.00
Table 3.8.1	100	2.70
	200	1.70
	300	1.30
	400	1.10
	500	1.00
	750	0.75

In order to permit lines to be loaded above the values shown in Table 3.8.1, various methods of compensation may be employed if the economics permit.

3.8.6   Series Capacitor Compensation

Under normal load conditions, the voltage regulation of a line section is determined principally by the series impedance of the line. For short- to moderate-length lines, voltage regulating transformers with the capability for tap-changing under load (TUL) are often sufficient to control the voltage regulation problem.

For longer lines, adequate range in tap positions may not be practical. In this case, series capacitors may be installed in the line, thereby reducing the effective series impedance of the line. If and when series capacitors are to be employed, it is common practice to introduce $-jX_C = 0.70jX_L$ (70% series compensation). The entire capacitance may be injected at the center of the line, or one-half may be injected at each end of the line. The decision is partly a matter of economics. There are other considerations, however. If we consider the line to suffer a continuous voltage drop along the length of the line, then at the discrete locations of the capacitor banks there will be a large voltage rise. Such injected series capacitor banks may limit the selection of relay protection that may be employed. Furthermore, fault and surge overvoltage conditions generally require disconnect, grounding, or bypass switches, and special gaps across the capacitors to protect them from overvoltages.

3.8.7   Shunt Reactor Compensation

According to the per unit data for the 100 km line in Equation 1.6.29, we observe that under no load conditions (i.e., $i_r = 0$), the sending end voltage is smaller than the receiving end voltage. Due to the line-charging capacitance there is therefore a voltage rise (Ferranti effect) towards the receiving end of the line.

In order to prevent excessive voltages at the receiving end of a long unloaded line, it may be necessary to install shunt (inductive) reactors. If the reactors are connected directly from line to ground, they are usually iron-cored reactors in oil-filled tanks. Low voltage reactors, connected to a transformer tertiary winding, are generally air-cored. Usually, shunt reactors are employed to compensate for 60% of the positive sequence line-charging MVA of the line section. Switches are provided to control the compensation as required. The "Near Resonance" problem associated with shunt reactor application will be taken up later.

3.8.8   Voltage Regulating and Phase-Shifting Transformers

Power transformers with tap-changing under load (TUL) capabilities are generally employed to maintain a reasonably constant voltage at the receiving end of the line. Invariably the tap changing is accomplished in such a fashion that only voltage magnitudes are varied, not phase angles.

A phase-shifting transformer is a power transformer provided with taps to permit a shift in phase from one side of the transformer to the other without changing the voltage magnitude.

The phase-shifting transformer controls the real power flow in the network, whereas the normal tap-changing transformer controls the var flow and voltage magnitude in the network.

3.8.9   Shunt Capacitors

A load, consisting in large part of induction motors, may have a relatively poor lagging power factor. That is to say, for a given line kVA, only a portion represents useful kW. The additional kVAR required by the load not only limits the total kVA that the line can carry, but may also lead to severe voltage regulation problems.

In such instances it may become technically and economically desirable to install a bank of shunt capacitors at or near the load. In this manner, the kVAR required by the load will be available at the load. At the sending end of the line, the load will appear to have a power factor much closer to unity.

3.8.10 Synchronous Condensers

In a synchronous machine, increasing the excitation (field current) will cause an increase in kVAR output. This characteristic is displayed graphically in Figure 3.8.7 in the form of so-called "V-curves."

A synchronous condenser is a special synchronous machine, with no capability for delivering either mechanical torque or electrical kW. The synchronous condenser is located at some appropriate point in a power system where there is a great demand for kVAR. By controlling the excitation of the machine, the output kVAR can be regulated. In such situations the output kVAR is regulated to maintain a given voltage at the point of installation (as opposed to maintaining a given power factor).

For improving the power factor of an otherwise highly inductive motor load, consideration should be given to employing some synchronous motors when the nature of at least part of the mechanical load will permit. By proper control of the excitation of synchronous motors, the kVAR required by the remaining induction motor load can be generated, thereby improving the net power factor of the total load.

3.9 Steady-State Synchronous Machine Characteristics

It is assumed that the reader is familiar with the essential principles underlying the performance of synchronous machines under balanced, steady-state, operation and will recognize the equivalent circuit shown in Figure 3.9.1. A review of Appendix I may be helpful.

The voltage E_i represents the 60 Hz open-circuit voltage of a synchronous machine and is directly proportional (neglecting saturation) to the field current (excitation). X_{ad} represents the magnetizing reactance (proportional to the permeance of the main or mutual flux path) of the machine, and X_ℓ represents reactance due to the leakage flux of the stator winding. For completeness, an element R_a, representing the stator (armature) winding resistance should be added in a series. Generally, R_a is negligibly small.

Shown on the diagram is e_g, which represents the voltage induced in the stator winding due to the net flux in the air gap of the machine. The voltage e_t will be somewhat different than this due to the leakage reactance drop through the stator winding.

The flux in the air gap of the machine is the net result of field current MMF_f and armature reaction MMF_a. The reduction in flux due to armature (load) current is reflected in the network by $jX_{ad}i_t$, and this voltage drop is often referred as the voltage due to armature reaction.

Figure 3.9.2 shows the phasor diagram for a round rotor synchronous generator. Often the synchronous impedance is given as:

$$Z_s = R_a + j(X_\ell + X_{ad}) \qquad (3.9.1)$$

On the basis of the phasor diagram, for a given e_t and i_t, at some power factor angle Φ, we may determine the required excitation. On the basis of what we learned in Section 3.8, we may write:

$$P = \frac{E_i e_t}{X_d} \sin \delta_{it} \qquad (3.9.2)$$

and recognize that the maximum real power transfer capability for the machine is synonymous with its steady-state stability limit.

We content ourselves, at this point, with this very brief and elementary representation for the synchronous machine. We postpone until a subsequent course a more detailed analysis of both non-salient and salient-pole machines as required for an understanding of steady-state, transient, and dynamic stability, and the impact of appropriate excitation systems upon such a performance.

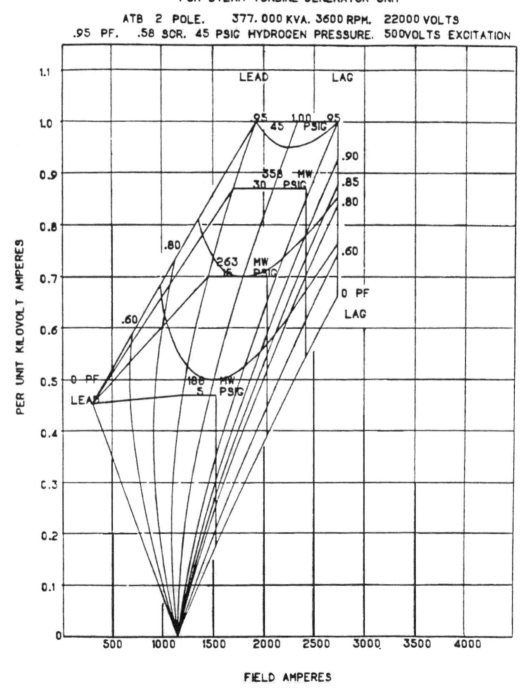

Figure 3.8.7: Compounding Curves (V-Curves) for a Synchronous Generator

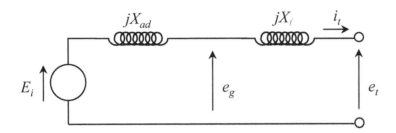

Figure 3.9.1: Equivalent Network for Synchronous Machines

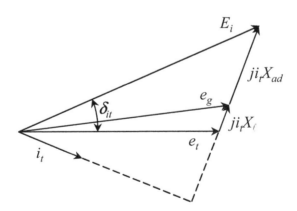

Figure 3.9.2: Synchronous Generator Phasor Diagram

3.10   Three-Phase Four-Wire Network

The series impedance of most three-phase power system components can be represented entirely by a 3 x 3 matrix. The series impedance for transmission line sections with one or more ground wires and/or subconductors is initially defined in terms of a much larger matrix. However, as described in Chapter 1, Section 1.2.6 the series impedance matrix can always be reduced to a 3 x 3 matrix (for a single circuit) representing a single equivalent conductor per phase.

In a three-phase four-wire network, the neutral return conductor (whatever form it may take) cannot, in general, be treated in the same fashion as a ground wire. Recall that a ground wire was assumed to be effectively grounded at every pole or tower, and for a line section consisting of a large number of spans, the approximation was made that the potential between ground wire and earth was uniformly zero along the length of the ground wire.

Here we define the fourth wire (neutral return) as a continuous conductor connecting only the voltage source neutral to the load neutral. In the most general analysis, this fourth wire will have a self-impedance and will also be mutually coupled magnetically with the phase conductors. On the basis of Figure 3.10.1, we wish to review the analytical approach required for this kind of configuration.

According to Kirchhoff's voltage Law, we may write the following equations:

$$
\begin{aligned}
\tilde{e}_a \;=\;& (\tilde{Z}_{aa} + \tilde{Z}_A)\tilde{i}_A + (\tilde{Z}_N + \tilde{Z}_{nn} + \tilde{Z}_n)\tilde{i}_n \\
&+ \tilde{Z}_{ab}\tilde{i}_b + \tilde{Z}_{ac}\tilde{i}_c - \tilde{Z}_{an}\tilde{i}_n - \tilde{Z}_{na}\tilde{i}_a - \tilde{Z}_{nb}\tilde{i}_b - \tilde{Z}_{nc}\tilde{i}_c
\end{aligned}
\tag{3.10.1}
$$

If we take:

$$
\tilde{i}_n = \tilde{i}_a + \tilde{i}_b + \tilde{i}_c
\tag{3.10.2}
$$

and substitute into Equation 3.10.1, we obtain:

$$
\begin{aligned}
\tilde{e}_a =\;& (\tilde{Z}_{aa} + \tilde{Z}_A + \tilde{Z}_N + \tilde{Z}_{nn} + \tilde{Z}_n - 2\tilde{Z}_{an})\tilde{i}_a \\
&+ (\tilde{Z}_{ab} + \tilde{Z}_N + \tilde{Z}_{nn} + \tilde{Z}_n - \tilde{Z}_{an} - \tilde{Z}_{bn})\tilde{i}_b \\
&+ (\tilde{Z}_{ac} + \tilde{Z}_N + \tilde{Z}_{nn} + \tilde{Z}_n - \tilde{Z}_{an} - \tilde{Z}_{cn})\tilde{i}_c
\end{aligned}
\tag{3.10.3}
$$

Similarly, for the other two phases we obtain:

$$
\begin{aligned}
\tilde{e}_b =\;& (\tilde{Z}_{ab} + \tilde{Z}_N + \tilde{Z}_{nn} + \tilde{Z}_n - \tilde{Z}_{bn} - \tilde{Z}_{an})\tilde{i}_a \\
&+ (\tilde{Z}_{bb} + \tilde{Z}_B + \tilde{Z}_N + \tilde{Z}_{nn} + \tilde{Z}_n - 2\tilde{Z}_{bn})\tilde{i}_b \\
&+ (\tilde{Z}_{bc} + \tilde{Z}_N + \tilde{Z}_{nn} + \tilde{Z}_n - \tilde{Z}_{bn} - \tilde{Z}_{cn})\tilde{i}_c
\end{aligned}
\tag{3.10.4}
$$

$$
\begin{aligned}
\tilde{e}_c =\;& (\tilde{Z}_{ac} + \tilde{Z}_N + \tilde{Z}_{nn} + \tilde{Z}_n - \tilde{Z}_{cn} - \tilde{Z}_{an})\tilde{i}_a \\
&+ (\tilde{Z}_{bc} + \tilde{Z}_N + \tilde{Z}_{nn} + \tilde{Z}_n - \tilde{Z}_{cn} - \tilde{Z}_{bn})\tilde{i}_b \\
&+ (\tilde{Z}_{cc} + \tilde{Z}_C + \tilde{Z}_N + \tilde{Z}_{nn} - \tilde{Z}_n - 2\tilde{Z}_{cn})\tilde{i}_c
\end{aligned}
\tag{3.10.5}
$$

In matrix format, Equations 3.10.3, 3.10.4, and 3.10.5 may be written as:

$$
\begin{bmatrix} \tilde{e}_a \\ \tilde{e}_b \\ \tilde{e}_c \end{bmatrix} =
\begin{bmatrix}
\tilde{Z}'_{aa} & \tilde{Z}'_{ab} & \tilde{Z}'_{ac} \\
\tilde{Z}'_{ba} & \tilde{Z}'_{bb} & \tilde{Z}'_{bc} \\
\tilde{Z}'_{ca} & \tilde{Z}'_{cb} & \tilde{Z}'_{cc}
\end{bmatrix}
\begin{bmatrix} \tilde{i}_a \\ \tilde{i}_b \\ \tilde{i}_c \end{bmatrix}
\tag{3.10.6}
$$

whereby $[\tilde{Z}'_\Phi]$ is a (3 x 3) matrix.

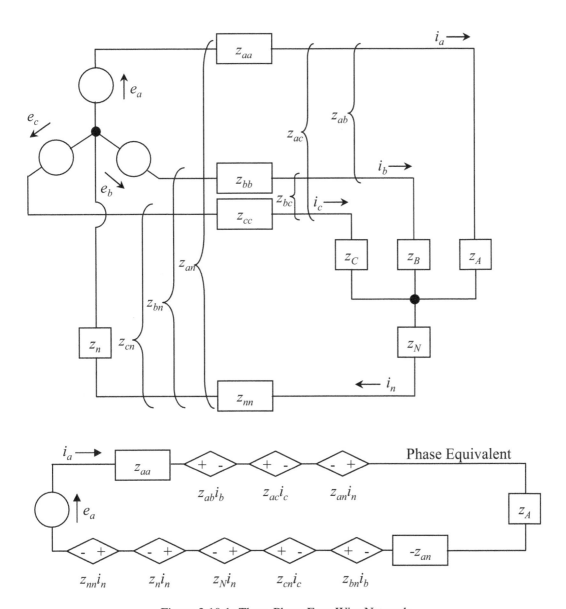

Figure 3.10.1: Three-Phase Four-Wire Network

3.11 Exercises

3.1 Three identical transformers have three windings that are rated at 65.8 kV, 38.1 kV, and 13.86 kV, respectively. The 13.86 kV windings are connected in delta and energized with rated voltage at rated frequency. The 38.1 kV windings are connected in wye and supply 9 MVA at 0 P.F. leading. The 65.8 kV windings are connected in wye and supply 15 MVA at 0.8 lagging P.F. Neglect the impedance of the transformers. Compute the secondary line voltages (L-L) and input line current.

3.2 Three identical 25 kVA transformers have three windings that are rated at 2200 V, 220 V, and 220 V, respectively. The high voltage windings are connected in delta and energized with rated voltage at rated frequency. Draw diagrams showing how to connect the low voltage windings so that balanced systems of three-phase voltages can be obtained that have the following line (L-L) voltages. All windings must be used in each connection.

(a) 220 V	(d) 582 V
(b) 381 V	(e) 762 V
(c) 440 V	(f) 762 V

3.3 Three transformers, with primaries connected in wye and secondaries connected in delta, are in parallel on the primary side with three others whose primaries and secondaries are connected in wye. If a secondary terminal of the first set is connected to a secondary terminal of the second set, what are the largest and smallest voltages that can exist between the remaining secondary terminals of the two sets? The line (L-L) voltages on the secondary sides are 1100 V for each set of transformers.

3.4 Three identical single-phase transformers are each rated 5 kVA, 2400 V, 120 V, 120 V. A three-phase source of 2400 V is available and a balanced three-phase load at 120 V is to be supplied. The high voltage windings are connected in delta.

(a) One proposal for the secondary connections is that the LV windings of each transformer be connected in a series. Then, with the LV windings connected in delta, and the load supplied from mid-taps, what maximum kVA may be supplied (to a single three-phase load) without exceeding the current rating of any of the windings?

(b) A second proposal is that only one-half of each low voltage winding be used, these being connected in delta. With these connections, what maximum kVA may be supplied without exceeding the current rating of any of the windings?

(c) With maximum permissible load supplied, compute the primary winding current in (a) and (b).

(d) In (a), a balanced three-phase load of 4 kVA at unity P.F. is being supplied at 120 V. Another three-phase load at 0.866 lagging P.F. is to be supplied from the ends (240 V) of the windings. How many kVA can be supplied to the second load?

3.5 Three identical single-phase transformers have 100-turn windings connected in wye to a 100 $\sqrt{3}$ V source. The 200-turn secondary windings are properly connected for delta, except that one corner is open. The rms voltage across the corner is 300 V. Compute the rms voltage across each secondary winding and that across two windings. The source voltage is sinusoidal at a fundamental frequency.

3.6 When a single-phase transformer is operated at rated sinusoidal voltage and frequency, the exciting current contains a fundamental frequency component and a third harmonic component (ignore higher frequency harmonics). Three of these transformers are connected in delta and connected to a balanced three-phase voltage source at a rated frequency. With rated voltage across each transformer, the rms line current is one ampere, and the rms winding current is one ampere. Determine the rms values of each of the components of the winding currents.

3.7 A single-phase 10 MVA, 66/12.7 kV transformer has an impedance of 7%. If a bank of three units is connected at 66/22 kV, what are the values of Z_{sc} in ohms on the 66 kV side and on the 22 kV side?

3.8 Three identical single-phase transformers are each rated 5 MVA, 66/13.2 kV. Each has 9% impedance and 0.65% resistance. They are connected delta-delta and deliver 10 MW at 0.8 lagging P.F. at an output voltage of 13.2 kV. At a rated frequency, compute the input voltage.

3.9 Show, by means of a phasor diagram, how to connect a bank of wye-delta transformers in parallel with a bank of delta-wye transformers.

3.10 Three identical single-phase transformers, each having 10% impedance, are to be connected in delta-delta. By an error, when the third winding was connected to close the secondary delta, its connections were reversed from what they should have been. With rated voltage applied to the primary windings, compute the secondary winding currents, the primary winding currents, and the primary line currents in terms of their rated values.

Chapter 4

Unbalanced Three-Phase Systems

Electric power systems are not always perfectly symmetrical nor do they always operate under balanced three-phase voltage and current conditions. One might, for example, wish to carry a single-phase load. We shall see that this is not always possible unless special provisions are made. Certainly most faults are unsymmetrical in nature.

We therefore need to review the performance characteristics of various network components from an unbalanced operation point of view, to which we devote the next two chapters. First, however, there are a few situations that can be analyzed most easily in a direct physical manner. Subsequently we will introduce the method of symmetrical components, an organized approach for handling most three-phase unbalances.

4.1 Open-Delta Connections

Consider the situation depicted in Figure 4.1.1, whereby the secondary winding in a delta connection has become inoperative. We wish to know whether the transformer bank can still continue to carry the load. If we ignore transformer leakage reactance and line impedance (the latter could be included in the load), we find that all three load impedances are subjected to a balanced three-phase voltage. Since the load currents are balanced, so are the line currents.

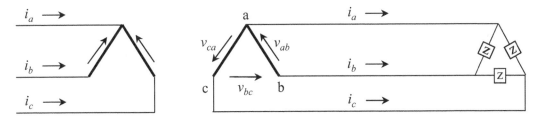

Figure 4.1.1: Open Delta Connection with Balanced Load

Since the open-delta windings now carry line current (which is times the normal delta current), we find that the windings are overloaded by 0.732 per unit. We conclude, therefore, that open-delta transformers can carry balanced three-phase load, but only 57.7% of that capable by a full delta bank. Observe also that the primary line currents are also balanced.

If allowance is made for balanced source and line impedances, all voltages and currents remain balanced. Therefore, as far as the conditions external to the transformer are concerned, we could

129

replace the transformer by a closed delta-delta or an equivalent ungrounded wye-wye transformer (see reference [1]).

If, however, we wish to include the effects of the transformer leakage reactance, which will generally be the case, then there will be a dissymmetry in the open-delta connection and the voltages and currents will no longer be balanced.

4.2   Single-Phase Load-Carrying Capability

We consider next the ability or inability of certain system configurations to carry unbalanced loads. The case of single-phase loading is relatively easy to analyze in terms of physical parameters. This physical approach also offers a better opportunity to study the conditions that may limit single-phase loading.

4.2.1   Line-to-Line Load

Figure 4.2.1(a) shows a wye-wye configuration with a line-to-line load. Ampere turn balance is attained and the load can be carried. Observe that whether one or both neutrals are grounded has no effect. Figure 4.2.1(b) shows a delta-wye connection, and the line-to-line load can be carried.

Figure 4.2.1(c) illustrates a wye-delta connection carrying line-to-line load. With the wye ungrounded, one delta winding carries twice the current in each of the other two windings. This situation is forced from the primary side by ampere turn balance considerations. What happens if the wye is grounded through some finite impedance? Nothing changes because there is no potential difference (in a perfectly symmetrical system) between the source neutral and the transformer neutral to cause current flow.

Figure 4.2.1(d) shows a delta-delta connection. The current in the secondary will divide as shown due to the leakage reactance of the transformers. The primary side exhibits the same current division due to ampere turn balance.

It is left to the reader to demonstrate that open-delta connections can also carry a single line-to-line load.

4.2.2   Line-to-Ground Load

Figure 4.2.2(a) shows a wye-wye connection with both neutrals ungrounded. Since there is no return path for the load current, no current can flow.

In Figure 4.2.2(b), with the secondary neutral grounded, there is now a return path for the load current. However, on the primary side, the current cannot flow since we cannot attain ampere turn balance in all three phases. This brings out some most important and fundamental considerations:

1. In order for currents to flow, there must exist a conductive closed path.

2. In addition, ampere-turn balance must be attained within the transformer (otherwise the net current must return via the magnetizing impedance path).

Example 4.2.1

Consider the configuration shown in Figure 4.2.2(b). Figure 4.2.3 contains additional figures to amplify this problem. Figure (a) shows the complete equivalent circuit for a single-phase transformer that reduces to (b) if we neglect losses and leakage flux. The appropriate three-phase interconnection is shown in Figure (c) with a balanced set of source voltages shown in Figure (d).

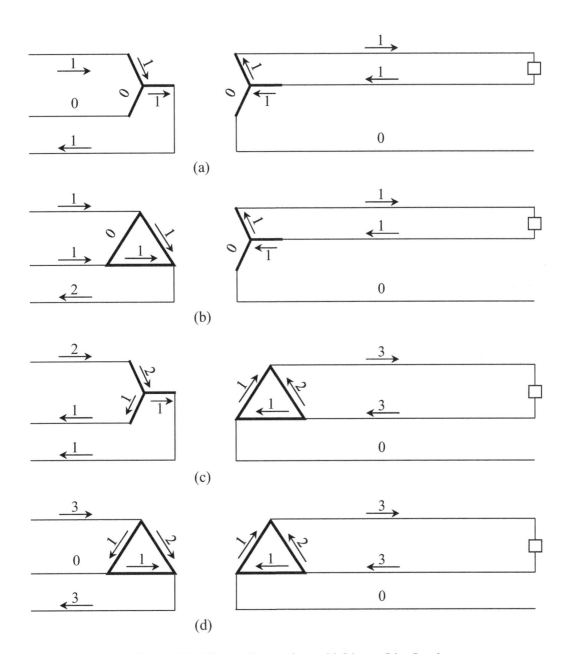

Figure 4.2.1: Various Connections with Line-to-Line Load

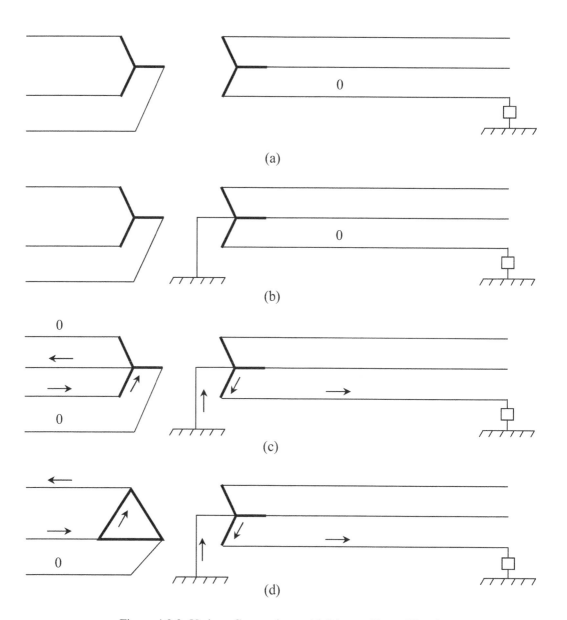

Figure 4.2.2: Various Connections with Line-to-Neutral Load

The center of the voltage triangle is identified as the geometric neutral (GN). This is also the source voltage neutral, even though this may not actually be accessible in the case of the delta-connected set of source voltages. For purposes of this example, we shall refer to $\frac{\tilde{E}_{aGN}}{jX_m}$ as the normal excitation current.

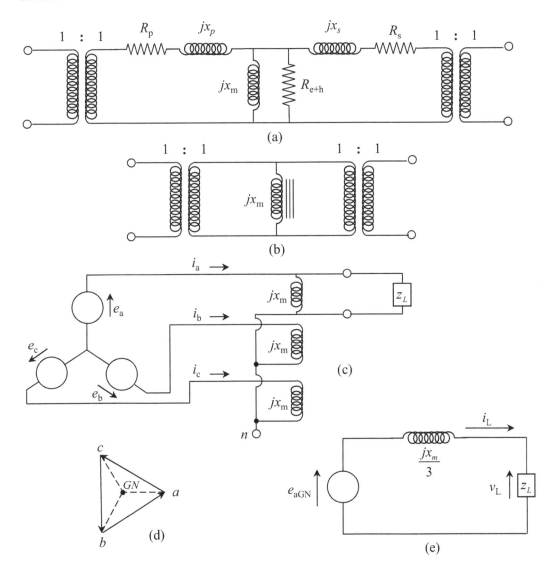

Figure 4.2.3: Figures for Example 4.2.1

We employ the Thevenin-Helmholtz theorem to obtain Figure (e). On this basis:

$$\tilde{I}_L = \frac{\tilde{E}_{aGN}}{\tilde{Z}_L + j\frac{X_m}{3}}$$

and it is apparent that X_m limits the current that can flow to the load.

Consider the limiting case for which:

$$\tilde{I}_L = \frac{3\tilde{E}_{aGN}}{jX_m}$$

and $\tilde{V}_L = \tilde{V}_{an} = 0$. This current returns and divides equally among the two X_m's. Thus, in phase "a," the net ampere turns are zero. Physically, the phase "a" primary winding carries three times normal excitation current, which is exactly balanced by the same current in the secondary winding. In the other two phases there is no ampere turn balance that accounts for the current limiting effect of the X_m's.

In phase "b," the primary excitation current is $\mid a^2 - 1 \mid = \sqrt{3}$ times normal, and in phase "c" the primary excitation current is $\mid a - 1 \mid = \sqrt{3}$ times normal. **The resultant main flux in phases "b" and "c" causes $\sqrt{3}\tilde{E}_{aGN}$ voltage magnitudes, and in phase "a" the flux is zero so that** $V_L = V_{an} = 0$.

Physically the system neutral n is not at the source neutral GN, it having been shifted to the phase "a" terminal. This results in:

$$\tilde{V}_{an} = 0$$

$$\tilde{V}_{bn} = -\tilde{V}_{ab} = \tilde{V}_{ba}$$

$$\tilde{V}_{cn} = \tilde{V}_{ca}$$

For other values of $\tilde{Z}_L$ there will of course be some current in the load, but only in the order of magnitude of magnetizing current, not in the order of magnitude of rated load current.

In Figure 4.2.2(c), both primary and secondary neutrals are grounded. Since both of the above conditions are met (assuming that the source neutral is grounded) the single-phase load can be carried.

In Figure 4.2.2(d), we observe that the delta-wye configuration can also carry a single-phase load.

If the secondary is connected in delta, no single-phase load can be carried since there exists no system neutral that can be connected to ground. In such an existing delta system, where it is desired to provide single-phase loading capability, we can accomplish this by employing grounding transformers. We shall describe two of the most common types.

Wye-Delta Grounding Transformer

Figure 4.2.4 illustrates the application of a wye-delta grounding transformer to establish a system **neutral that may be connected to ground. Conductive paths exist for the load current to flow as** shown. Additionally, ampere turn balance is attained in both the main transformer bank and in the grounding transformer bank. The delta secondary of the grounding transformer may or may not be loaded; generally, it is not. In the latter case, the grounding transformer reacts according to magnetizing reactance if the system load is perfectly balanced.

Observe that the rated MVA of the grounding transformer windings need only be one-third of the MVA of the single-phase load that they are to carry.

Zig-Zag Grounding Transformer

Figure 4.2.5(a) illustrates the application of a zig-zag grounding transformer. Again, conductive **paths exist for the load current to flow as shown. Additionally, the required ampere turn balance is** attained in both the main transformer bank and in the zig-zag transformer bank.

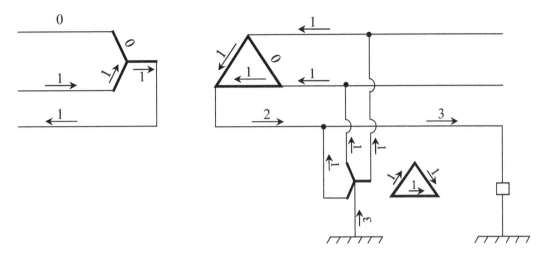

Figure 4.2.4: Application of Wye-Delta Grounding Transformer

As shown in Figure 4.2.5(b), the zig-zag is actually a two-winding transformer, with equal turns in the primary and secondary windings. The bottom end of a primary winding is connected to the bottom end of a secondary winding in an adjacent phase. When the three currents entering the primary winding are equal in magnitude and in phase, the primary and secondary ampere-turns are equal and opposite in each leg. This ampere turn balance provides an easy path for such currents to flow through the (leakage reactance) of the transformer. It is left for the reader to verify that for a balanced set of three-phase currents, ampere turn balance is not attained and the zig-zag transformer reacts according to magnetizing reactance to such currents.

Observe that while the currents in the windings are one-third of the single-phase load current, the voltage rating of each winding is only 57.7% of the system line to neutral voltage. Consequently, for grounding transformer purposes, a zig-zag transformer requires only 57.7% of the MVA rating of a corresponding wye-delta grounding transformer.

4.3   Symmetrical Components

Although several simple cases of system unbalance can be analyzed quite easily in a direct manner in terms of physical quantities, most system unbalances do not lend themselves to such an easy analysis. It has therefore been found desirable to develop a systematic approach for analyzing a much broader variety of system unbalances.

4.3.1   History

Historically, it was L.G. Stokvis in 1912 who indicated the possibility of resolving an unbalanced system of currents into positive and negative sequence components of current. In 1918, Dr. C.L. Fortescue presented a paper that introduced the zero sequence current and voltage concept. Even more generally, Dr. Fortescue proved that "a system of N vectors or quantities may be re-solved, when N is prime, into N different symmetrical groups or systems, one of which consists of N equal vectors and the remaining (N-1) systems consist of N equi-spaced vectors which, with the first mentioned groups of equal vectors, forms an equal number of symmetrical N-phase systems."

The method of symmetrical components is a general one, applicable to many polyphase systems. In this chapter, however, only the symmetrical component equations applicable to three-phase power

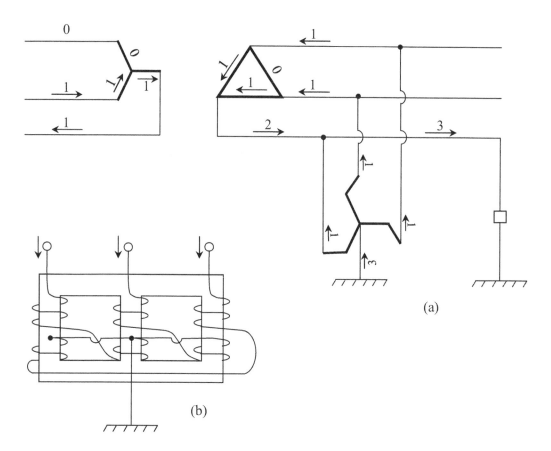

(a)

(b)

Figure 4.2.5: Application of Zig-Zag Grounding Transformer

systems will be developed.

Dr. Fortescue's systems of symmetrical components afford the possibility of resolving an unsymmetrical voltage system into three sets of symmetrical voltage systems. When the network is linear, the superposition principle may be applied to determine the phase currents from the solutions for symmetrical component currents. However, the network must also be symmetrical in order that the analysis can be based on independent or uncoupled single-phase networks.

In view of this last statement, it is perhaps somewhat surprising that symmetrical components have been applied so successfully to an unbalanced load or unbalanced fault problems. The answer lies in the fortuitous (rather than designed) ability to establish relatively simple interconnections between sequence networks, at the point of unbalance, for a limited number of unbalance conditions. As a matter of fact, if an untransposed transmission line is considered with $\tilde{Z}_{ab} \neq \tilde{Z}_{bc} \neq \tilde{Z}_{ca}$, the method of symmetrical components offers almost no simplification of the analysis.

Nonetheless, a sufficiently large number of studies can be based on the assumption of a symmetrical network, at least for practical purposes, so that the method of symmetrical components has become widely accepted. Other systems of components have also been proposed and applied to advantage, and we shall consider one of these, the α, β, 0 (or Clarke) components in a subsequent chapter.

In some instances it will not be permissible to neglect dissymmetries introduced, for example, by the untransposed transmission line without introducing a significant error into the results. Such is the case in the analysis of radio noise propagation and traveling waves. Here the emphasis is upon finding an appropriate transformation that will allow the three-phase network to be replaced by single-phase networks that are not mutually coupled to one another. In the language of matrix algebra, we seek a transformation that will diagonalize a given physical impedance matrix, no matter how unsymmetrical this matrix might be. The complex mathematical process required to determine this transformation is the nature of the eigenvalue and eigenvector problem. In terms of engineering language we speak of modal components and the modal transformation. While we shall not pursue this subject further in this book, it should be noted that both the symmetrical component transformation and the α, β, 0 component transformation can be shown to be special cases of modal transformations when the three-phase networks are inherently symmetrical.

Finally, looking both backward and forward, it should be pointed out that symmetrical components, and related systems, were developed several decades ago when network solutions were obtained by hand calculations or, at best, by analog simulations as, for example, the alternating current network analyzer. Today, where high-speed digital computer programs are available for solving network problems, some of the advantage of the component systems is lost. Digital computers can handle matrix algebra operations quite easily and very often it may be found more convenient to perform the entire analysis in terms of physical or three-phase quantities directly rather than to transform to other quantities first.

4.3.2 Unbalanced Voltage Decomposition

Consider the three-phase configuration shown in Figure 4.3.1. The three-phase source voltages consist of the superposition of three different voltages. Let the source and line impedance be negligible, and the load impedance balanced.

Let one three-phase set of voltages be defined by:

$$\begin{bmatrix} \tilde{E}_{a1} \\ \tilde{E}_{b1} \\ \tilde{E}_{c1} \end{bmatrix} = (\tilde{E}_{a1}) \begin{bmatrix} 1 \\ a^2 \\ a \end{bmatrix} \tag{4.3.1}$$

and the second by:

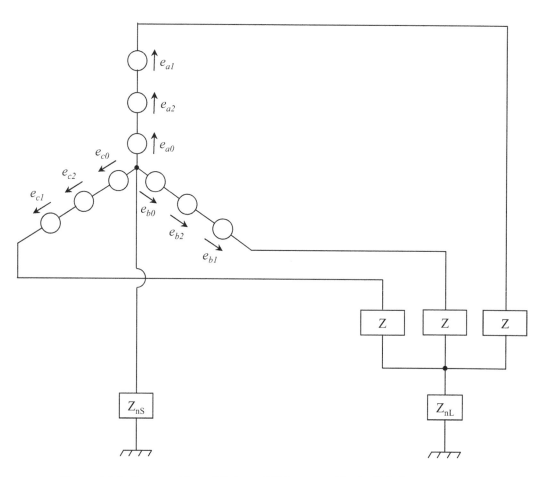

Figure 4.3.1: Superposition of Balanced Voltages to Obtain Unbalanced Voltages

$$\begin{bmatrix} \tilde{E}_{a2} \\ \tilde{E}_{b2} \\ \tilde{E}_{c2} \end{bmatrix} = (\tilde{E}_{a2}) \begin{bmatrix} 1 \\ a \\ a^2 \end{bmatrix} \tag{4.3.2}$$

and the third by:

$$\begin{bmatrix} \tilde{E}_{a0} \\ \tilde{E}_{b0} \\ \tilde{E}_{c0} \end{bmatrix} = (\tilde{E}_{a0}) \begin{bmatrix} 1 \\ 1 \\ 1 \end{bmatrix} \tag{4.3.3}$$

Then:

$$\begin{aligned}
\tilde{E}_a &= \tilde{E}_{a0} + \tilde{E}_{a1} + \tilde{E}_{a2} = \tilde{E}_{a0} + \tilde{E}_{a1} + \tilde{E}_{a2} & (4.3.4) \\
\tilde{E}_b &= \tilde{E}_{b0} + \tilde{E}_{b1} + \tilde{E}_{b2} = \tilde{E}_{a0} + a^2 \tilde{E}_{a1} + a \tilde{E}_{a2} & (4.3.5) \\
\tilde{E}_c &= \tilde{E}_{c0} + \tilde{E}_{c1} + \tilde{E}_{c2} = \tilde{E}_{a0} + a \tilde{E}_{a1} + a^2 \tilde{E}_{a2} & (4.3.6)
\end{aligned}$$

and:

$$\begin{bmatrix} \tilde{E}_a \\ \tilde{E}_b \\ \tilde{E}_c \end{bmatrix} = \begin{bmatrix} 1 & 1 & 1 \\ 1 & a^2 & a \\ 1 & a & a^2 \end{bmatrix} \begin{bmatrix} \tilde{E}_{a0} \\ \tilde{E}_{a1} \\ \tilde{E}_{a2} \end{bmatrix} \tag{4.3.7}$$

Example 4.3.1

Let $\tilde{E}_{a0} = a$, $\tilde{E}_{a1} = 1$, and $\tilde{E}_{a2} = -a$. Then:

$$\begin{bmatrix} \tilde{E}_a \\ \tilde{E}_b \\ \tilde{E}_c \end{bmatrix} = \begin{bmatrix} 1 & 1 & 1 \\ 1 & a^2 & a \\ 1 & a & a^2 \end{bmatrix} \begin{bmatrix} a \\ 1 \\ -a \end{bmatrix} = \begin{bmatrix} 1 \\ a \\ -2 + j\sqrt{3} \end{bmatrix}$$

The source voltages are very unbalanced as shown in Figure 4.3.2.

With different choices for $\tilde{E}_{a0}$, $\tilde{E}_{a1}$, and $\tilde{E}_{a2}$, a variety of different unbalanced source voltages could be created. As a matter of fact, besides the particular sets of voltage sources selected in Equations 4.3.1, 4.3.2, and 4.3.3, we could have selected many other kinds. We shall show, shortly, why our selection happens to be a particularly convenient one.

First we shall determine the nature of the currents that flow through the loads. Since we have a linear network, we may employ the principle of superposition, taking one set of voltage sources at a time, with the others set to zero.

Consider first the set of voltage sources $\tilde{E}_{a1}$, $\tilde{E}_{b1}$, and $\tilde{E}_{c1}$. These are a balanced set of voltages with phase rotation a-b-c. In Chapter 3, Section 3.1, we examined just this kind of configuration and concluded that the phase currents were balanced. We also found that the analysis could be reduced to a single-phase analysis on the basis of Figure 3.1.3.

Consider now the second set of voltage sources $\tilde{E}_{a2}$, $\tilde{E}_{b2}$, and $\tilde{E}_{c2}$. These are again a balanced set of voltages with, however, phase rotation a-c-b. For the particular configuration in Figure 4.3.1, it is apparent that a reversal in phase rotation of the source voltages will merely result in a phase

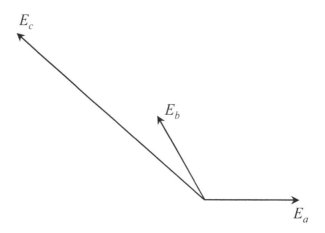

Figure 4.3.2: Resultant Unbalanced Voltages for Example 4.3.1

rotation of the currents. Therefore, the single-phase impedance network in Figure 3.1.3 is also applicable for this set of source voltages.

We consider now the final set of source voltages $\tilde{E}_{a0}$, $\tilde{E}_{b0}$, and $\tilde{E}_{c0}$. These voltages are of equal magnitude, and also in phase with one another. We need to determine, now, just what the line currents will be under this condition. We shall begin with Equation 3.1.2, and for $\tilde{E}_{a0}$, $\tilde{E}_{b0}$, and $\tilde{E}_{c0}$ we find that $\tilde{I}_{a0}$, $\tilde{I}_{b0}$, and $\tilde{I}_{c0}$. Thus, we have:

$$\tilde{I}_{a0} = \frac{\tilde{E}_{a0} - (\tilde{V}_{nS} + \tilde{V}_{nL}}{\tilde{Z}_L} \tag{4.3.8}$$

and on the basis of Equation 3.1.3:

$$(\tilde{V}_{nS} + \tilde{V}_{nL} = 3\tilde{I}_{a0}(\tilde{Z}_{nS} + \tilde{Z}_{nL}) \tag{4.3.9}$$

Substituting Equation 4.3.9 into Equation 4.3.8, we obtain:

$$\tilde{I}_{a0} = \frac{\tilde{E}_{a0}}{\tilde{Z}_L + 3(\tilde{Z}_{nS} + \tilde{Z}_{nL})} \tag{4.3.10}$$

Again the analysis may be carried out on a single-phase basis. Figure 4.3.3 summarizes the appropriate equivalent single-phase networks.

Thus, by our judicious selection of three sets of source voltages defined by Equations 4.3.1, 4.3.2, and 4.3.3, we find that the currents associated with these voltages can be determined on the basis of single-phase network representations. The net line currents are simply the superposition of these individual solutions. We also observe, at this point, that the network response, while the same for $\tilde{E}_{a1}$ and $\tilde{E}_{a2}$ (in this case), is different for $\tilde{E}_{a0}$.

In Example 4.3.1 we chose certain values for our selected sets of source voltages to create an unbalanced set of three-phase source voltages. In actual practice, we are given a set of three-phase unbalanced voltages or currents and are asked to predict system performance. If, indeed, we could take a set of three-phase unbalanced voltages, and determine three sets of voltages defined by Equations 4.3.1 through 4.3.3, then we would be able to perform three single-phase analyses per Figure 4.3.3, and by superposition obtain the desired results. In short, if we can perform the inverse of Equation 4.3.7, then we will have attained our objective. The reader may verify that:

$$
\begin{bmatrix} \tilde{E}_{a0} \\ \tilde{E}_{a1} \\ \tilde{E}_{a2} \end{bmatrix} = \left(\frac{1}{3}\right) \begin{bmatrix} 1 & 1 & 1 \\ 1 & a & a^2 \\ 1 & a^2 & a \end{bmatrix} \begin{bmatrix} \tilde{E}_a \\ \tilde{E}_b \\ \tilde{E}_c \end{bmatrix} \tag{4.3.11}
$$

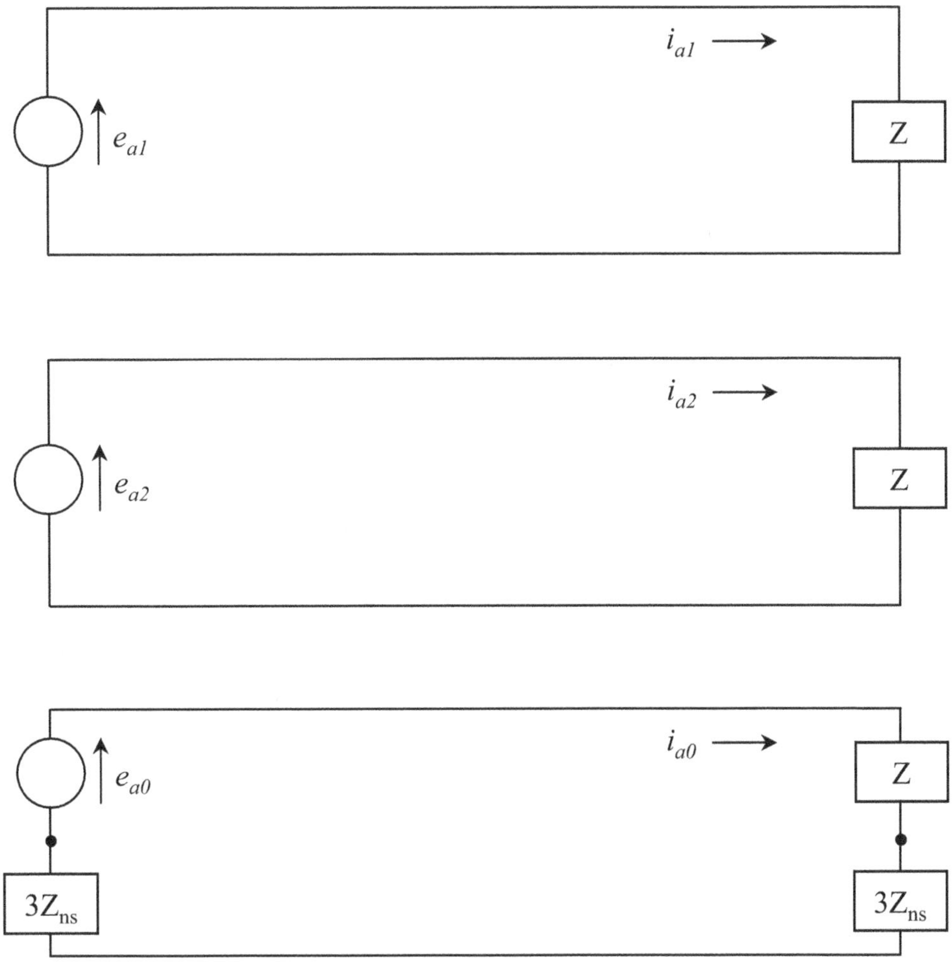

Figure 4.3.3: Equivalent Network Appropriate to Each Set of Balanced Voltages

Equation 4.3.11 states, that, given an unbalanced set of three-phase voltages, three sets of balanced voltages according to Equations 4.3.1, 4.3.2, and 4.3.3 can always be evaluated.

For the particular configuration in Figure 4.3.1, we have shown that the analysis may be carried out on the basis of Figure 4.3.3. By superposition it follows that:

$$
\begin{bmatrix} \tilde{I}_a \\ \tilde{I}_b \\ \tilde{I}_c \end{bmatrix} = \begin{bmatrix} 1 & 1 & 1 \\ 1 & a^2 & a \\ 1 & a & a^2 \end{bmatrix} \begin{bmatrix} \tilde{I}_{a0} \\ \tilde{I}_{a1} \\ \tilde{I}_{a2} \end{bmatrix} \tag{4.3.12}
$$

and:

$$\begin{bmatrix} \tilde{I}_{a0} \\ \tilde{I}_{a1} \\ \tilde{I}_{a2} \end{bmatrix} = (\frac{1}{3}) \begin{bmatrix} 1 & 1 & 1 \\ 1 & a & a^2 \\ 1 & a^2 & a \end{bmatrix} \begin{bmatrix} \tilde{I}_a \\ \tilde{I}_b \\ \tilde{I}_c \end{bmatrix} \tag{4.3.13}$$

4.3.3 Graphical Decomposition

According to Equations 4.3.11 and 4.3.13, if we are given an unbalanced set of voltages (currents), we can find three different sets of balanced voltages (currents) which, by superposition, will give the unbalanced set of voltages (currents). This decomposition can be performed graphically as illustrated in Figure 4.3.4.

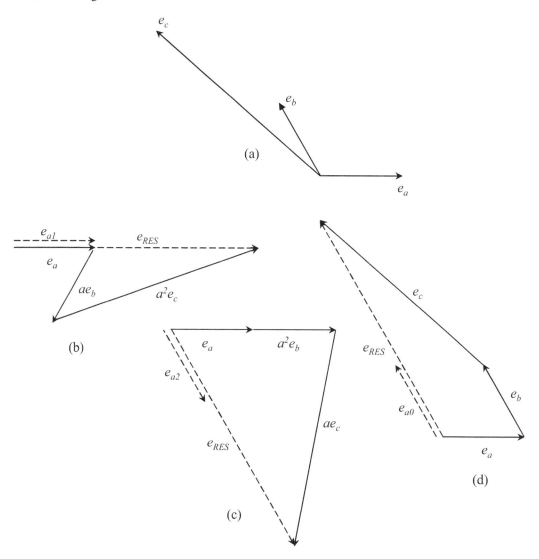

Figure 4.3.4: Graphical Decomposition of Unbalanced Three-Phase Voltages into Symmetrical Components

Figure 4.3.4(a) illustrates a set of unbalanced voltages. In Figure 4.3.4(d) we simply perform phasor addition of these unbalanced voltages. The resultant phasor, divided by three, gives the magnitude and phase of $\tilde{E}_{a0}$.

In Figure 4.3.4(b), to phasor $\tilde{E}_a$ we add $\tilde{E}_b$ rotated forward by 120 degrees, and to this we add $\tilde{E}_c$ rotated forward by 240 degrees. The resultant phasor, divided by three, gives the magnitude and phase of $\tilde{E}_{a1}$.

In Figure 4.3.4(c), to phasor $\tilde{E}_a$ we add $\tilde{E}_b$ rotated backward by 120 degrees, and to this we add $\tilde{E}_c$ rotated backward by 240 degrees. The resultant phasor, divided by three, gives the magnitude and phase of $\tilde{E}_{a2}$.

4.3.4 Matrix Transformation

From another point of view, we might consider Equations 4.3.11 and 4.3.13 as a transformation from a set of physical variables to a more convenient (though perhaps nonphysical) set of variables. Based upon the definitions in Appendix E, Section E.17, we may rewrite Equations 4.3.11 and 4.3.13 in terms of symbolic matrix notation:

$$[\tilde{E}_c] = [S][\tilde{E}_\Phi] \tag{4.3.14}$$
$$[\tilde{I}_c] = [S][\tilde{I}_\Phi] \tag{4.3.15}$$

Now, given:

$$[\tilde{E}_\Phi] = [\tilde{Z}_\Phi][\tilde{I}_\Phi] \tag{4.3.16}$$

we may write:

$$[S][\tilde{E}_\Phi] = [S][\tilde{Z}_\Phi][S]^{-1}[S][\tilde{I}_\Phi] \tag{4.3.17}$$

and:

$$[\tilde{E}_c] = [\tilde{Z}_c][\tilde{I}_c] \tag{4.3.18}$$

At this point we define the set of balanced voltages $\tilde{E}_{a1}$, $\tilde{E}_{b1}$, and $\tilde{E}_{c1}$ as the positive sequence set of voltages. We define the set of balanced voltages $\tilde{E}_{a2}$, $\tilde{E}_{b2}$, and $\tilde{E}_{c2}$ as the negative sequence set of voltages. We define the set of balanced voltages $\tilde{E}_{a0}$, $\tilde{E}_{b0}$, and $\tilde{E}_{c0}$ as the zero sequence set of voltages. They are explicitly defined by Equations 4.3.1, 4.3.2, and 4.3.3.

4.3.5 General Symmetrical Component Impedance Matrix

The characteristics of most three-phase network components can be defined in terms of a 3 x 3 matrix. In the formative stage, a transmission line section with several subconductors and ground wires requires a considerably larger impedance matrix. However, as described in Chapter 1, Sections 1.1.7.3 and 1.1.7.4, every single-circuit transmission line section can be reduced to an equivalent three-conductor line.

Thus, for a single-circuit three-phase network element we can always write:

$$[\tilde{Z}_\Phi] = \begin{bmatrix} \tilde{Z}_{aa} & \tilde{Z}_{ab} & \tilde{Z}_{ac} \\ \tilde{Z}_{ba} & \tilde{Z}_{bb} & \tilde{Z}_{bc} \\ \tilde{Z}_{ca} & \tilde{Z}_{cb} & \tilde{Z}_{cc} \end{bmatrix} \tag{4.3.19}$$

Then, in terms of symmetrical components we have:

$$\begin{bmatrix} \tilde{Z}_{00} & \tilde{Z}_{01} & \tilde{Z}_{02} \\ \tilde{Z}_{10} & \tilde{Z}_{11} & \tilde{Z}_{12} \\ \tilde{Z}_{20} & \tilde{Z}_{21} & \tilde{Z}_{22} \end{bmatrix} = (\frac{1}{3}) \begin{bmatrix} 1 & 1 & 1 \\ 1 & a & a^2 \\ 1 & a^2 & a \end{bmatrix} \begin{bmatrix} \tilde{Z}_{aa} & \tilde{Z}_{ab} & \tilde{Z}_{ac} \\ \tilde{Z}_{ba} & \tilde{Z}_{bb} & \tilde{Z}_{bc} \\ \tilde{Z}_{ca} & \tilde{Z}_{cb} & \tilde{Z}_{cc} \end{bmatrix} \begin{bmatrix} 1 & 1 & 1 \\ 1 & a^2 & a \\ 1 & a & a^2 \end{bmatrix} \quad (4.3.20)$$

If we wish to evaluate only $\tilde{Z}_{12}$, for example, which lies in the second row and third column, we have:

$$(\frac{1}{3})[1 \; a \; a^2] \begin{bmatrix} \tilde{Z}_{aa} & \tilde{Z}_{ab} & \tilde{Z}_{ac} \\ \tilde{Z}_{ba} & \tilde{Z}_{bb} & \tilde{Z}_{bc} \\ \tilde{Z}_{ca} & \tilde{Z}_{cb} & \tilde{Z}_{cc} \end{bmatrix} \begin{bmatrix} 1 \\ a \\ a^2 \end{bmatrix}$$

$$= (\frac{1}{3})[1 \; 1 \; 1] \begin{bmatrix} \tilde{Z}_{aa} & \tilde{Z}_{ab} & \tilde{Z}_{ac} \\ a\tilde{Z}_{ba} & a\tilde{Z}_{bb} & a\tilde{Z}_{bc} \\ a^2\tilde{Z}_{ca} & a^2\tilde{Z}_{cb} & a^2\tilde{Z}_{cc} \end{bmatrix} \begin{bmatrix} 1 \\ a \\ a^2 \end{bmatrix}$$

$$= (\frac{1}{3})[1 \; 1 \; 1] \begin{bmatrix} \tilde{Z}_{aa} & a\tilde{Z}_{ab} & a^2\tilde{Z}_{ac} \\ a\tilde{Z}_{ba} & a^2\tilde{Z}_{bb} & \tilde{Z}_{bc} \\ a^2\tilde{Z}_{ca} & \tilde{Z}_{cb} & a\tilde{Z}_{cc} \end{bmatrix} \begin{bmatrix} 1 \\ 1 \\ 1 \end{bmatrix}$$

$$= (\frac{1}{3})[(\tilde{Z}_{aa} + a^2\tilde{Z}_{bb} + a\tilde{Z}_{cc}) + 2(\tilde{Z}_{bc} + a^2\tilde{Z}_{ca} + a\tilde{Z}_{ab})] \quad (4.3.21)$$

With only a little practice, one can go directly from Equation 4.3.20 to the second last step in Equation 4.3.21. Table 4.3.1 defines each $[\tilde{Z}_c]$ element for a three-phase, three-wire transmission line, and Table 4.3.2 defines each $[\tilde{Z}_c]$ element for a three-phase, four-wire transmission line described in Chapter 3, Section 3.10. It is apparent that in general, $[\tilde{Z}_c]$ is not a symmetrical matrix. This is verified by Equation E.17.7 in Appendix E. Table 4.3.2 shows quite clearly that neutral self-impedances influence only those symmetrical component impedances associated with the zero sequence network.

4.3.6   Diagonal Symmetrical Component Impedance Matrix

Fortunately, practical power systems are inherently symmetrical with perhaps only minor dissymmetries (except at fault points), which may be neglected in most major types of studies. Thus, if we may assume that:

$$\tilde{Z}_{aa} = \tilde{Z}_{bb} = \tilde{Z}_{cc} \quad (4.3.22)$$

and:

$$\tilde{Z}_{ab} = \tilde{Z}_{bc} = \tilde{Z}_{ca} \quad (4.3.23)$$

Table 4.3.1
Three-phase, Three-wire Transmission Line

$$3\tilde{Z}_{00} = \quad = (\tilde{Z}_{aa} + \tilde{Z}_{bb} + \tilde{Z}_{cc}) + 2(\tilde{Z}_{bc} + \tilde{Z}_{ca} + \tilde{Z}_{ab})$$
$$3\tilde{Z}_{11} = 3\tilde{Z}_{22} = (\tilde{Z}_{aa} + \tilde{Z}_{bb} + \tilde{Z}_{cc}) - (\tilde{Z}_{bc} + \tilde{Z}_{ca} + \tilde{Z}_{ab})$$
$$3\tilde{Z}_{01} = 3\tilde{Z}_{20} = (\tilde{Z}_{aa} + a^2\tilde{Z}_{bb} + a\tilde{Z}_{cc}) - (\tilde{Z}_{bc} + a^2\tilde{Z}_{ca} + a\tilde{Z}_{ab})$$
$$3\tilde{Z}_{10} = 3\tilde{Z}_{02} = (\tilde{Z}_{aa} + a\tilde{Z}_{bb} + a^2\tilde{Z}_{cc}) - (\tilde{Z}_{bc} + a\tilde{Z}_{ca} + a^2\tilde{Z}_{ab})$$
$$3\tilde{Z}_{12} = \quad = (\tilde{Z}_{aa} + a^2\tilde{Z}_{bb} + a\tilde{Z}_{cc}) + 2(\tilde{Z}_{bc} + a^2\tilde{Z}_{ca} + a\tilde{Z}_{ab})$$
$$3\tilde{Z}_{21} = \quad = (\tilde{Z}_{aa} + a\tilde{Z}_{bb} + a^2\tilde{Z}_{cc}) + 2(\tilde{Z}_{bc} + a\tilde{Z}_{ca} + a^2\tilde{Z}_{ab})$$

Table 4.3.2
Three-phase, Four-wire Transmission Line

$$3\tilde{Z}_{00} = (\tilde{Z}_{aa} + \tilde{Z}_{bb} + \tilde{Z}_{cc} + \tilde{Z}_A + \tilde{Z}_B + \tilde{Z}_C)$$
$$2(\tilde{Z}_{bc} + \tilde{Z}_{ca} + \tilde{Z}_{ab})$$
$$-6(\tilde{Z}_{an} + \tilde{Z}_{bn} + \tilde{Z}_{cn}) + 9(\tilde{Z}_N + \tilde{Z}_{nn} + \tilde{Z}_n)$$

$$3\tilde{Z}_{11} = 3\tilde{Z}_{22} = (\tilde{Z}_{aa} + \tilde{Z}_{bb} + \tilde{Z}_{cc} + \tilde{Z}_A + \tilde{Z}_B + \tilde{Z}_C)$$
$$-(\tilde{Z}_{bc} + \tilde{Z}_{ca} + \tilde{Z}_{ab})$$

$$3\tilde{Z}_{01} = 3\tilde{Z}_{20} = (\tilde{Z}_{aa} + a^2\tilde{Z}_{bb} + a\tilde{Z}_{cc} + \tilde{Z}_A + a^2\tilde{Z}_B + a\tilde{Z}_C)$$
$$-(\tilde{Z}_{bc} + a^2\tilde{Z}_{ca} + a\tilde{Z}_{ab})$$
$$-3(\tilde{Z}_{an} + a^2\tilde{Z}_{bn} + a\tilde{Z}_{cn})$$

$$3\tilde{Z}_{10} = 3\tilde{Z}_{02} = (\tilde{Z}_{aa} + a\tilde{Z}_{bb} + a^2\tilde{Z}_{cc} + \tilde{Z}_A + a\tilde{Z}_B + a^2\tilde{Z}_C)$$
$$-(\tilde{Z}_{bc} + a\tilde{Z}_{ca} + a^2\tilde{Z}_{ab})$$
$$-3(\tilde{Z}_{an} + a\tilde{Z}_{bn} + a^2\tilde{Z}_{cn})$$

$$3\tilde{Z}_{12} = (\tilde{Z}_{aa} + a^2\tilde{Z}_{bb} + a\tilde{Z}_{cc} + \tilde{Z}_A + a^2\tilde{Z}_B + a\tilde{Z}_C)$$
$$2(\tilde{Z}_{bc} + a^2\tilde{Z}_{ca} + a\tilde{Z}_{ab})$$

$$3\tilde{Z}_{21} = (\tilde{Z}_{aa} + a\tilde{Z}_{bb} + a^2\tilde{Z}_{cc} + \tilde{Z}_A + a\tilde{Z}_B + a^2\tilde{Z}_C)$$
$$2(\tilde{Z}_{bc} + a\tilde{Z}_{ca} + a^2\tilde{Z}_{ab})$$

---

the equations in Table 4.3.1 reduce to:

$$\tilde{Z}_{00} = \tilde{Z}_{aa} + 2\tilde{Z}_{ab} \tag{4.3.24}$$

$$\tilde{Z}_{11} = \tilde{Z}_{22} = \tilde{Z}_{aa} - \tilde{Z}_{ab} \tag{4.3.25}$$

$$\tilde{Z}_{01} = \tilde{Z}_{02} = \tilde{Z}_{12} = \tilde{Z}_{21} = \tilde{Z}_{10} = \tilde{Z}_{20} = 0 \tag{4.3.26}$$

There then results:

$$\begin{bmatrix} \tilde{E}_0 \\ \tilde{E}_1 \\ \tilde{E}_2 \end{bmatrix} = \begin{bmatrix} \tilde{Z}_{00} & 0 & 0 \\ 0 & \tilde{Z}_{11} & 0 \\ 0 & 0 & \tilde{Z}_{22} \end{bmatrix} \begin{bmatrix} \tilde{I}_0 \\ \tilde{I}_1 \\ \tilde{I}_2 \end{bmatrix} \tag{4.3.27}$$

and, under the stipulated assumptions, thereby obtaining three uncoupled single-phase networks representing a three-phase mutually coupled physical network as illustrated in Figure 4.3.5.

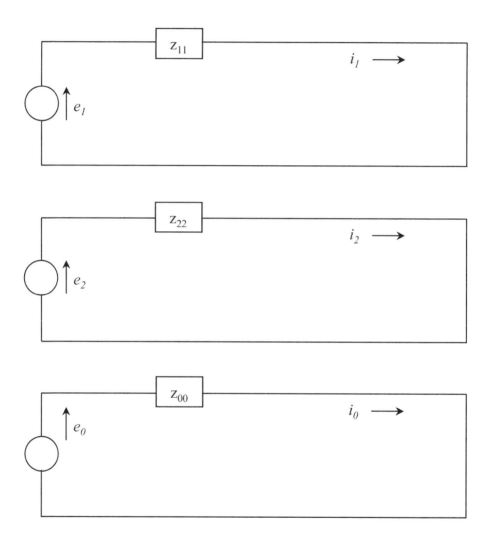

Figure 4.3.5: Symmetrical Component Equivalent Networks

For the three-phase four-wire configuration, the equations in Table 4.3.2 reduce to:

$$\tilde{Z}_{00} = (\tilde{Z}_{aa} + \tilde{Z}_A) + 2\tilde{Z}_{ab} + 3(\tilde{Z}_N + \tilde{Z}_{nn} + \tilde{Z}_n) - 6\tilde{Z}_{an} \tag{4.3.28}$$

$$\tilde{Z}_{11} = \tilde{Z}_{22} = (\tilde{Z}_{aa} + \tilde{Z}_A) - \tilde{Z}_{ab} \tag{4.3.29}$$

$$\tilde{Z}_{01} = \tilde{Z}_{10} = \tilde{Z}_{12} = \tilde{Z}_{21} = \tilde{Z}_{02} = \tilde{Z}_{20} = 0 \tag{4.3.30}$$

4.4 Elementary Fault Interconnections

In the following two chapters we shall review the performance characteristics of transmission lines and three-phase transformers in terms of symmetrical components. In Chapter 7, we shall employ symmetrical components in the analysis of a variety of system unbalanced fault conditions.

Before closing this chapter, it might be helpful to consider at least two elementary system faults in order to present a preview as to how symmetrical components will be employed later.

Given the configuration shown in Figure 4.4.1(a). With a balanced source voltage (no internal impedance) we have:

$$\tilde{E}_1 = \tilde{E}_a \ with \ \tilde{E}_2 = \tilde{E}_0 = 0 \tag{4.4.1}$$

Let the transmission line section be fully transposed so that Equations 4.3.24, 4.3.25, and 4.3.26 apply. The respective single-phase symmetrical component networks are shown in Figure 4.4.1(b). At the end of the line (singular point of unbalance) let there be a single-line-to-ground fault (SLGF) with no fault impedance. Then we have:

$$\tilde{I}_a = \tilde{I}_a \ with \ \tilde{I}_b = \tilde{I}_c \tag{4.4.2}$$

On the basis of Equation 4.3.13, we find:

$$\tilde{I}_0 = \tilde{I}_1 = \tilde{I}_2 = \frac{\tilde{I}_a}{3} \tag{4.4.3}$$

At the point of fault we also have:

$$\tilde{V}_{ag} = 0 \tag{4.4.4}$$

On the basis of Equation 4.3.12, but substituting $\tilde{V}$ for $\tilde{I}$, we have:

$$\tilde{V}_{ag} = 0 = \tilde{V}_{0g} + \tilde{V}_{1g} + \tilde{V}_{2g} \tag{4.4.5}$$

The series interconnections of the single-phase symmetrical component networks at the point of fault, as shown in Figure 4.4.1(b), satisfy these constraints. Thus, what could be a complex analysis in terms of physical (mutually coupled) three-phase quantities, is now a very simple single-phase circuit analysis. For example:

$$\tilde{I}_0 = \tilde{I}_1 = \tilde{I}_2 = \frac{\tilde{E}_1}{\tilde{Z}_{11} + \tilde{Z}_{22} + \tilde{Z}_{00}} \tag{4.4.6}$$

and:

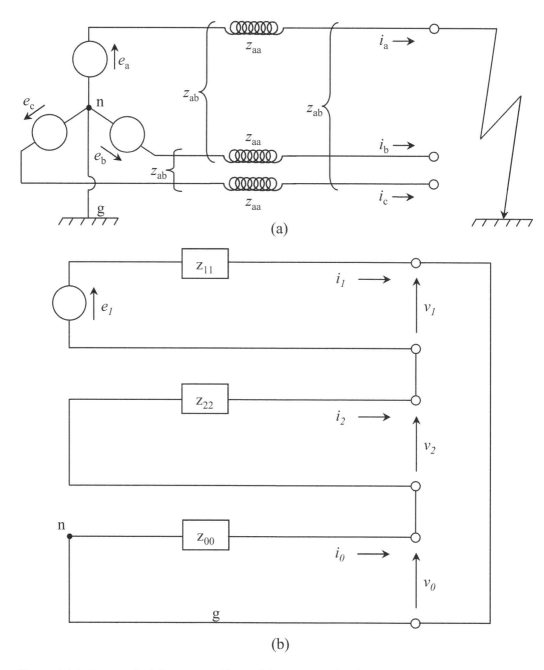

(a)

(b)

Figure 4.4.1: Symmetrical Component Network Interconnection for Representing a Single-Line-to-Ground (SLGF) Fault

$$\tilde{V}_1 = \tilde{E}_1 - \tilde{Z}_{11}\tilde{I}_1 = (\frac{\tilde{Z}_{22} + \tilde{Z}_{00}}{\tilde{Z}_{11} + \tilde{Z}_{22} + \tilde{Z}_{00}})\tilde{E}_1 \qquad (4.4.7)$$

$$\tilde{V}_2 = -\tilde{Z}_{22}\tilde{I}_2 = (\frac{-\tilde{Z}_{22}}{\tilde{Z}_{11} + \tilde{Z}_{22} + \tilde{Z}_{00}})\tilde{E}_1 \qquad (4.4.8)$$

$$\tilde{V}_0 = -\tilde{Z}_{00}\tilde{I}_0 = (-\frac{\tilde{Z}_{00}}{\tilde{Z}_{11} + \tilde{Z}_{22} + \tilde{Z}_{00}})\tilde{E}_1 \qquad (4.4.9)$$

Then, in terms of phase quantities:

$$
\begin{aligned}
\tilde{V}_a &= \tilde{V}_0 + \tilde{V}_1 + \tilde{V}_2 = 0 & (4.4.10) \\
\tilde{V}_b &= \tilde{V}_0 + a^2\tilde{V}_1 + a\tilde{V}_2 \\
&= (\frac{a^2\tilde{Z}_{22} + a^2\tilde{Z}_{00} - a\tilde{Z}_{22} - \tilde{Z}_{00}}{\tilde{Z}_{11} + \tilde{Z}_{22} + \tilde{Z}_{00}})\tilde{E}_1 \\
&= -j\sqrt{3}\tilde{E}_1(\frac{\tilde{Z}_{22} - a\tilde{Z}_{00}}{\tilde{Z}_{11} + \tilde{Z}_{22} + \tilde{Z}_{00}}) & (4.4.11) \\
\tilde{V}_c &= \tilde{V}_0 + a\tilde{V}_1 + a^2\tilde{V}_2 \\
&= (\frac{a\tilde{Z}_{22} + a\tilde{Z}_{00} - a^2\tilde{Z}_{22} - \tilde{Z}_{00}}{\tilde{Z}_{11} + \tilde{Z}_{22} + \tilde{Z}_{00}})\tilde{E}_1 \\
&= j\sqrt{3}\tilde{E}_1(\frac{\tilde{Z}_{22} - a^2\tilde{Z}_{00}}{\tilde{Z}_{11} + \tilde{Z}_{22} + \tilde{Z}_{00}}) & (4.4.12)
\end{aligned}
$$

and:

$$\tilde{I}_a = 3\tilde{I}_1 = \frac{3\tilde{E}_1}{\tilde{Z}_{11} + \tilde{Z}_{22} + \tilde{Z}_{00}} \qquad (4.4.13)$$

Consider now the single open-conductor (SOC) configuration shown in Figure 4.4.2(a). We have:

$$\Delta\tilde{V}_a = \Delta\tilde{V}_a \text{ with } \Delta\tilde{V}_b = \Delta\tilde{V}_c \qquad (4.4.14)$$

Therefore:

$$\Delta\tilde{V}_0 = \Delta\tilde{V}_1 = \Delta\tilde{V}_2 = \frac{\Delta\tilde{V}_a}{3} \qquad (4.4.15)$$

Also:

$$\tilde{I}_a = 0 = \tilde{I}_0 + \tilde{I}_1 + \tilde{I}_2 \qquad (4.4.16)$$

The interconnection of the symmetrical component networks shown in Figure 4.4.2(b) satisfies these constraints. We observe that:

$$\tilde{V}_{anL} = \tilde{V}_{0nL} + \tilde{V}_1 + \tilde{V}_2 = (\tilde{I}_0 + \tilde{I}_1 + \tilde{I}_2)\tilde{Z}_L = 0 \qquad (4.4.17)$$

as befits the physical configuration. Also:

$$\Delta\tilde{V}_1 = \tilde{E}_1 - (\tilde{Z}_{11} + \tilde{Z}_L)\tilde{I}_1 = -(\tilde{Z}_{22} + \tilde{Z}_L)\tilde{I}_2 = -(\tilde{Z}_{00} + 3\tilde{Z}_{nS} + 3\tilde{Z}_{nL} + \tilde{Z}_L)\tilde{I}_0 \qquad (4.4.18)$$

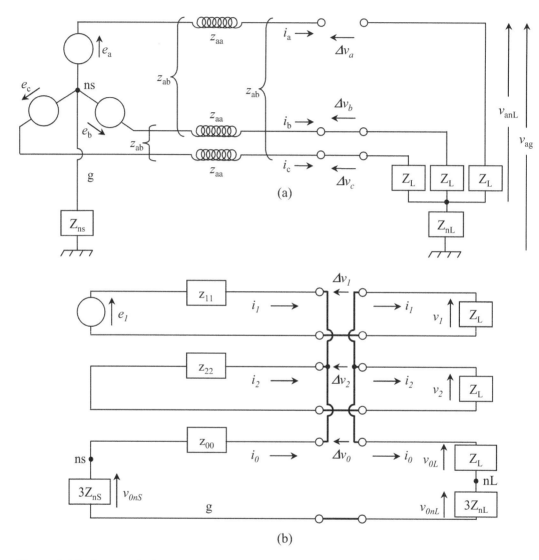

Figure 4.4.2: Symmetrical Component Network Interconnection for Representing a Single Open-Conductor Fault

$$
\begin{aligned}
\Delta \tilde{V}_a &= \tilde{E}_1 - \tilde{Z}_{11}\tilde{I}_1 - \tilde{Z}_{22}\tilde{I}_2 - (\tilde{Z}_{00} + 3\tilde{Z}_{nS} + 3\tilde{Z}_{nL})\tilde{I}_0 \\
&= \tilde{E}_1 - \tilde{Z}_{11}(\tilde{I}_1 + \tilde{I}_2) - (\tilde{Z}_{00} + 3\tilde{Z}_{nS} + 3\tilde{Z}_{nL})\tilde{I}_0 \\
&= \tilde{E}_1 + (\tilde{Z}_{11} - \tilde{Z}_{00} - 3\tilde{Z}_{nS} - 3\tilde{Z}_{nL})\tilde{I}_0 \\
&= \tilde{E}_{anS} + 3\tilde{I}_0(\tilde{Z}_{ab} + \tilde{Z}_{nS} + \tilde{Z}_{nL}) \\
&= \tilde{E}_{anS} - (\tilde{Z}_{ab} + \tilde{Z}_{nS} + \tilde{Z}_{nL})\tilde{I}_n
\end{aligned}
\tag{4.4.19}
$$

If either neutral (or both neutrals) are not grounded, then $\tilde{I}_0 = 0$ and $\Delta \tilde{V}_a = \tilde{E}_{anS}$ as expected.

It is to be especially observed that in the positive and negative sequence networks there is no voltage difference between the source neutral (nS) and the load neutral (nL) and between each of them and ground. Thus, the reference bus in the positive sequence and negative sequence networks represents all three (nS, nL, and G).

Only in the zero sequence network do these three points have separate and distinct identities as shown. Therefore, only if there exists a zero sequence current (neutral current) may we expect neutral shifts (with respect to ground) at the voltage source and at the load.

4.5   References

[1] Members of the Massachusetts Institute of Technology Staff. *Magnetic Circuits and Transformers*. John Wiley and Sons, Inc., New York, New York, 1943.

4.6   Exercises

4.1 It is desired to supply a single-phase load as shown in the system in Figure 4.6.1. For each transformer connection given in Figure 4.6.2, find whether or not such a connection is possible.

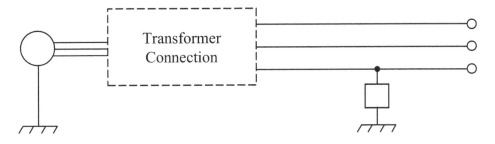

Figure 4.6.1: Single-Phase Load for Exercise 4.1

4.2 A three-phase load of two 50 HP, 440 V, 3Φ motors is to be supplied by transformers connected in V (open delta) on a 2300 volt line. Assume a motor efficiency of 90% and a lagging 85% P.F.

(a) Determine the kVA capacity of each of the transformers and their transformation ratio.

(b) What would be the available kVA, for a delta-delta configuration, if a third transformer of the same rating as the first were added?

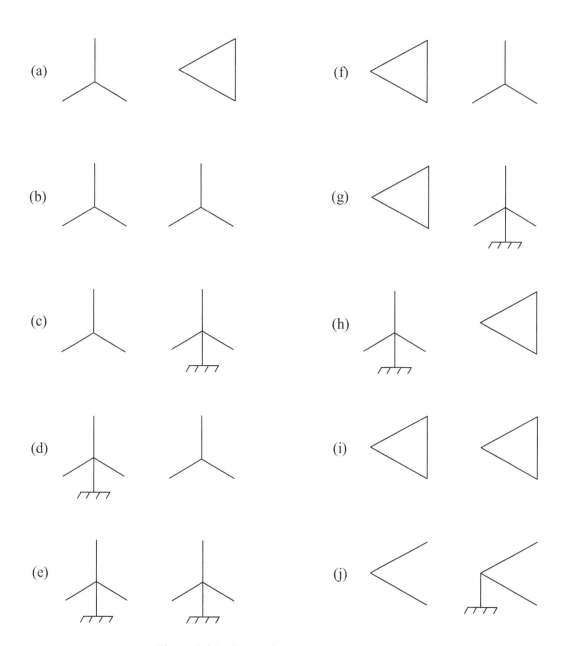

Figure 4.6.2: Connection Types for Exercise 4.1

4.3 Three identical single-phase transformers are connected in delta-delta and energized with balanced input voltages. A single-phase load drawing a current of 100 amperes is connected line-to-line. Compute the current in each of the three secondary windings.

4.4 Four identical 10 kVA single-phase transformers are to be used for a three-phase transformation.

   (a) If they are connected as two V-banks in parallel and operated at rated voltage and frequency, what is the maximum kVA that may be delivered to a balanced three-phase load without exceeding rated current in any winding?

   (b) Solve (a) if three of the transformers form a delta-delta bank and a fourth is connected in parallel with one of them.

4.5 Calculate the symmetrical component voltages of the following unbalanced line-to-neutral voltages and check your results by calculating the phase voltages from the symmetrical component voltages.

$$\tilde{V}_a = 100\angle 90° \quad \tilde{V}_b = 116\angle 0° \quad \tilde{V}_c = 71\angle 225°$$
$$\tilde{V}_a = 100\angle 90° \quad \tilde{V}_b = 32.4\angle -94.8° \quad \tilde{V}_c = 37.4\angle 176.5°$$

4.6 Solve Exercise 4.5 graphically.

4.7 A generator has the following symmetrical component internal impedances:

$$\tilde{Z}_1 = j0.15 \quad \tilde{Z}_2 = j0.17 \quad \tilde{Z}_0 = j0.09$$

We wish to compare a three-phase short-circuit current with a single-line-to-ground fault (SLGF) current. What value of generator neutral impedance must be used to limit the SLGF current to not more than the 3 times the fault current if:

   (a) the neutral impedance is a resistance?

   (b) the neutral impedance is an inductive reactance?

4.8 Given the configuration shown in Figure 4.4.2(a) with:

$$\tilde{Z}_{aa} = j0.15 \quad \tilde{Z}_{ab} = j0.05$$

$$\tilde{Z}_L = 0.8 + j0.6 \quad \tilde{Z}_{nS} = \tilde{Z}_{nL} = 0.04 + j0.01$$

and with a balanced set of source voltages such that:

   (a) Determine the symmetrical component impedances for the symmetrical component networks.

   (b) Evaluate $\tilde{I}_0, \tilde{I}_1, \tilde{I}_2$ and $\tilde{I}_a, \tilde{I}_b, \tilde{I}_c$.

   (c) Evaluate $\tilde{V}_0, \tilde{V}_1, \tilde{V}_2$ and $\tilde{V}_a, \tilde{V}_b, \tilde{V}_c$.

   (d) Evaluate $\tilde{V}_{0L}, \tilde{V}_{0nL}, \tilde{V}_{0nS}$.

   (e) Evaluate $\tilde{V}_{anL}, \tilde{V}_{bnL}, \tilde{V}_{cnL}$.

   (f) Evaluate the neutral shift at the load.

   (g) Evaluate the neutral shift at the source.

Chapter 5

Symmetrical Component Representation of Transmission Lines

The basic electromagnetic characteristics of transmission lines were described in Chapter 1 where considerable emphasis was placed upon the flux-linkage concept. It was established that, in a most rigorous sense, the net self and mutual flux-linkages for a general section of untransposed line including ground wire and earth effects were not equal among the three phases. The consequent unbalances in voltage and currents, interpreted in terms of physical quantities, are somewhat difficult to recognize and define.

In this chapter we shall review transmission line performance again, this time with emphasis not only upon the more practical terms of resistance and reactance, but also in terms of symmetrical components. We shall see that the effects of untransposed line sections, when viewed in terms of symmetrical components, can more readily be comprehended and defined.

On this basis, engineering judgment can be introduced in deciding if and when such unbalances are sufficiently important to be retained in the analysis. In many instances it will be sufficiently accurate to ignore these unbalances and to consider the line section to be (even if not in fact) completely transposed.

We shall proceed to see how transmission line sections perform under positive, negative, and zero sequence currents and voltages, and to present appropriate equivalent networks.

5.1 Series Impedance

For the series impedance of a transmission line we define:

$$\tilde{Z}_{ii} = R_{ii} + \frac{R_e}{3} + j(X_{ii} + \frac{X_e}{3}) \qquad [\Omega/km] \qquad (5.1.1)$$

$$\tilde{Z}_{ik} = \frac{R_e}{3} + j(X_{ik} + \frac{X_e}{3}) \qquad [\Omega/km] \qquad (5.1.2)$$

where:

155

$R_{ii} =$ resistance of the conductor $[\Omega/km]$ (5.1.3)

$X_{ii} = 0.0754(f/60)ln(1/GMRi) =$ inductive reactance

of the conductor to one-meter spacing $[\Omega/km]$ (5.1.4)

$X_{ik} = 0.0754(f/60)ln(1/dik) =$ inductive reactance

between conductors i and k $[\Omega/km]$ (5.1.5)

$GMR_i =$ geometric mean radius of

the conductor $[m]$ (5.1.6)

$d_{ik} =$ center-to-center spacing

between conductors i and k $[m]$ (5.1.7)

$\rho_e =$ earth resistivity $[\Omega - m]$ (5.1.8)

$R_e = 0.17766$ $[\Omega/km]$ for $f = 60Hz$, $\rho_e = 100$ $[\Omega - m]$ (5.1.9)

$X_e = 1.5259$ $[\Omega/km]$ for $f = 60Hz$, $\rho_e = 100$ $[\Omega - m]$ (5.1.10)

The appropriate voltage equation, in symbolic matrix notation, for a transmission line section is:

$$[\Delta \tilde{V}_\Phi] = [\tilde{Z}_\Phi][\tilde{I}_\Phi] \qquad [V/km] \tag{5.1.11}$$

where $[\Delta \tilde{V}_\Phi]$ has as many rows and columns as there are conductors, subconductors, and ground wires in the configuration.

It was shown in Sections 1.1.7.3 and 1.1.7.4 in Chapter 1 that this impedance matrix could always be reduced to one representing a single net conductor per phase per circuit. On this basis we have, for a single-circuit line:

$$\begin{bmatrix} \Delta \tilde{V}_a \\ \Delta \tilde{V}_b \\ \Delta \tilde{V}_c \end{bmatrix} = \begin{bmatrix} \tilde{Z}_{aa} & \tilde{Z}_{ab} & \tilde{Z}_{ac} \\ \tilde{Z}_{ba} & \tilde{Z}_{bb} & \tilde{Z}_{bc} \\ \tilde{Z}_{ca} & \tilde{Z}_{cb} & \tilde{Z}_{cc} \end{bmatrix} \begin{bmatrix} \tilde{I}_a \\ \tilde{I}_b \\ \tilde{I}_c \end{bmatrix} \qquad [V/km] \tag{5.1.12}$$

This system of voltage equations can be expressed in terms of symmetrical components in accordance with the principles discussed in Chapter 4, Section 4.3.4. Thus:

$$[\Delta \tilde{V}_c] = [\tilde{S}][\Delta \tilde{V}_\Phi] = [\tilde{S}][\tilde{Z}_\Phi][\tilde{S}]^{-1}[\tilde{I}_c] \tag{5.1.13}$$

and:

$$[\Delta \tilde{V}_c] = [\tilde{Z}_c][\tilde{I}_c] \tag{5.1.14}$$

with:

$$\begin{bmatrix} \Delta \tilde{V}_0 \\ \Delta \tilde{V}_1 \\ \Delta \tilde{V}_2 \end{bmatrix} = \begin{bmatrix} \tilde{Z}_{00} & \tilde{Z}_{01} & \tilde{Z}_{02} \\ \tilde{Z}_{10} & \tilde{Z}_{11} & \tilde{Z}_{12} \\ \tilde{Z}_{20} & \tilde{Z}_{21} & \tilde{Z}_{22} \end{bmatrix} \begin{bmatrix} \tilde{I}_0 \\ \tilde{I}_1 \\ \tilde{I}_2 \end{bmatrix} \qquad [V/km] \tag{5.1.15}$$

where the individual symmetrical component impedance elements are defined in Tables 4.3.1 and 4.3.3 in Section 4.3.5 of Chapter 4.

5.2   Numerical Example

Consider the configuration shown in Figure 5.2.1. Without ground wires we have:

$$[\tilde{Z}_\Phi] = \begin{bmatrix} 0.12135 + j0.83905 & 0.05922 + j0.30445 & 0.05922 + j0.30445 \\ 0.05922 + j0.30445 & 0.12135 + j0.83905 & 0.05922 + j0.25219 \\ 0.05922 + j0.30445 & 0.05922 + j0.25219 & 0.12135 + j0.83905 \end{bmatrix}$$

$$[\Omega/km] \tag{5.2.1}$$

and in terms of symmetrical components:

$$[\tilde{Z}_c] = \begin{bmatrix} 0.23979 + j1.41311 & 0 + j0.01742 & 0 + j0.01742 \\ 0 + j0.01742 & 0.06214 + j0.55202 & 0 - j0.03484 \\ 0 + j0.01742 & 0 - j0.03484 & 0.06214 + j0.55202 \end{bmatrix}$$

$$[\Omega/km] \tag{5.2.2}$$

For the configuration shown in Figure 5.2.1, but with ground wires, we have for the equivalent single circuit:

$$[\tilde{Z}_\Phi] = \begin{bmatrix} 0.08856 + j0.68628 & 0.02662 + j0.16210 & 0.02662 + j0.16210 \\ 0.02662 + j0.16210 & 0.08929 + j0.70506 & 0.02673 + j0.12096 \\ 0.02662 + j0.16210 & 0.02673 + j0.12096 & 0.08929 + j0.70506 \end{bmatrix}$$

$$[\Omega/km] \tag{5.2.3}$$

and in terms of symmetrical components:

$$[\tilde{Z}_c] =$$
$$\begin{bmatrix} 0.14236 + j0.99557 & -0.00029 + j0.00745 & -0.00029 + j0.00745 \\ -0.00029 + j0.00745 & 0.06239 + j0.55041 & -0.00017 + -j0.03368 \\ -0.00029 + j0.00745 & -0.00017 - j0.03368 & 0.06239 + j0.55041 \end{bmatrix}$$

$$[\Omega/km] \tag{5.2.4}$$

We make several observations:

1. The symmetrical component impedance matrices in Equations 5.2.2 and 5.2.4 have a strong diagonal (i.e. relatively small off-diagonal elements).

2. $\tilde{Z}_{11} \equiv \tilde{Z}_{22}$.

3. For no ground wires, $\tilde{Z}_{00} \approx 3\tilde{Z}_{11}$.

4. Ground wires have a strong (reducing) influence on $\tilde{Z}_{00}$, but negligible effect on $\tilde{Z}_{12}$ and $\tilde{Z}_{21}$.

5. Ground wires have some effect on $\tilde{Z}_{01}$, $\tilde{Z}_{02}$, $\tilde{Z}_{10}$, and $\tilde{Z}_{20}$, but almost none on $\tilde{Z}_{12}$ and $\tilde{Z}_{21}$.

6. For no ground wires, $\tilde{Z}_{aa} = \tilde{Z}_{bb} = \tilde{Z}_{cc}$ in our case. With ground wires, however, $\tilde{Z}_{aa} \neq \tilde{Z}_{bb} = \tilde{Z}_{cc}$.

7. Both symmetrical component impedance matrices happen to be symmetrical, although this is generally not the case. It happened here because we selected phase "a" for the centrally located conductor. Had we selected a phase designation with the conductor on the right as "b," the conductor in the center as "c," and the conductor on the left as "a," it can be shown that we would have obtained:

$$[\tilde{Z}_c] = \begin{bmatrix} \tilde{Z}_{00} & \tilde{a}\tilde{Z}_{01} & \tilde{a}^2\tilde{Z}_{02} \\ \tilde{a}^2\tilde{Z}_{10} & \tilde{Z}_{11} & \tilde{a}\tilde{Z}_{12} \\ \tilde{a}\tilde{Z}_{20} & \tilde{a}^2\tilde{Z}_{22} & \tilde{Z}_{22} \end{bmatrix} \quad (5.2.5)$$

whereby the various $\tilde{Z}_{ik}$ elements are the same as those obtained in Equations 5.2.2 and 5.2.4. Another phase rotation would produce a form similar to that in Equation 5.2.5, except that the a's and a^2's would be interchanged. The main diagonal terms remain unaffected by phase rotations. It becomes apparent that for a completely transposed transmission line section we would obtain ultimately:

$$[\tilde{Z}_c] = \begin{bmatrix} \tilde{Z}_{00} & 0 & 0 \\ 0 & \tilde{Z}_{11} & 0 \\ 0 & 0 & \tilde{Z}_{22} \end{bmatrix} \quad (5.2.6)$$

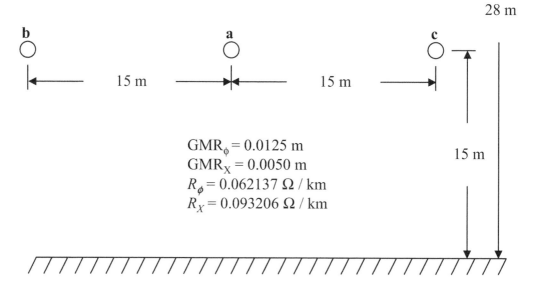

Figure 5.2.1: Single-Circuit Untransposed Line Section

5.3 Single-Circuit Untransposed Line – Electromagnetic Unbalance

Let an untransposed line section be energized at the source end by a balanced set of three-phase voltages. At the receiving end let there be a three-phase-to-ground short circuit as shown in Figure 5.3.1. The voltage drop equations are in terms of symmetrical components:

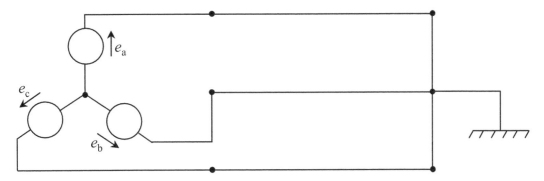

Figure 5.3.1: Short-Circuit Test on Untransposed Transmission Line Section

$$\begin{bmatrix} 0 \\ \Delta \tilde{V}_1 \\ 0 \end{bmatrix} \begin{bmatrix} \tilde{Z}_{00} & \tilde{Z}_{01} & \tilde{Z}_{02} \\ \tilde{Z}_{10} & \tilde{Z}_{11} & \tilde{Z}_{12} \\ \tilde{Z}_{20} & \tilde{Z}_{21} & \tilde{Z}_{22} \end{bmatrix} \begin{bmatrix} \tilde{I}_0 \\ \tilde{I}_1 \\ \tilde{I}_2 \end{bmatrix} \qquad [V/km] \qquad (5.3.1)$$

We recognize that the off-diagonal terms in the impedance matrix are much smaller than the diagonal terms, and we anticipate that I_0 and I_2 will be much smaller than I_1. Consequently we shall neglect products of such small terms to obtain:

$$\begin{bmatrix} 0 \\ \Delta \tilde{V}_1 \\ 0 \end{bmatrix} \begin{bmatrix} \tilde{Z}_{00} & \tilde{Z}_{01} & 0 \\ 0 & \tilde{Z}_{11} & 0 \\ 0 & \tilde{Z}_{21} & \tilde{Z}_{22} \end{bmatrix} \begin{bmatrix} \tilde{I}_0 \\ \tilde{I}_1 \\ \tilde{I}_2 \end{bmatrix} \qquad [V/km] \qquad (5.3.2)$$

We define electromagnetic unbalance factors as:

$$\tilde{m}_0 = \frac{\tilde{I}_0}{\tilde{I}_1} \quad \tilde{m}_2 = \frac{\tilde{I}_2}{\tilde{I}_1} \qquad (5.3.3)$$

Then, on the basis of the above approximate equations we obtain:

$$\tilde{m}_0 \approx -\frac{\tilde{I}_{01}}{\tilde{I}_{00}} \quad \tilde{m}_2 \approx -\frac{\tilde{I}_{21}}{\tilde{I}_{22}} \qquad (5.3.4)$$

According to the numerical example considered in Section 5.2 we have, for the case of no ground wires:

$$\tilde{m}_0 \approx \frac{-j0.01742}{0.23979 + j1.41311} = 0.01215\angle -170.37^o \qquad (5.3.5)$$

$$\tilde{m}_2 \approx \frac{j0.03484}{0.06214 + j0.55202} = 0.06272\angle 6.42^o \qquad (5.3.6)$$

For the same line, but with ground wires we have

$$\tilde{m}_0 \approx \frac{0.00029 - j0.00745}{0.14236 + j0.99557} = 0.00741\angle - 169.63^o \tag{5.3.7}$$

$$\tilde{m}_2 \approx \frac{0.00017 + j0.03368}{0.06239 + j0.55041} = 0.06080\angle 6.18 \tag{5.3.8}$$

It is apparent that the electromagnetic unbalance is small and that the ground wires have a significant effect upon $\tilde{m}_0$ but not upon $\tilde{m}_2$. It is somewhat dangerous to generalize beyond this, but as a rule-of-thumb, $\tilde{Z}_{00}$ is about three times $\tilde{Z}_{11}$ for many practical line configurations without ground wires, and for the horizontal line configuration with no ground wires, $\tilde{Z}_{21}$ is about twice $\tilde{Z}_{01}$ [see also Tables 4.3.1 and 4.3.2]. On this basis we see that $\tilde{m}_2$ is approximately six times $\tilde{m}_0$ for the single-circuit horizontal line configuration with no ground wires. References [5], [7], and [4] contain results of a parametric study of electromagnetic unbalance factors for a number of different line configurations.

For comparison purposes we have the inverses of Equations 5.2.2 and 5.2.4, as obtained by digital computer. For the no-ground wire case we have:

$$\left[\tilde{Z}_c\right]^{-1} =$$

$$\begin{bmatrix} 0.11697 + j0.68836 & -0.00663 + j0.02239 & -0.00663 + j0.02239 \\ -0.00663 + j0.02239 & 0.20402 - j1.79629 & 0.02560 - j0.11121 \\ -0.00663 + j0.02239 & 0.02560 - j0.11121 & 0.20402 - j1.79629 \end{bmatrix}$$

$$[S \cdot km] \tag{5.3.9}$$

and for the two-ground wire case:

$$\left[\tilde{Z}_c\right]^{-1} =$$

$$\begin{bmatrix} 0.14085 - j0.98452 & -0.00422 + j0.01362 & -0.00422 + j0.01362 \\ -0.00422 + j0.01362 & 0.20559 - j1.80031 & 0.02429 - j0.10767 \\ -0.00422 + j0.01362 & 0.02429 - j0.10767 & 0.20559 - j1.80031 \end{bmatrix}$$

$$[S \cdot km] \tag{5.3.10}$$

The unbalance factors for the no-ground wire case are:

$$\tilde{m}_0 \approx \frac{-0.00663 - j0.02239}{0.20402 - j1.79629} = 0.01292\angle - 169.99^o \tag{5.3.11}$$

$$\tilde{m}_2 \approx \frac{0.02560 - j0.11121}{0.20402 - j1.79629} = 0.06312\angle 6.48^o \tag{5.3.12}$$

and for the two-ground wire case:

$$\tilde{m}_0 \approx \frac{-0.00422 + j0.01362}{0.20559 - j1.80031} = 0.00787\angle - 169.30^o \tag{5.3.13}$$

$$\tilde{m}_2 \approx \frac{0.02429 - j0.10767}{0.20559 - j1.80031} = 0.06091\angle 6.20 \tag{5.3.14}$$

Thus, we conclude that the approximate method, reflected by Equations 5.3.5 through 5.3.8, is quite accurate. Employing the approximate method avoids the necessity of inverting the $[\tilde{Z}_c]$ matrix.

The approximate method has a further advantage in analyzing the more realistic situation in which source and load impedance should be considered in estimating expected electromagnetic current unbalance. Since source and load impedances are generally represented as having no mutual impedances amongst the phases (i.e. $\tilde{Z}_{AB} = \tilde{Z}_{BC} = \tilde{Z}_{CA} = 0$), their respective symmetrical component impedances are diagonal matrices containing only $\tilde{Z}_{00}$, $\tilde{Z}_{11}$, and $\tilde{Z}_{22}$. For load impedances, $\tilde{Z}_{00} = \tilde{Z}_{11} = \tilde{Z}_{22}$. For source impedances representing rotating machinery we shall show later that $\tilde{Z}_{22} < \tilde{Z}_{11}$ in the steady state. Adding the source and load symmetrical component impedance matrices to the transmission line impedance matrix in Equation 5.3.1 does not affect the elements $\tilde{Z}_{01}$ and $\tilde{Z}_{21}$. Thus, we have the more realistic estimate:

$$\tilde{m}_0 \approx \frac{-\tilde{Z}_{01}}{\tilde{Z}_{0-source} + \tilde{Z}_{00-line} + \tilde{Z}_{0-load}} \tag{5.3.15}$$

$$\tilde{m}_2 \approx \frac{-\tilde{Z}_{21}}{\tilde{Z}_{2-source} + \tilde{Z}_{22-line} + \tilde{Z}_{2-load}} \tag{5.3.16}$$

Thus, such inductive series impedances tend to reduce the net electromagnetic current unbalance (see also Reference [5]). On the other hand, 70% series capacitor line compensation tends to increase this unbalance.

5.4 Transposed Line Sections

When a transmission line section is completely transposed, the off-diagonal terms in the symmetrical component impedance matrix become exactly zero and only the main diagonal terms remain per Tables 4.3.1 and 4.3.2. Then we have:

$$\begin{bmatrix} \Delta \tilde{V}_0 \\ \Delta \tilde{V}_1 \Delta \tilde{V}_2 \end{bmatrix} = \begin{bmatrix} \tilde{Z}_{00} & 0 & 0 \\ 0 & \tilde{Z}_{11} & 0 \\ 0 & 0 & \tilde{Z}_{22} \end{bmatrix} \begin{bmatrix} \tilde{I}_0 \\ \tilde{I}_1 \\ \tilde{I}_2 \end{bmatrix} \qquad [V/km] \tag{5.4.1}$$

where:

$$\tilde{Z}_{00} = \tilde{Z}_{aa} + 2\tilde{Z}_{ab} = R_a + R_e + j(X_{aa} + 2X_{ab} + X_e) \tag{5.4.2}$$

$$\tilde{Z}_{11} = \tilde{Z}_{22} = \tilde{Z}_{aa} - \tilde{Z}_{ab} = R_a + j(X_{aa} - X_{ab}) \tag{5.4.3}$$

with:

$$R_a = \left(\frac{R_{aa} + R_{bb} + R_{cc}}{3}\right) \qquad [\Omega/km] \tag{5.4.4}$$

$$X_{aa} = 0.0754\left(\frac{f}{60}\right) \ln \frac{1}{GMR} \qquad [\Omega/km] \tag{5.4.5}$$

$$GMR = (GMR_a \, GMR_b \, GMR_c)^{\frac{1}{3}} \qquad [m] \tag{5.4.6}$$

$$X_{ab} = 0.0754\left(\frac{f}{60}\right) \ln \frac{1}{GMD} \qquad [\Omega/km] \tag{5.4.7}$$

$$GMD = (D_{ab}D_{bc}D_{ca})^{\frac{1}{3}} \qquad [m] \tag{5.4.8}$$

with R_e and X_e defined by Equations 5.1.9 and 5.1.10, respectively.

In most load flow studies, short-circuit studies, and stability studies, the unbalances introduced by untransposed line sections are sufficiently small as to be negligible.

Unless specifically stated to the contrary, all transmission line sections shall be considered to be completely transposed for purposes of various system analysis

5.5   Double-Circuit Lines

We consider now two single-circuit lines, which are parallel to one another, perhaps on the same right-of-way. After "bundling" and "ground wire removal," we obtain the voltage equations for the "equivalent double-circuit line":

$$
\begin{bmatrix} \delta\tilde{V}_a \\ \delta\tilde{V}_b \\ \delta\tilde{V}_c \\ \delta\tilde{V}_{a'} \\ \delta\tilde{V}_{b'} \\ \delta\tilde{V}_{c'} \end{bmatrix} = \begin{bmatrix} \tilde{Z}_{aa} & \tilde{Z}_{ab} & \tilde{Z}_{ac} & \tilde{Z}_{aa'} & \tilde{Z}_{ab'} & \tilde{Z}_{ac'} \\ \tilde{Z}_{ba} & \tilde{Z}_{bb} & \tilde{Z}_{bc} & \tilde{Z}_{ba'} & \tilde{Z}_{bb'} & \tilde{Z}_{bc'} \\ \tilde{Z}_{ca} & \tilde{Z}_{cb} & \tilde{Z}_{cc} & \tilde{Z}_{ca'} & \tilde{Z}_{cb'} & \tilde{Z}_{cc'} \\ \tilde{Z}_{a'a} & \tilde{Z}_{a'b} & \tilde{Z}_{a'c} & \tilde{Z}_{a'a'} & \tilde{Z}_{a'b'} & \tilde{Z}_{a'c'} \\ \tilde{Z}_{b'a} & \tilde{Z}_{b'b} & \tilde{Z}_{b'c} & \tilde{Z}_{b'a'} & \tilde{Z}_{b'b'} & \tilde{Z}_{b'c'} \\ \tilde{Z}_{c'a} & \tilde{Z}_{c'b} & \tilde{Z}_{c'c} & \tilde{Z}_{c'a'} & \tilde{Z}_{c'b'} & \tilde{Z}_{c'c'} \end{bmatrix} \begin{bmatrix} \tilde{I}_a \\ \tilde{I}_b \\ tildeI_c \\ \tilde{I}_{a'} \\ \tilde{I}_{b'} \\ \tilde{I}_{c'} \end{bmatrix} \quad [V/km] \quad (5.5.1)
$$

In terms of symmetrical components we have:

$$
\begin{bmatrix} \delta\tilde{V}_0 \\ \delta\tilde{V}_1 \\ \delta\tilde{V}_2 \\ \delta\tilde{V}_{0'} \\ \delta\tilde{V}_{1'} \\ \delta\tilde{V}_{2'} \end{bmatrix} = \begin{bmatrix} \tilde{Z}_{00} & \tilde{Z}_{01} & \tilde{Z}_{02} & \tilde{Z}_{00'} & \tilde{Z}_{01'} & \tilde{Z}_{02'} \\ \tilde{Z}_{10} & \tilde{Z}_{11} & \tilde{Z}_{12} & \tilde{Z}_{10'} & \tilde{Z}_{11'} & \tilde{Z}_{12'} \\ \tilde{Z}_{20} & \tilde{Z}_{21} & \tilde{Z}_{22} & \tilde{Z}_{20'} & \tilde{Z}_{21'} & \tilde{Z}_{22'} \\ \tilde{Z}_{0'0} & \tilde{Z}_{0'1} & \tilde{Z}_{0'2} & \tilde{Z}_{0'0'} & \tilde{Z}_{0'1'} & \tilde{Z}_{0'2'} \\ \tilde{Z}_{1'0} & \tilde{Z}_{1'1} & \tilde{Z}_{1'2} & \tilde{Z}_{1'0'} & \tilde{Z}_{1'1'} & \tilde{Z}_{1'2'} \\ \tilde{Z}_{2'0} & \tilde{Z}_{2'1} & \tilde{Z}_{2'2} & \tilde{Z}_{2'0'} & \tilde{Z}_{2'1'} & \tilde{Z}_{2'2'} \end{bmatrix} \begin{bmatrix} \tilde{I}_0 \\ \tilde{I}_1 \\ tildeI_2 \\ \tilde{I}_{0'} \\ \tilde{I}_{1'} \\ \tilde{I}_{2'} \end{bmatrix} \quad [V/km] \quad (5.5.2)
$$

whereby the various elements of the matrix can be evaluated on the basis of the definitions in Tables 4.3.1 and 4.3.2. For example:

$$
\begin{aligned}
\tilde{Z}_{21'} &= \frac{1}{3}(\tilde{Z}_{aa'} + \tilde{a}\tilde{Z}_{bb'} + \tilde{a}^2\tilde{Z}_{cc'}) \\
&+ \frac{1}{3}(\tilde{Z}_{bc'} + \tilde{a}\tilde{Z}_{ca'} + \tilde{a}^2\tilde{Z}_{ab'}) \\
&+ \frac{1}{3}(\tilde{Z}_{cb'} + \tilde{a}\tilde{Z}_{ac'} + \tilde{a}^2\tilde{Z}_{ba'}) \quad [\Omega/km]
\end{aligned} \quad (5.5.3)
$$

whereby we must use the expanded form since $\tilde{Z}_{bc'}$ may not be equal to $\tilde{Z}_{cb'}$.

5.6   Numerical Example

Consider the double-circuit configuration shown in Figure 5.6.1(a). The phase impedance matrix and the symmetrical component impedance matrix for the double-circuit line with no ground wires are shown in Figure 5.6.2 and for the double-circuit line with four ground wires (equivalent line) in Figure 5.6.3.

Looking first at Figure 5.6.2 and comparing the self-impedance matrices with Equations 5.2.1 and 5.2.2, we observe no change, as expected. In Figure 5.6.2, we again observe that all off-diagonal terms in the symmetrical component impedance matrix are significantly smaller than the diagonal terms with the exception of $\tilde{Z}_{00'} = \tilde{Z}_{0'0}$. We conclude that there is negligible positive and negative sequence coupling between the two circuits. However, there is a strong zero sequence coupling between the two circuits.

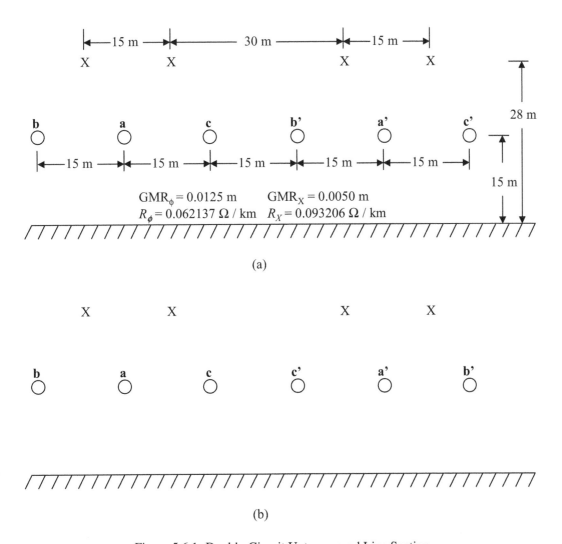

Figure 5.6.1: Double-Circuit Untransposed Line Section

$$[\tilde{Z}_\varphi] =$$

		I			II		
		a	b	c	a'	b'	c'
I	a	0.12136 +j0.83905	0.05922 +j0.30445	0.05922 +j0.30445	0.05922 +j0.22161	0.05922 +j0.25218	0.05922 +j0.19992
	b	0.05922 +j0.30445	0.12136 +j0.83905	0.05922 +j0.25219	0.05922 +j0.19992	0.05922 +j0.22161	0.05922 +j0.18310
	c	0.05922 +j0.30445	0.05922 +j0.25219	0.12136 +j0.83905	0.05922 +j0.25218	0.05922 +j0.30443	0.05922 +j0.22161
II	a'	0.05922 +j0.22161	0.05922 +j0.19992	0.05922 +j0.25218	0.12136 +j0.83905	0.05922 +j0.30445	0.05922 +j0.30445
	b'	0.05922 +j0.25218	0.05922 +j0.22161	0.05922 +j0.30443	0.05922 +j0.30445	0.12136 +j0.83905	0.05922 +j0.25219
	c'	0.05922 +j0.19992	0.05922 +j0.18310	0.05922 +j0.22161	0.05922 +j0.30445	0.05922 +j0.25219	0.12136 +j0.83905

$$[\tilde{Z}_c] =$$

		I			II		
		0	1	2	0'	1'	2'
I	0	0.23979 +j1.41311	0.00000 +j0.01742	0.00000 +j0.01742	0.17766 +j0.68552	0.05011 -j0.00591	-0.05011 -j0.00591
	1	0.00000 +j0.01742	0.06214 +j0.55202	0.00000 -j0.03484	0.05011 -j0.00591	-0.00486 -j0.01034	0.00000 +j0.01181
	0'	0.00000 +j0.01742	0.00000 -j0.03484	0.06214 +j0.55202	-0.05011 -j0.00591	0.00000 +j0.01181	0.00486 -j0.01034
II	0'	0.17766 +j0.68552	-0.05011 -j0.00591	0.05011 -j0.00591	0.23979 +j1.41311	0.00000 +j0.01742	0.00000 +j0.01742
	1'	-0.05011 -j0.00591	0.00486 -j0.01034	0.00000 +j0.01181	0.00000 +j0.01742	0.06214 +j0.55202	0.00000 -j0.03484
	2'	0.05011 -j0.00591	0.00000 +j0.01181	-0.00486 -j0.01034	0.00000 +j0.01742	0.00000 -j0.03484	0.06214 +j0.55202

Figure 5.6.2: Impedance Matrix for Figure 5.6.1(a) with No Ground Wires [Ω/km]

$$[\tilde{Z}_{\phi}] =$$

		I			II		
		a	b	c	a'	b'	c'
I	a	0.08060 +j0.66445	0.01881 +j0.14283	0.01777 +j0.13301	0.01618 +j0.06587	0.01682 +j0.08860	0.01646 +j0.05770
	b	0.01881 +j0.14283	0.08170 +j0.68807	0.01789 +j0.09525	0.01646 +j0.05770	0.01700 +j0.07219	0.01698 +j0.05310
	c	0.01777 +j0.13301	0.01789 +j0.09525	0.07978 +j0.66639	0.01682 +j0.08860	0.01720 +j0.13536	0.01700 +j0.07218
II	a'	0.01618 +j0.06587	0.01646 +j0.05770	0.01682 +j0.08860	0.08060 +j0.66445	0.01777 +j0.13301	0.01881 +j0.14283
	b'	0.01682 +j0.08860	0.01700 +j0.07219	0.01720 +j0.13536	0.01777 +j0.13301	0.07978 +j0.66639	0.01789 +j0.09525
	c'	0.01646 +j0.05770	0.01698 +j0.05310	0.01700 +j0.07218	0.01881 +j0.14283	0.01789 +j0.09525	0.08170 +j0.68807

$$[\tilde{Z}_{c}] =$$

		I			II		
		0	1	2	0'	1'	2'
I	0	0.11701 +j0.92037	0.00918 +j0.00911	-0.00901 +j0.01081	0.05031 +j0.23043	0.03224 -j0.00929	-0.03309 -j0.00897
	1	-0.00901 +j0.01081	0.06254 +j0.54927	0.00027 -j0.03266	0.03224 -j0.00929	-0.00597 -j0.01024	0.00003 +j0.01195
	0'	0.00918 +j0.00911	-0.00090 -j0.03276	0.06254 +j0.54927	-0.03309 -j0.00897	0.00003 +j0.01195	0.00584 -j0.00995
II	0'	0.05031 +j0.23043	-0.03309 -j0.00897	0.03224 -j0.00929	0.11701 +j0.92037	-0.00900 +j0.01081	0.00918 +j0.00911
	1'	-0.03309 -j0.00897	0.00584 -j0.00995	0.00003 +j0.01195	0.00918 +j0.00911	0.06254 +j0.54927	-0.00090 -j0.03276
	2'	0.03224 -j0.00929	0.00003 +j0.01195	-0.00597 -j0.01024	-0.00900 +j0.01081	0.00027 -j0.03266	0.06254 +j0.54927

Figure 5.6.3: Impedance Matrix for Figure 5.6.1(a) with All Ground Wires [Ω/km]

$$[\tilde{Z}_\phi] =$$

		I a	I b	I c	II a'	II b'	II c'
I	a	0.08060 +j0.66445	0.01881 +j0.14283	0.01777 +j0.13301	0.01618 +j0.06587	0.01646 +j0.05770	0.01682 +j0.08860
	b	0.01881 +j0.14283	0.08170 +j0.68807	0.01789 +j0.09525	0.01646 +j0.05770	0.01698 +j0.05310	0.01700 +j0.07219
	c	0.01777 +j0.13301	0.01789 +j0.09525	0.07978 +j0.66639	0.01682 +j0.08860	0.01700 +j0.07218	0.01720 +j0.13536
II	a'	0.01618 +j0.06587	0.01646 +j0.05770	0.01682 +j0.08860	0.08060 +j0.66445	0.01881 +j0.14283	0.01777 +j0.13301
	b'	0.01646 +j0.05770	0.01698 +j0.05310	0.01700 +j0.07218	0.01881 +j0.14283	0.08170 +j0.68807	0.01789 +j0.09525
	c'	0.01682 +j0.08860	0.01700 +j0.07219	0.01720 +j0.13536	0.01777 +j0.13301	0.01789 +j0.09525	0.07978 +j0.66639

$$[\tilde{Z}_c] =$$

		I 0	I 1	I 2	II 0'	II 1'	II 2'
I	0	0.11701 +j0.92037	0.00918 +j0.00911	-0.00901 +j0.01081	0.05031 +j0.23043	0.03224 -j0.00929	-0.03309 -j0.00897
	1	-0.00901 +j0.01081	0.06254 +j0.54927	0.00027 -j0.03266	0.03224 -j0.00929	-0.00597 -j0.01024	0.00003 +j0.01195
	0'	0.00918 +j0.00911	-0.00090 -j0.03276	0.06254 +j0.54927	-0.03309 -j0.00897	0.00003 +j0.01195	0.00584 -j0.00995
II	0'	0.05031 +j0.23043	-0.03309 -j0.00897	0.03224 -j0.00929	0.11701 +j0.92037	0.00918 +j0.00911	-0.00900 +j0.01081
	1'	0.03224 -j0.00929	0.00003 +j0.01195	-0.00597 -j0.01024	-0.00900 +j0.01081	0.06254 +j0.54927	0.00027 -j0.03266
	2'	-0.03309 -j0.00897	0.00584 -j0.00995	0.00003 +j0.01195	0.00918 +j0.00911	-0.00090 -j0.03276	0.06254 +j0.54927

Figure 5.6.4: Impedance Matrix for Figure 5.6.1(b) with All Ground Wires [Ω/km]

Figure 5.6.3 shows the strong influence of the ground wires upon the zero sequence impedances and negligible effect upon all the other terms.

5.7 Double-Circuit Untransposed Line – Electromagnetic Unbalance

In order to determine the electromagnetic unbalanced currents for the untransposed double-circuit lines, we may proceed in a manner similar to that described in Section 5.3 for the single-circuit line. Performing on each circuit a short-circuit test as illustrated in Figure 5.3.1, we have (approximately):

$$
\begin{bmatrix} 0 \\ \delta\tilde{V}_1 \\ 0 \\ 0 \\ \delta\tilde{V}_{1'} \\ 0 \end{bmatrix} = \begin{bmatrix} \tilde{Z}_{00} & \tilde{Z}_{01} & 0 & \tilde{Z}_{00'} & \tilde{Z}_{01'} & 0 \\ 0 & \tilde{Z}_{11} & 0 & 0 & \tilde{Z}_{11'} & 0 \\ 0 & \tilde{Z}_{21} & \tilde{Z}_{22} & 0 & \tilde{Z}_{21'} & \tilde{Z}_{22'} \\ \tilde{Z}_{0'0} & \tilde{Z}_{0'1} & 0 & \tilde{Z}_{0'0'} & \tilde{Z}_{0'1'} & 0 \\ 0 & \tilde{Z}_{1'1} & 0 & 0 & \tilde{Z}_{1'1'} & 0 \\ 0 & \tilde{Z}_{2'1} & 0 & \tilde{Z}_{2'0'} & \tilde{Z}_{2'1'} & \tilde{Z}_{2'2'} \end{bmatrix} \begin{bmatrix} \tilde{I}_0 \\ \tilde{I}_1 \\ \tilde{I}_2 \\ \tilde{I}_{0'} \\ \tilde{I}_{1'} \\ \tilde{I}_{2'} \end{bmatrix} \quad [V/km] \quad (5.7.1)
$$

$$
\begin{bmatrix} \Delta\tilde{V}_1 \\ \Delta\tilde{V}_{1'} \end{bmatrix} \approx \begin{bmatrix} \tilde{Z}_{11} & \tilde{Z}_{11'} \\ \tilde{Z}_{1'1} & \tilde{Z}_{1'1'} \end{bmatrix} \begin{bmatrix} \tilde{I}_1 \\ \tilde{I}_{1'} \end{bmatrix} \quad [V/km] \quad (5.7.2)
$$

$$
\begin{bmatrix} 0 \\ 0 \end{bmatrix} \approx \begin{bmatrix} \tilde{Z}_{00} & \tilde{Z}_{01} & \tilde{Z}_{00'} & \tilde{Z}_{01'} \\ \tilde{Z}_{0'0} & \tilde{Z}_{0'1} & \tilde{Z}_{0'0'} & \tilde{Z}_{0'1'} \end{bmatrix} \begin{bmatrix} \tilde{I}_0 \\ \tilde{I}_1 \\ \tilde{I}_{0'} \\ \tilde{I}_{1'} \end{bmatrix}
$$
$$
= \begin{bmatrix} \tilde{Z}_{00} & \tilde{Z}_{00'} \\ \tilde{Z}_{0'0} & \tilde{Z}_{0'0'} \end{bmatrix} \begin{bmatrix} \tilde{I}_0 \\ \tilde{I}_{0'} \end{bmatrix} + \begin{bmatrix} \tilde{Z}_{01} & \tilde{Z}_{01'} \\ \tilde{Z}_{0'1} & \tilde{Z}_{0'1'} \end{bmatrix} \begin{bmatrix} \tilde{I}_1 \\ \tilde{I}_{1'} \end{bmatrix} \quad (5.7.3)
$$

$$
\begin{bmatrix} 0 \\ 0 \end{bmatrix} \approx \begin{bmatrix} \tilde{Z}_{21} & \tilde{Z}_{22} & \tilde{Z}_{21'} & \tilde{Z}_{22'} \\ \tilde{Z}_{2'1} & \tilde{Z}_{2'2} & \tilde{Z}_{2'1'} & \tilde{Z}_{2'2'} \end{bmatrix} \begin{bmatrix} \tilde{I}_1 \\ \tilde{I}_2 \\ \tilde{I}_{1'} \\ \tilde{I}_{2'} \end{bmatrix}
$$
$$
= \begin{bmatrix} \tilde{Z}_{22} & \tilde{Z}_{22'} \\ \tilde{Z}_{2'2} & \tilde{Z}_{2'2'} \end{bmatrix} \begin{bmatrix} \tilde{I}_2 \\ \tilde{I}_{2'} \end{bmatrix} + \begin{bmatrix} \tilde{Z}_{21} & \tilde{Z}_{21'} \\ \tilde{Z}_{2'1} & \tilde{Z}_{2'1'} \end{bmatrix} \begin{bmatrix} \tilde{I}_1 \\ \tilde{I}_{1'} \end{bmatrix} \quad (5.7.4)
$$

Of particular interest is the special case in which the two circuits have common sending-end and receiving-end buses (i.e. the two circuits are electric-ally in parallel) so that $\Delta\tilde{V}_1 = \Delta\tilde{V}_1$. On the basis of Equation 5.7.2, we have:

$$
\begin{bmatrix} \tilde{I}_1 \\ \tilde{I}_{1'} \end{bmatrix} = \frac{\Delta\tilde{V}_1}{(\tilde{Z}_{11}\tilde{Z}_{1'1'} - \tilde{Z}_{11'}\tilde{Z}_{1'1})} \begin{bmatrix} \tilde{Z}_{1'1'} & -\tilde{Z}_{11'} \\ -\tilde{Z}_{1'1} & \tilde{Z}_{11} \end{bmatrix} \begin{bmatrix} 1 \\ 1 \end{bmatrix} \quad (5.7.5)
$$

$$
\begin{bmatrix} \tilde{I}_0 \\ \tilde{I}_{0'} \end{bmatrix} = \frac{-1}{(\tilde{Z}_{00}\tilde{Z}_{0'0'} - \tilde{Z}_{00'}\tilde{Z}_{0'0})} \begin{bmatrix} \tilde{Z}_{0'0'} & -\tilde{Z}_{00'} \\ -\tilde{Z}_{0'0} & \tilde{Z}_{00} \end{bmatrix} \begin{bmatrix} \tilde{Z}_{01} & \tilde{Z}_{01'} \\ \tilde{Z}_{0'1} & \tilde{Z}_{0'1'} \end{bmatrix} \begin{bmatrix} \tilde{I}_1 \\ \tilde{I}_{1'} \end{bmatrix} \quad (5.7.6)
$$

$$
\begin{bmatrix} \tilde{I}_2 \\ \tilde{I}_{2'} \end{bmatrix} = \frac{-1}{(\tilde{Z}_{22}\tilde{Z}_{2'2'} - \tilde{Z}_{22'}\tilde{Z}_{2'2})} \begin{bmatrix} \tilde{Z}_{2'2'} & -\tilde{Z}_{22'} \\ -\tilde{Z}_{2'2} & \tilde{Z}_{22} \end{bmatrix} \begin{bmatrix} \tilde{Z}_{21} & \tilde{Z}_{21'} \\ \tilde{Z}_{2'1} & \tilde{Z}_{2'1'} \end{bmatrix} \begin{bmatrix} \tilde{I}_1 \\ \tilde{I}_{1'} \end{bmatrix} \quad (5.7.7)
$$

For Figure 5.6.2	For Figure 5.6.3
$\dfrac{\tilde{I}_1}{\Delta \tilde{V}_1} = 0.19372 - j1.82577$	$\dfrac{\tilde{I}_1}{\Delta \tilde{V}_1} = 0.19341 - j1.83512$
$= 1.83601\angle -83.944^o$	$= 1.84528\angle -83.984^o$
$\dfrac{\tilde{I}_{1'}}{\Delta \tilde{V}_1} = 0.22444 - j1.11876$	$\dfrac{\tilde{I}_{1'}}{\Delta \tilde{V}_1} = 0.23087 - j1.82550$
$= 1.83255\angle -82.965^o$	$= 1.84004\angle -82.792^o$
$\dfrac{\tilde{I}_0}{\Delta \tilde{V}_1} = 0.12609 + j0.03482$	$\dfrac{\tilde{I}_0}{\Delta \tilde{V}_1} = 0.10857 + j0.02236$
$= 0.12561\angle 16.093^o$	$= 0.11085\angle 11.639^o$
$\dfrac{\tilde{I}_{0'}}{\Delta \tilde{V}_1} = -0.12701 - j0.01536$	$\dfrac{\tilde{I}_{0'}}{\Delta \tilde{V}_1} = -0.11060 - j0.01938$
$= 0.12794\angle 186.894^o$	$= 0.11228\angle 189.939^o$
$\dfrac{\tilde{I}_2}{\Delta \tilde{V}_1} = 0.01694 - j0.07566$	$\dfrac{\tilde{I}_2}{\Delta \tilde{V}_1} = 0.01258 - j0.06965$
$= 0.07754\angle -78.237^o$	$= 0.07078\angle -80.649^o$
$\dfrac{\tilde{I}_{2'}}{\Delta \tilde{V}_1} = 0.01814 - j0.07525$	$\dfrac{\tilde{I}_{2'}}{\Delta \tilde{V}_1} = 0.01773 - j0.06791$
$= 0.07741\angle -77.294^o$	$= 0.07019\angle -76.208^o$

For Figure 5.6.4
$\dfrac{\tilde{I}_1}{\Delta \tilde{V}_1} = 0.19622 - j1.75996 = 1.77086\angle -83.638^o$
$\dfrac{\tilde{I}_{1'}}{\Delta \tilde{V}_1} = 0.19622 - j1.75996 = 1.77086\angle -83.638^o$
$\dfrac{\tilde{I}_0}{\Delta \tilde{V}_1} = -0.03528 - j0.03482 = 0.03641\angle 194.299^o$
$\dfrac{\tilde{I}_0}{\Delta \tilde{V}_1} = -0.03528 - j0.03482 = 0.03641\angle 194.299^o$
$\dfrac{\tilde{I}_2}{\Delta \tilde{V}_1} = 0.04461 - j0.12724 = 0.13483\angle -71.464^o$
$\dfrac{\tilde{I}_2}{\Delta \tilde{V}_1} = 0.04461 - j0.12724 = 0.13483\angle -71.464^o$

Figure 5.7.1: Approximate Electromagnetic Unbalanced Currents for an Electrically Parallel Double-Circuit Transmission Line

Consider the double-circuit line with no ground wires for which the data in Figure 5.6.2, 5.6.3, and 5.6.4 apply. By this approximate method we obtain the results shown in Figure 5.7.1.

For accurate evaluations we write:

$$
\begin{bmatrix}
\tilde{I}_0 \\
\tilde{I}_1 \\
\tilde{I}_2 \\
\tilde{I}_{0'} \\
\tilde{I}_{1'} \\
\tilde{I}_{2'}
\end{bmatrix}
=
\begin{bmatrix}
\tilde{Y}_{00} & \tilde{Y}_{01} & \tilde{Y}_{02} & \tilde{Y}_{00'} & 0\tilde{1}' & \tilde{Y}_{02'} \\
\tilde{Y}_{10} & \tilde{Y}_{11} & \tilde{Y}_{12} & \tilde{Y}_{10'} & 1\tilde{1}' & \tilde{Y}_{12'} \\
\tilde{Y}_{20} & \tilde{Y}_{21} & \tilde{Y}_{22} & \tilde{Y}_{20'} & 2\tilde{1}' & \tilde{Y}_{22'} \\
\tilde{Y}_{0'0} & \tilde{Y}_{0'1} & \tilde{Y}_{0'2} & \tilde{Y}_{0'0'} & 0'\tilde{1}' & \tilde{Y}_{0'2'} \\
\tilde{Y}_{1'0} & \tilde{Y}_{1'1} & \tilde{Y}_{1'2} & \tilde{Y}_{1'0'} & 1'\tilde{1}' & \tilde{Y}_{1'2'} \\
\tilde{Y}_{2'0} & \tilde{Y}_{2'1} & \tilde{Y}_{2'2} & \tilde{Y}_{2'0'} & 2'\tilde{1}' & \tilde{Y}_{2'2'}
\end{bmatrix}
\begin{bmatrix}
\Delta\tilde{V}_0 \\
\Delta\tilde{V}_1 \\
\Delta\tilde{V}_2 \\
\Delta\tilde{V}_{0'} \\
\Delta\tilde{V}_{1'} \\
\Delta\tilde{V}_{2'}
\end{bmatrix}
\tag{5.7.8}
$$

Performing on each circuit a short-circuit test as illustrated in Figure 5.3.1, we have:

$$
\begin{bmatrix}
\tilde{I}_0 \\
\tilde{I}_1 \\
\tilde{I}_2 \\
\tilde{I}_{0'} \\
\tilde{I}_{1'} \\
\tilde{I}_{2'}
\end{bmatrix}
=
\begin{bmatrix}
\tilde{Y}_{01} & \tilde{Y}_{01'} \\
\tilde{Y}_{11} & \tilde{Y}_{11'} \\
\tilde{Y}_{21} & \tilde{Y}_{21'} \\
\tilde{Y}_{0'1} & \tilde{Y}_{0'1'} \\
\tilde{Y}_{1'1} & \tilde{Y}_{1'1'} \\
\tilde{Y}_{2'1} & \tilde{Y}_{2'1'}
\end{bmatrix}
\begin{bmatrix}
\Delta\tilde{V}_1 \\
\Delta\tilde{V}_{1'}
\end{bmatrix}
\tag{5.7.9}
$$

Again consider the special case in which the two circuits have common sending-end and receiving-end buses (i.e. the two circuits are electrically in parallel) so that $\Delta\tilde{V}_1 = \Delta\tilde{V}_{1'}$. For the double-circuit line with no ground wires for which the data in Figure 5.6.2 applies, we took the inverse of $[Z_c]$ by means of a digital computer to obtain:

$$
\begin{aligned}
\tilde{Y}_{11} &= 0.20693 - j1.80497 & \tilde{Y}_{11'} &= -0.01491 - j0.03667 \\
\tilde{Y}_{1'1} &= 0.02822 - j0.02533 & \tilde{Y}_{1'1'} &= 0.20693 - j1.80497 \\
\tilde{Y}_{01} &= 0.03367 + j0.04261 & \tilde{Y}_{01'} &= 0.08370 - j0.00819 \\
\tilde{Y}_{0'1} &= -0.07381 - j0.04381 & \tilde{Y}_{0'1'} &= -0.05029 + j0.02971 \\
\tilde{Y}_{21} &= 0.01803 - j0.10581 & \tilde{Y}_{21'} &= -0.00878 + j0.03878 \\
\tilde{Y}_{2'1} &= -0.00878 + j0.03878 & \tilde{Y}_{2'1'} &= 0.02833 - j0.10257
\end{aligned}
\tag{5.7.10}
$$

Then, on the basis of Equation 5.7.9, we obtain:

$$
\begin{aligned}
\frac{\tilde{I}_1}{\Delta\tilde{V}_1} &= 0.19202 - j1.84164 = 1.85162\angle -84.05^o \\[2ex]
\frac{\tilde{I}_{1'}}{\Delta\tilde{V}_1} &= 0.23515 - j1.83030 = 1.84534\angle -82.68^o \\[2ex]
\frac{\tilde{I}_0}{\Delta\tilde{V}_1} &= 0.11737 + j0.03442 = 0.12231\angle 16.34^o \\[2ex]
\frac{\tilde{I}_{0'}}{\Delta\tilde{V}_1} &= -0.12410 - j0.01410 = 0.12490\angle 186.48_o \\[2ex]
\frac{\tilde{I}_2}{\Delta\tilde{V}_1} &= 0.00925 - j0.06703 = 0.06767\angle -82.14^o \\[2ex]
\frac{\tilde{I}_{2'}}{\Delta\tilde{V}_1} &= 0.01955 - j0.06379 = 0.06672\angle -72.96^o
\end{aligned}
\tag{5.7.11}
$$

Comparing the exact results in (5.7.11) with the approximate results in Figure 5.7.1, we note that the approximate method gives a good estimate of the current unbalance. More importantly, if we wish to estimate the impact of a series capacitor line compensation, this is most easily done by accordingly modifying $\tilde{Z}_{11}$, $\tilde{Z}_{1'1'}$, $\tilde{Z}_{00}$, $\tilde{Z}_{0'0'}$, $\tilde{Z}_{22}$ and $\tilde{Z}_{2'2'}$ in Equations 5.7.5, 5.7.6, and 5.7.7, respectively.

The electromagnetic current unbalance factors for each circuit are:

$$
\begin{aligned}
\tilde{m}_0 &= \frac{\tilde{I}_0}{\tilde{I}_1} = 0.06606\angle 100.39^o \quad \tilde{m}_{0'} = \frac{\tilde{I}_{0'}}{\tilde{I}_{1'}} = 0.06768\angle 269.16^o \\
\tilde{m}_2 &= \frac{\tilde{I}_2}{\tilde{I}_1} = 0.03655\angle 1.91^o \quad \tilde{m}_{2'} = \frac{\tilde{I}_{2'}}{\tilde{I}_{1'}} = 0.03616\angle 9.72^o
\end{aligned}
\tag{5.7.12}
$$

Because we have two circuits, we must now contend with two unbalanced current components. In Equation 5.7.12, we observe that the negative sequence currents are nearly in phase in both circuits. Consequently they will be forced to flow through the source and load impedances and will thereby be reduced significantly as before. However, the zero sequence current components are nearly out of phase. These components will tend to circulate around the two circuits and will be limited in magnitude only by the transmission line impedances.

It becomes convenient to separate each symmetrical component current into two parts; a "through current" and a "circulating current." Thus, we have:

Through Current   Circulating Current

$$\frac{\tilde{I}_0 + \tilde{I}_{0'}}{2\Delta\tilde{V}_1} = -0.00337 + j0.01016 \qquad \frac{\tilde{I}_0 - \tilde{I}_{0'}}{2\Delta\tilde{V}_1} = 0.12074 + j0.02426$$

$$\frac{\tilde{I}_2 + \tilde{I}_{2'}}{2\Delta\tilde{V}_1} = 0.01440 - j0.06541 \qquad \frac{\tilde{I}_2 - \tilde{I}_{2'}}{2\Delta\tilde{V}_1} = -0.00515 - j0.00162$$

$$\frac{\tilde{I}_1 + \tilde{I}_{1'}}{2\Delta\tilde{V}_1} = 0.21359 - j1.83597 \qquad \frac{\tilde{I}_1 - \tilde{I}_{1'}}{2\Delta\tilde{V}_1} = -0.02157 - j0.00567$$

$$\mid m_{0t} \mid = \frac{\mid I_{0t} \mid}{\mid I_{1t} \mid} = 0.0058 \qquad \mid m_{0c} \mid = \frac{\mid I_{0c} \mid}{\mid I_{1t} \mid} = 0.0666$$

$$\mid m_{2t} \mid = \frac{\mid I_{2t} \mid}{\mid I_{1t} \mid} = 0.0362 \qquad \mid m_{2c} \mid = \frac{\mid I_{2c} \mid}{\mid I_{1t} \mid} = 0.0029$$

$$\mid m_{1t} \mid = 1.0 \qquad \mid m_{1c} \mid = \frac{\mid I_{1c} \mid}{\mid I_{1t} \mid} = 0.0121 \qquad (5.7.13)$$

For the particular phasing arrangement chosen in Figure 5.6.1(a) it becomes apparent that we can expect a significant zero sequence circulating current between the two circuits if they are electrically in parallel. The negative sequence current is predominantly a "through current" and will be reduced by source and load impedances.

The "circulating current" component becomes significantly larger when a 70% series capacitor line compensation is employed. Reference [14] discusses this in greater detail. It was found, for example, with one capacitor bank out of service in only one line, differences between phase current magnitudes up to 42% were recorded. This led to relay operation even at moderate line loads.

It was pointed out in Section 5.2 that a phase rotation of a single-circuit untransposed line causes the symmetrical component impedances to be modified as shown in Equation 5.2.5. By changing the phasing arrangement of the six conductors in a double-circuit line, we would expect to see phase angle changes in the symmetrical component mutual impedance elements. Figures 5.6.3, 5.6.4, and 5.7.1 clearly illustrate the effect of changing the phasing arrangement. Figures 5.7.2 and 5.7.3 show the $[\tilde{Z}_c]$'s for the twelve significantly different phasing arrangements, and Figures 5.7.4 and 5.7.5 show the corresponding $[\tilde{Z}_c]^{-1}$'s.

On the basis of Equations 5.7.9 and 5.7.13, we may write:

$$\tilde{I}_{0c} = (\tilde{Y}_{01} + \tilde{Y}_{01'} - \tilde{Y}_{0'1} - \tilde{Y}_{0'1'})\Delta\tilde{V}_1$$
$$\tilde{I}_{2c} = (\tilde{Y}_{21} + \tilde{Y}_{21'} - \tilde{Y}_{2'1} - \tilde{Y}_{2'1'})\Delta\tilde{V}_1 \qquad (5.7.14)$$

In Figure 5.7.4, we observe that the phasing arrangement bac-cab leads to zero circulating current for both $\tilde{I}_{0c}$ and $\tilde{I}_{2c}$. In Figure 5.7.5 we observe that the phasing arrangement cab-bac also leads to zero circulating current for both $\tilde{I}_{0c}$ and $\tilde{I}_{2c}$.

We conclude that, for the line configuration shown in Figure 5.6.1, zero circulating current for both $\tilde{I}_{0c}$ and $\tilde{I}_{2c}$ can be attained when the phasing arrangement is selected such that there exists a mirror image about the centerline between the two circuits of the selected phasing.

This conclusion may apply for some other geometric conductor arrangements, but not necessarily to all other configurations.

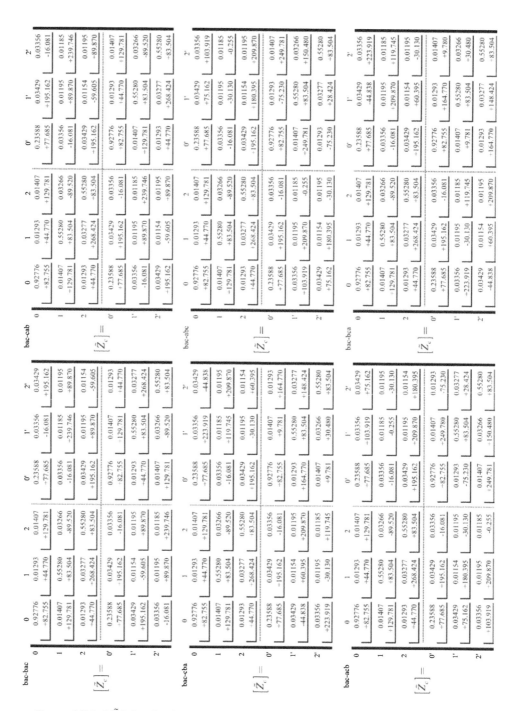

Figure 5.7.2: $[\tilde{Z}_c]$ for Configuration in Figure 5.6.1 with Various Phase Arrangements

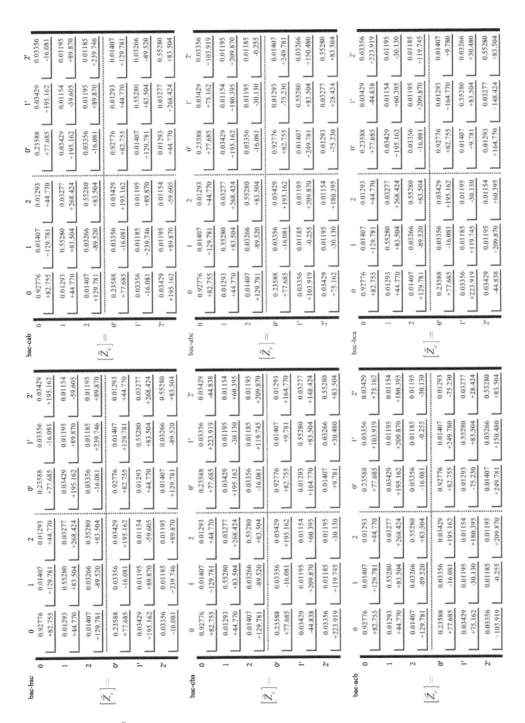

Figure 5.7.3: $[\tilde{Z}_c]$ for Configuration in Figure 5.6.1 with Various Phase Arrangements

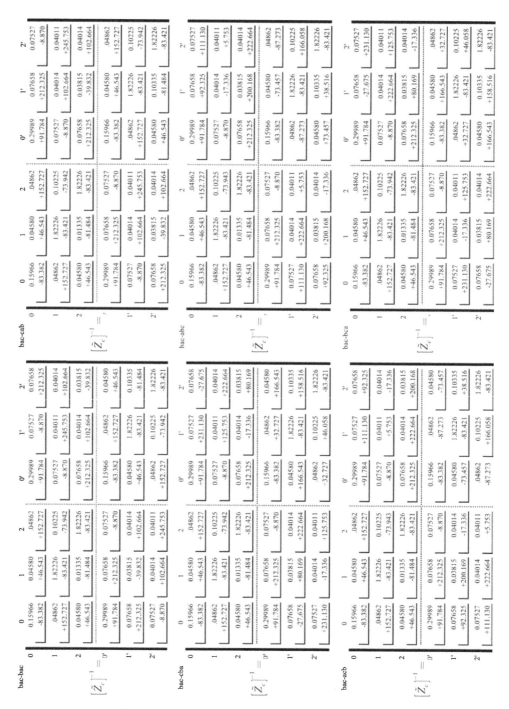

Figure 5.7.4: $[\tilde{Z}_c^{-1}]$ for Configuration in Figure 5.6.1 with Various Phase Arrangements

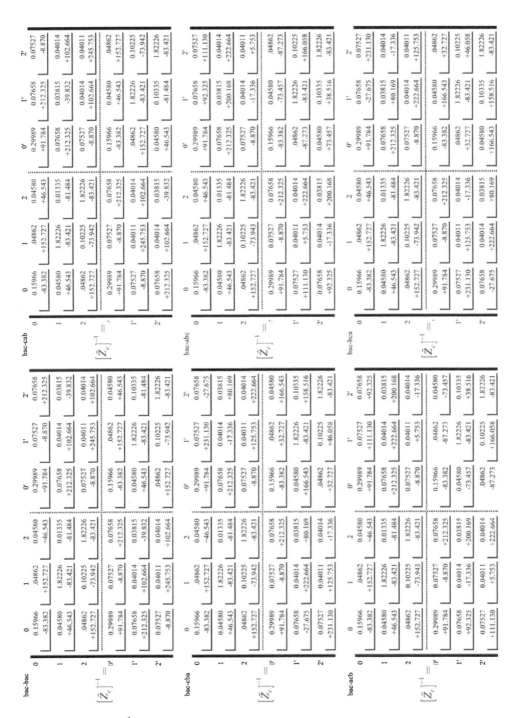

Figure 5.7.5: $[\tilde{Z}_c^{-1}]$ for Configuration in Figure 5.6.1 with Various Phase Arrangements

5.8   Shunt Capacitive Reactance

For the shunt capacitive reactance of a transmission line we define:

$$X'_{i-g} = \frac{P_{ii}}{\omega} = 0.047746(\frac{60}{f})\ln\frac{2h_i}{r_i} \qquad [M\Omega - km] \qquad (5.8.1)$$

$$X'_{ik} = \frac{P_{ik}}{\omega} = 0.047746(\frac{f}{60})\ln\frac{D_{ik}}{d_{ik}} \qquad [M\Omega - km] \qquad (5.8.2)$$

where:

P_{ii} = self-potential coefficient of the conductor $[km/\mu F]$ (5.8.3)

P_{ik} = mutual-potential coefficient between

conductors i and k $[km/\mu F]$ (5.8.4)

h_i = height of conductor i above the earth $[m]$ (5.8.5)

r_i = radius of conductor i $[m]$ (5.8.6)

d_{ik} = center-to-center spacing between

conductors i and k $[m]$ (5.8.7)

D_{ik} = spacing between conductor i and the image

of conductor k $[m]$ (5.8.8)

The appropriate voltage equation, in symbolic matrix notation, for a transmission line section is:

$$[\tilde{V}_\Phi] = -\frac{j}{\omega}[P_\Phi][\Delta \tilde{I}_\Phi] \qquad [V] \qquad (5.8.9)$$

where $[\Delta \tilde{I}_\Phi]$ represents the line-charging current in microamperes per km.

Now [P] has as many rows and columns as there are conductors, subconductors, and ground wires. It was shown in Chapter 1 that such a matrix could always be reduced to one representing a single net conductor per phase per circuit. On this basis we have, for a single-circuit line:

$$\begin{bmatrix} \tilde{V}_a \\ \tilde{V}_b \\ \tilde{V}_c \end{bmatrix} = -\frac{j}{\omega} \begin{bmatrix} P_{aa} & P_{ab} & P_{ac} \\ P_{ba} & P_{bb} & P_{bc} \\ P_{ca} & P_{cb} & P_{cc} \end{bmatrix} \begin{bmatrix} \Delta \tilde{I}_a \\ \Delta \tilde{I}_b \\ \Delta \tilde{I}_c \end{bmatrix} \qquad [V] \qquad (5.8.10)$$

This system of voltage equations can be expressed in terms of symmetrical components in accordance with the principles discussed in Chapter 4. Thus:

$$[\tilde{V}_c] = -fracj\omega[P_c][\Delta \tilde{I}_c] \qquad [V] \qquad (5.8.11)$$

with:

$$\begin{bmatrix} \tilde{V}_0 \\ \tilde{V}_1 \\ \tilde{V}_2 \end{bmatrix} = \frac{1}{\omega} \begin{bmatrix} \tilde{P}_{00} & \tilde{P}_{01} & \tilde{P}_{02} \\ \tilde{P}_{10} & \tilde{P}_{11} & \tilde{P}_{12} \\ \tilde{P}_{20} & \tilde{P}_{21} & \tilde{P}_{22} \end{bmatrix} \begin{bmatrix} \Delta \tilde{I}_0 \\ \Delta \tilde{I}_1 \\ \Delta \tilde{I}_2 \end{bmatrix} \qquad [V] \qquad (5.8.12)$$

where, in general, the symmetrical components are complex numbers.

5.9   Numerical Example

Consider the configuration shown in Figure 5.2.1. Without ground wires we have:

$$\frac{1}{\omega}[P_\Phi] = \begin{array}{ccc} 0.37163 & 0.03842 & 0.03842 \\ 0.03842 & 0.37163 & 0.01655 \\ 0.03842 & 0.01655 & 0.37163 \end{array} \quad [M\Omega - km] \qquad (5.9.1)$$

and in terms of symmetrical components:

$$\frac{1}{\omega}[P_c] = \begin{array}{ccc} 0.43390 & 0.00729 & 0.00729 \\ 0.00729 & 0.34050 & -0.01458 \\ 0.00729 & -0.01458 & 0.34050 \end{array} \quad [M\Omega - km] \qquad (5.9.2)$$

For the configuration shown in Figure 5.2.1, but with ground wires, we have for the equivalent single circuit:

$$\frac{1}{\omega}[P_\Phi] = \begin{bmatrix} 0.36144 & 0.03034 & 0.03034 \\ 0.03034 & 0.36464 & 0.01073 \\ 0.03034 & 0.01073 & 0.36464 \end{bmatrix} \quad [M\Omega - km] \qquad (5.9.3)$$

and in terms of symmetrical components:

$$\frac{1}{\omega}[P_c] = \begin{bmatrix} 0.41119 & 0.00547 & 0.00547 \\ 0.00547 & 0.33977 & -0.01414 \\ 0.00547 & -0.01414 & 0.33977 \end{bmatrix} \quad [M\Omega - km] \qquad (5.9.4)$$

The $\tilde{P}_{ij}$'s happen to be real numbers here due to our selection of phasing arrangement.

Since this analysis parallels that in Section 1.2, we shall merely note that many (not all) of the observations made there, also apply here. However, since considerations of line-charging current are more commonly modeled and analyzed in terms of capacitances rather than shunt reactances, we shall show these appropriate values also.

For the configuration shown in Figure 5.2.1, and without ground wires, we have:

$$\omega[C_\Phi] = \begin{bmatrix} 2.74704 & -0.27191 & -0.27191 \\ -0.27191 & 2.72307 & -0.09314 \\ -0.27191 & -0.09314 & 2.72307 \end{bmatrix} \quad [\mu S/km] \qquad (5.9.5)$$

and in terms of symmetrical components:

$$\omega[C_c] = \begin{bmatrix} 2.30642 & -0.05160 & -0.05160 \\ -0.05160 & 2.94339 & 0.12717 \\ -0.05160 & 0.12717 & 2.94339 \end{bmatrix} \quad [\mu S/km] \qquad (5.9.6)$$

For the equivalent single circuit with ground wires we have:

$$\omega[C_\Phi] = \begin{bmatrix} 2.80478 & -0.22673 & -0.22673 \\ -0.22673 & 2.76311 & -0.06245 \\ -0.22673 & -0.06245 & 2.76311 \end{bmatrix} \quad [\mu S/km] \qquad (5.9.7)$$

and in terms of symmetrical components:

$$\omega[C_c] = \begin{bmatrix} 2.43306 & -0.04087 & -0.04087 \\ -0.04087 & 2.94897 & 0.12341 \\ -0.04087 & 0.12341 & 2.94897 \end{bmatrix} \quad [\mu S/km] \qquad (5.9.8)$$

For a completely transposed line section we would have, for the configuration with no ground wires:

$$\omega C_{aa} = (1/3)(2.74704 + 2.72307 + 2.72307) = 2.73106 \quad [\mu S/km]$$
$$-\omega C_{ab} = (1/3)(0.09314 + 0.27191 + 0.27191) = 0.21232 \quad [\mu S/km]$$

$$(5.9.9)$$

so that in terms of the equivalent capacitance model described in Section 1.2.6 we have:

$$\omega C_{a-g} = 2.30642 \quad [\mu S/km]$$
$$\omega C_{a-b} = 0.21232 \quad [\mu S/km]$$

$$(5.9.10)$$

Similarly, for the fully transposed single-circuit line with ground wires we have:

$$\omega C_{aa} = 2.777 \quad [\mu/km]$$
$$-\omega C_{ab} = 0.17197 \quad [\mu S/km]$$

$$(5.9.11)$$

so that:

$$\omega C_{a-g} = 2.43306 \quad [\mu S/km]$$
$$\omega C_{a-b} = 0.17197 \quad [\mu S/km]$$

$$(5.9.12)$$

Figure 5.9.1(a) shows the physical equivalent capacitance model, and Figure 5.9.1(b) shows the symmetrical component network representation. The line-to-ground capacitances appear directly in all three symmetrical component networks, while the line-to-line capacitances appear only in the positive and negative sequence networks. Observe that on a line-to-neutral representation we must use three times the value of C_{a-b} given in Equations 5.9.10 and 5.9.12.

5.10   Single-Circuit Untransposed Line – Electrostatic Unbalance

In the United States, where electric power system neutrals are generally effectively grounded, the electrostatic unbalance due to untransposed single-circuit transmission lines is of little concern. However, in some other countries the electric power systems at distribution and subtransmission voltage levels are operated with ungrounded neutral or with neutrals connected to ground by a tuned inductance (Peterson coil, resonant neutral grounding). In these instances electrostatic unbalance can be a serious problem.

Let an untransposed line section be energized at the source end by a set of positive sequence voltages with neutral grounded through with the possibility of a zero sequence voltage due to a neutral shift (but no negative sequence voltage). At the receiving end let there be an open circuit as shown in Figure 5.10.1. The line-charging current equations are, in terms of symmetrical components:

$$\begin{bmatrix} \Delta \tilde{I}_0 \\ \Delta \tilde{I}_1 \\ \Delta \tilde{I}_2 \end{bmatrix} = \omega \begin{bmatrix} \tilde{C}_{00} & \tilde{C}_{01} & tildeC_{02} \\ \tilde{C}_{10} & \tilde{C}_{11} & tildeC_{12} \\ \tilde{C}_{20} & \tilde{C}_{21} & tildeC_{22} \end{bmatrix} \begin{bmatrix} \tilde{V}_0 \\ \tilde{V}_1 \\ \tilde{V}_2 \end{bmatrix} \quad [A/m] \qquad (5.10.1)$$

We consider first the equation:

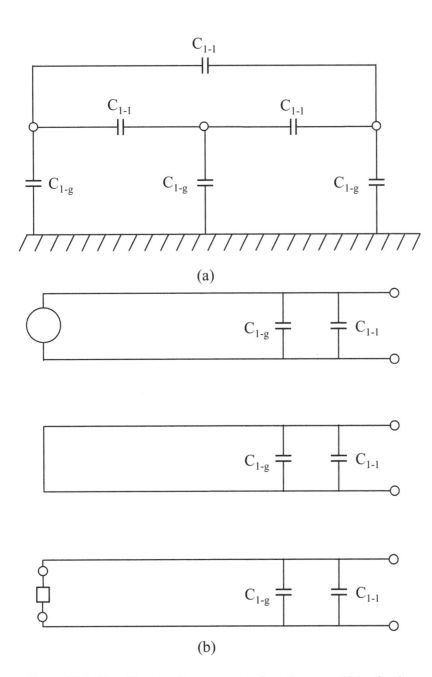

(a)

(b)

Figure 5.9.1: Line-Charging Representation for a Transposed Line Section

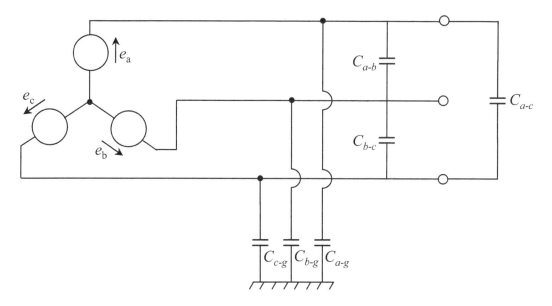

Figure 5.10.1: Open-Circuit Test on Untransposed Transmission Line Section

$$\Delta \tilde{I}_0 = j\omega C_{00}\tilde{V}_0 + \omega \tilde{C}_{01}\tilde{V}_1 \qquad [A/m] \qquad (5.10.2)$$

where $\Delta \tilde{I}_0$ represents the zero sequence charging current supplied by the source. If the source neutral is ungrounded, then $\Delta \tilde{I}_0 = 0$ and we have for the open-circuit voltage at the source neutral:

$$\tilde{V}_{0-o.c.} = j\left(\frac{\tilde{C}_{01}}{C_{00}}\right)\tilde{V}_1 \qquad (5.10.3)$$

If the source neutral is solidly grounded, then $= 0$ and:

$$\Delta \tilde{I}_{0-s.c.} = \omega \tilde{C}_{01}\tilde{V}_1 = -j\omega C_{00}\left\{j\left(\frac{\tilde{C}_{01}}{C_{00}}\right)\tilde{V}_1\right\} = \frac{-1}{-jX_{C00}}\left\{j\left(\frac{\tilde{C}_{01}}{C_{00}}\right)\tilde{V}_1\right\} \qquad (5.10.4)$$

Figure 5.10.2 shows the introduction of the electrostatic unbalance effects into the zero sequence network.

Now let:

$$\tilde{Z}_n \quad = \quad R_n + jX_n \qquad (5.10.5)$$

$$Q_n \quad = \quad \frac{X_n}{R_n} \qquad (5.10.6)$$

$$\tilde{Z}_n \quad = \quad (1 + jQ_n)\frac{X_n}{Q_n} \qquad (5.10.7)$$

Then, according to Figure 5.10.2 we have:

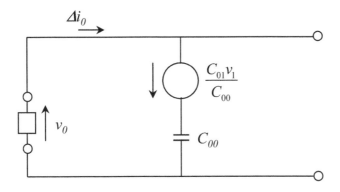

Figure 5.10.2: Zero Sequence Electrostatic Unbalance Representation

$$
\begin{aligned}
\tilde{V}_0 &= \frac{3\tilde{Z}_n}{3\tilde{Z}_n + \frac{1}{j\omega C_{00}}}(\frac{j\tilde{C}_{01}}{C_{00}})\tilde{V}_1 \\[2ex]
&= \frac{(1 + jQ_n)}{1 + jQ_n(1 - \frac{1}{3\omega C_{00}X_n})}(\frac{j\tilde{C}_{01}}{C_{00}})\tilde{V}_1 \\[2ex]
&= \frac{(1 + jQ_n)}{1 + jQ_n(1 - \frac{X_{00}}{3X_n})}(\frac{j\tilde{C}_{01}}{C_{00}})\tilde{V}_1 \\[2ex]
&= \tilde{k}\,|\,\frac{C_{01}}{C_{00}}\,||\,V_1\,|
\end{aligned}
\tag{5.10.8}
$$

Table 5.10.1 provides values of k as a function of Q and $(X_{00}/3X_n)$.

Table 5.10.1				
$(X_{00}/3X_n)$	Q = 50	100	150	200
0.95	18.57	19.61	19.83	19.90
0.96	22.37	24.25	24.66	24.81
0.97	27.74	31.62	32.54	32.88
0.98	35.36	44.72	47.44	48.51
0.99	44.73	70.71	83.21	89.44
1.00	50.01	100.01	150.00	200.00
1.01	44.73	70.71	83.21	89.44
1.02	35.36	44.72	47.44	48.51
1.03	27.74	31.62	32.54	32.88
1.04	22.37	24.25	24.66	24.81
1.05	18.57	19.61	19.83	19.90
$|\,k\,|$ As a Function of Q and $(X_{00}/3X_n)$.				

Because we have assumed balanced source voltages with = 0, then:

$$
\Delta\tilde{I}_2 = \omega\tilde{C}_{21}\tilde{V}_1 = -j\omega C_{22}\{j(\frac{\tilde{C}_{21}}{C_{22}})\tilde{V}_1\} = \frac{-1}{-jX_{C22}}\{j(\frac{\tilde{C}_{21}}{C_{22}})\tilde{V}_1\}
\tag{5.10.9}
$$

Figure 5.10.3 shows the appropriate negative sequence network representation.

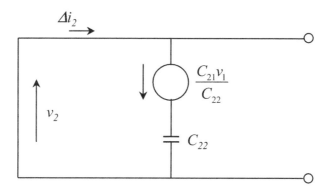

Figure 5.10.3: Negative Sequence Electrostatic Unbalance Representation

5.11   Double-Circuit Lines

We consider now two single-circuit lines that are parallel to one another, perhaps on the same right-of-way. After "bundling" and "ground wire removal," we obtain the voltage equations for the "equivalent double-circuit line:"

$$
\begin{bmatrix} \tilde{V}_a \\ \tilde{V}_b \\ \tilde{V}_c \\ \tilde{V}_{a'} \\ \tilde{V}_{b'} \\ \tilde{V}_{c'} \end{bmatrix} = -\frac{j}{\omega} \begin{bmatrix} P_{aa} & P_{ab} & P_{ac} & P_{aa'} & P_{ab'} & P_{ac'} \\ P_{ba} & P_{bb} & P_{bc} & P_{ba'} & P_{bb'} & P_{bc'} \\ P_{ca} & P_{cb} & P_{cc} & P_{ca'} & P_{cb'} & P_{cc'} \\ P_{a'a} & P_{a'b} & P_{a'c} & P_{a'a'} & P_{a'b'} & P_{a'c'} \\ P_{b'a} & P_{b'b} & P_{b'c} & P_{b'a'} & P_{b'b'} & P_{b'c'} \\ P_{c'a} & P_{c'b} & P_{c'c} & P_{c'a'} & P_{c'b'} & P_{c'c'} \end{bmatrix} \begin{bmatrix} \Delta\tilde{I}_a \\ \Delta\tilde{I}_b \\ \Delta\tilde{I}_c \\ \Delta\tilde{I}_{a'} \\ \Delta\tilde{I}_{b'} \\ \Delta\tilde{I}_{c'} \end{bmatrix} \quad [V] \quad (5.11.1)
$$

In terms of symmetrical components, we have:

$$
\begin{bmatrix} \tilde{V}_0 \\ \tilde{V}_1 \\ \tilde{V}_2 \\ \tilde{V}_{0'} \\ \tilde{V}_{1'} \\ \tilde{V}_{2'} \end{bmatrix} = \frac{1}{\omega} \begin{bmatrix} P_{00} & P_{01} & P_{02} & P_{00'} & P_{01'} & P_{02'} \\ P_{10} & P_{11} & P_{12} & P_{10'} & P_{11'} & P_{12'} \\ P_{20} & P_{21} & P_{22} & P_{20'} & P_{21'} & P_{22'} \\ P_{0'0} & P_{0'1} & P_{0'2} & P_{0'0'} & P_{0'1'} & P_{0'2'} \\ P_{1'0} & P_{1'1} & P_{1'2} & P_{1'0'} & P_{1'1'} & P_{1'2'} \\ P_{2'0} & P_{2'1} & P_{2'2} & P_{2'0'} & P_{2'1'} & P_{2'2'} \end{bmatrix} \begin{bmatrix} \Delta\tilde{I}_0 \\ \Delta\tilde{I}_1 \\ \Delta\tilde{I}_2 \\ \Delta\tilde{I}_{0'} \\ \Delta\tilde{I}_{1'} \\ \Delta\tilde{I}_{2'} \end{bmatrix} \quad [V] \quad (5.11.2)
$$

where some of the $\tilde{P}$ elements in the symmetrical component matrix may be complex numbers.

5.12   Numerical Example

Consider the double-circuit configuration shown in Figure 5.6.1(a). The phase shunt reactance matrix and the symmetrical component shunt reactance matrix for the double-circuit line with no ground wires are shown in Figure 5.12.1 and for the double-circuit line with four ground wires (equivalent line) in Figures 5.12.2 and 5.12.3.

$$\frac{1}{\omega}\left[\tilde{P}_{\phi}\right] = -\frac{j}{\omega}$$

		I			II		
		a	b	c	a'	b'	c'
I	a	0.37163	0.03842	0.03842	0.00878	0.01655	0.00533
	b	0.03842	0.37163	0.01655	0.00533	0.00878	0.00354
	c	0.03842	0.01655	0.37163	0.01655	0.03842	0.00878
II	a'	0.00878	0.00533	0.01655	0.37163	0.03842	0.03842
	b'	0.01655	0.00878	0.03842	0.03842	0.37163	0.01655
	c'	0.00533	0.00354	0.00878	0.03842	0.01655	0.37163

$$\frac{1}{\omega}\left[\tilde{P}_{c}\right] = \frac{1}{\omega}$$

		I			II		
		0	1	2	0'	1'	2'
I	0	0 -j0.43390	0 -j0.00729	0 -j0.00729	0 -j0.03735	-0.01331 +j0.00335	0.01331 +j0.00335
	1	0 -j0.00729	0 -j0.34050	0 +j0.01458	-0.01331 +j0.00335	0.00359 +j0.00551	0 -j0.00670
	2	0 -j0.00729	0 +j0.01458	0 -j0.34050	0.01331 +j0.00335	0 -j0.00670	-0.00359 +j0.00551
II	0'	0 -j0.03735	0.01331 +j0.00335	-0.01331 +j0.00335	0 -j0.43390	0 -j0.00729	0 -j0.00729
	1'	0.01331 +j0.00335	-0.00359 +j0.00551	0 -j0.00670	0 -j0.00729	0 -j0.34050	0 +j0.01458
	2'	-0.01331 +j0.00335	0 -j0.00670	0.00359 +j0.00551	0 -j0.00729	0 +j0.01458	0 -j0.34050

Figure 5.12.1: Shunt Reactance Matrices for Figure 5.6.1(a) with No Ground Wires $[M\Omega - km]$

$$\frac{1}{\omega}[\tilde{P}_\phi] = -\frac{j}{\omega}$$

	c	0.02956	0.01027	0.36298	0.01031	0.03111	0.00450
	a'	0.00414	0.00219	0.01031	0.36107	0.02956	0.03013
II	b'	0.01031	0.00450	0.03111	0.02956	0.36298	0.01027
	c'	0.00219	0.00140	0.00450	0.03013	0.01027	0.36451

$$\frac{1}{\omega}[\tilde{P}_c] = \frac{1}{\omega}$$

		I			II		
		0	1	2	0'	1'	2'
I	0	0 -j0.40950	-0.00061 -j0.00563	0.00061 -j0.00563	0 -j0.02354	-0.01092 +j0.00345	0.01092 +j0.00345
	1	0.00061 -j0.00563	0 -j0.33954	-0.00012 +j0.01394	-0.01092 +j0.00345	0.00389 +j0.00521	0 -j0.00655
	2	-0.00061 -j0.00563	0.00012 +j0.01394	0 -j0.33954	0.01092 +j0.00345	0 -j0.00655	-0.00389 +j0.00521
II	0'	0 -j0.02354	0.01092 +j0.00345	-0.01092 +j0.00345	0 -j0.40950	0.00060 -j0.00563	-0.00060 -j0.00563
	1'	0.01092 +j0.00345	-0.00389 +j0.00521	0 -j0.00655	-0.00060 -j0.00563	0 -j0.33954	0.00012 +j0.01394
	2'	-0.01092 +j0.00345	0 -j0.00655	0.00389 +j0.00521	0.00060 -j0.00563	-0.00012 +j0.01394	0 -j0.33954

Figure 5.12.2: Shunt Reactance Matrices for Figure 5.6.1(a) with All Ground Wires $[M\Omega - km]$

$$\frac{1}{\omega}\left[\tilde{P}_\phi\right]=-\frac{j}{\omega}$$

		I			II		
	c	0.02956	0.01027	0.36298	0.01031	0.00450	0.03111
	a'	0.00414	0.00219	0.01031	0.36107	0.03013	0.02956
II b'		0.00219	0.00140	0.00450	0.03013	0.36451	0.01027
	c'	0.01031	0.00450	0.03111	0.02956	0.01027	0.36298

$$\frac{1}{\omega}\left[\tilde{P}_c\right]=\frac{1}{\omega}$$

		I			II		
		0	1	2	0'	1'	2'
I	0	0 -j0.40950	-0.00061 -j0.00563	0.00061 -j0.00563	0 -j0.02354	0.01092 +j0.00345	-0.01092 +j0.00345
	1	0.00061 -j0.00563	0 -j0.33954	-0.00012 +j0.01394	-0.01092 +j0.00345	0 -j0.00655	0.00389 +j0.00521
	2	-0.00061 -j0.00563	0.00012 +j0.01394	0 -j0.33954	0.01092 +j0.00345	-0.00389 +j0.00521	0 -j0.00655
II	0'	0 -j0.02354	0.01092 +j0.00345	-0.01092 +j0.00345	0 -j0.40950	-0.00060 -j0.00563	0.00060 -j0.00563
	1'	-0.01092 +j0.00345	0 -j0.00655	0.00389 +j0.00521	0.00060 -j0.00563	0 -j0.33954	-0.00012 +j0.01394
	2'	0.01092 +j0.00345	-0.00389 +j0.00521	0 -j0.00655	-0.00060 -j0.00563	0.00012 +j0.01394	0 -j0.33954

Figure 5.12.3: Shunt Reactance Matrices for Figure 5.6.1(b) with All Ground Wires $[M\Omega - km]$

5.13   References

[1] D. Coleman, F. Watts, and R.B. Shipley. *Digital Calculation of Overhead Transmission Line Constants*. AIEE Transactions, Vol 78, February, 1959, pgs. 1266-1270.

[2] Eric T.B. Gross. *Unbalances of Untransposed Overhead Lines*. Journal of the Franklin Institute, Vol 254, Philadelphia, PA, 1952, pgs. 487-497.

[3] Eric T.B. Gross and Wing Chin. *Electrostatic Unbalance of Untransposed Single-Circuit Lines*. AIEE Transactions, Vol 87, 1968, pgs. 24-34.

[4] Eric T.B. Gross, John H. Drinnan, and Erich Jochum. *Electromagnetic Unbalance of Untransposed Transmission Lines: III - Double Circuit Lines*. AIEE Transactions, Vol 78, December, 1959, pgs. 1362-1371.

[5] Eric T.B. Gross and M. Harry Hesse. *Electromagnetic Unbalance of Untransposed Transmission Lines*. AIEE Transactions, PAS-72, 1953, pgs. 1323-1336.

[6] Eric T.B. Gross and W.J. McNutt. *Electrostatic Unbalance to Ground of Twin Conductor Lines*. AIEE Transactions, PAS-72, 1953, pgs. 1288-1297.

[7] Eric T.B. Gross and S.W. Nelson. *Electromagnetic Unbalance of Untransposed Transmission Lines: II - Single Circuit Lines With Horizontal Conductor Arrangement*. AIEE Transactions, Vol 74, October, 1955, pgs. 887-893.

[8] Eric T.B. Gross and Andrew H. Westo. *Transposition and Unbalance of High Voltage Lines*. Electrical Engineering, Vol 71, July, 1952, pg 606.

[9] Eric T.B. Gross and Andrew H. Weston. *Transposition of High-Voltage Overhead Lines and Elimination of Electrostatic Unbalance to Ground*. AIEE Transactions, Vol 70, Part II, 1951, pgs. 1837-1844.

[10] M.H. Hesse. *Circulating Currents in Paralleled Untransposed Multi-Circuit Lines. Pt. I - Numerical Evaluation*. AIEE Transactions, Vol 85, July, 1966, pgs. 802-811.

[11] M.H. Hesse. *Circulating Currents in Paralleled Untransposed Multi-Circuit Lines. Pt. II - Methods for Estimating Current Unbalance*. AIEE Transactions, Vol 85, July, 1966, pgs. 812-820.

[12] M.H. Hesse. *Electromagnetic and Electrostatic Transmission Line Parameters by Digital Computer*. AIEE Transactions, Vol 81, June, 1963, pgs. 282-291.

[13] M.H. Hesse. *Simplified Approach for Estimating Current Unbalances in EHV Loop Circuits*. Proceedings, IEE (London), Vol 119, No 11, November, 1972, pgs. 1621-1627.

[14] M.H. Hesse and J. Sabath. *EHV Double-Circuit Untransposed Transmission Line: Analysis and Tests*. AIEE Transactions, Vol 90, 1971, pgs. 984-992.

[15] M.H. Hesse and D.D. Wilson. *Near Resonant Coupling on EHV Circuits, Pt. II - Methods of Analysis*. AIEE Transactions, Vol 87, 1968, pgs. 326-333.

[16] H. Holley, D. Coleman, and R.B. Shipley. *Untransposed EHV Line Computations*. AIEE Transactions, Vol 82, March, 1964, pgs. 291-296.

[17] R.F. Lawrence and D.J. Povejsil. *Determination of Inductive and Capacitive Unbalance for Untransposed Transmission Lines*. AIEE Transactions, Vol 71, Part III, 1952, pgs. 548-555.

[18] M.J. Pickett, H.L. Manning, and H.N. VanGeem. *Near Resonant Coupling on EHV Circuits, Pt. I - Field Investigation*. AIEE Transactions, Vol 87, 1968, pgs. 322-325.

5.14 Exercises

5.1 Consider the configuration shown in Figure 5.2.1. For no ground wires, we have given a phasing arrangement (from left to right):

Case I   b · a · c

The remaining significantly different phasing arrangements are:

Case II   c · b · a
Case III   a · c · b
Case IV   c · a · b
Case V   a · b · c
Case VI   b · c · a

(a) Determine $[\tilde{Z}_c]$ for Cases I through VI, with the elements in $[\tilde{Z}_c]$ expressed in polar form.

(b) For which cases are there only two significantly different off-diagonal elements? Why is this so? [Note: See Tables 4.3.1 and 4.3.2]

(c) Determine $\tilde{m}_2$, $\tilde{m}_0$, and $\frac{\tilde{m}_2}{\tilde{m}_0}$ for the six cases employing impedances.

5.2 Figures 5.7.2 through 5.7.5 give the $[\tilde{Z}_c]$ (Ω/km) and associated $[\tilde{Z}_c]^{-1}$ matrices for the double-circuit line shown in Figure 5.6.1, for all significantly different phasing arrangements. Consider the two circuits to be electrically in parallel.

(a) Determine $\tilde{m}_{0t}$, $\tilde{m}_{2t}$, $\tilde{m}_{0c}$, and $\tilde{m}_{2c}$ by the approximate method employing the $[\tilde{Z}_c]^{-1}$ matrices.

(b) For the cab-cab arrangement, determine $\tilde{m}_{0t}$, $\tilde{m}_{2t}$, $\tilde{m}_{0c}$, and $\tilde{m}_{2c}$ by the approximate method employing the $[\tilde{Z}_c]$ matrix.

(c) Repeat (b) but with the addition of a balanced source impedance $z_{00} = z_{11} = z_{22} = j0.35$ and with the addition of a balanced load impedance $z_{00} = z_{11} = z_{22} = j0.35 = 2.5$ at 30 degrees.

(d) Repeat (c) but with the addition of a 70% series capacitor compensation in each circuit.

Chapter 6

Symmetrical Component Representation of Transformers

In Chapter 2, the physical characteristics of single-phase transformers were investigated and interpreted in terms of equivalent circuits. In Chapter 3, the balanced three-phase system was considered and it was found that for symmetrically constructed three-phase transformers, a single-phase representation per phase could be used for the wye-wye or delta-delta connections. For wye-delta connections, an appropriate phase shift has to be introduced, in general. Therefore, we have already established what the transformer equivalent circuits look like in the positive sequence network.

It remains, therefore, to establish the proper equivalent circuits for three-phase transformers in the negative and zero sequence networks.

6.1 Phase Shift through Y-Δ Transformers

Since both positive and negative sequence voltage sources are balanced sets of three-phase voltages, one rotating in the forward sense, the other in a backward sense, the arrangements developed in Chapter 3 apply also to the negative sequence equivalent network, i.e., positive and negative sequence networks are simply balanced three-phase cases for the wye-wye and delta-delta connections. The same applies essentially for the wye-delta connection, except for the phase shift through the transformer.

Figure 6.1.1 shows a wye-delta transformer connection, together with the phasor diagram for the case where positive sequence voltage is applied to the wye terminals. Figure 6.1.2 shows the complete per unit equivalent circuit for this transformer, which is the appropriate representation for this transformer in the positive sequence network, provided, of course, that the substitution of a 90 degree phase shift for the -30 degree phase shift is acceptable.

Figure 6.1.3 shows the phasor diagram of the transformer connection of Figure 6.1.1 with negative sequence voltages applied to the wye terminals. Note that in this case, the phase shift across the transformer is in the opposite direction from that for the positive sequence equivalent. The labeling of the per unit equivalent circuit of Figure 6.1.3 includes this difference in the phase shift, and is the appropriate equivalent for the negative sequence network.

In the alternative wye-delta connection that was shown in Chapter 3, the -90 degree phase shift seen going from wye to delta windings in the positive sequence would become +90 degree phase shift for the negative sequence voltages applied to the wye winding.

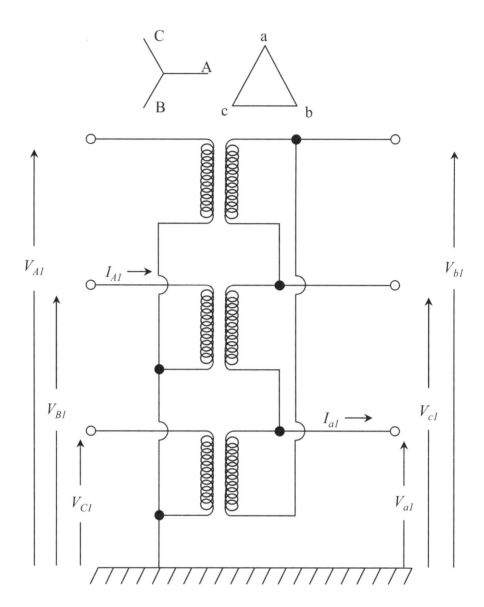

Figure 6.1.1: Wye-Delta Transformer with Positive Sequence Voltages Applied

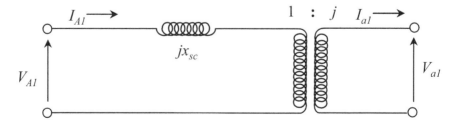

Figure 6.1.2: Positive Sequence Equivalent Circuit of the Transformer Connection of Figure 6.1.1

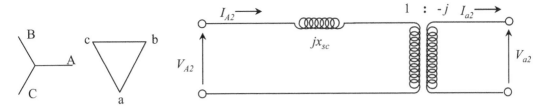

Figure 6.1.3: Phasor Diagram and Negative Sequence Equivalent Circuit of the Transformer Connection of Figure 6.1.1

Example 6.1.1
Assume that no zero sequence current is present in either side of the connection of Figure 6.1.1, but that positive and negative sequence currents are:

$$\tilde{I}_{A1} = 0 - j0.60$$

$$\tilde{I}_{A2} = 0 - j0.40$$

To find the secondary current in each phase, note that:

$$-j\tilde{I}_{a1} = -j0.60$$

$$\tilde{I}_{a1} = +0.60 = j\tilde{I}_{A1}$$

Similarly:

$$j\tilde{I}_{a2} = -j0.40$$

$$\tilde{I}_{a2} = -0.40 = -j\tilde{I}_{A2}$$

From the given vector diagrams of Figures 6.1.1 and 6.1.3, we can see that for this particular example, the positive and negative sequence phase A currents are in phase on the wye side of the transformer, but 180 degrees out of phase on the delta side. Then:

$$\tilde{I}_a = \tilde{I}_{a1} + \tilde{I}_{a2} = +0.20$$

$$\tilde{I}_b = \tilde{a}^2 \tilde{I}_{a1} + \tilde{a} \tilde{I}_{a2} = -0.10 - j0.866$$

$$\tilde{I}_c = \tilde{a} \tilde{I}_{a1} + \tilde{a}^2 \tilde{I}_{a2} = -0.10 + j0.866$$

Notice that in the foregoing example, the wye-side currents were initially given. In many solutions, the currents initially solved would be on the delta side, i.e., $\tilde{I}_{a0}$, $\tilde{I}_{a1}$, and $\tilde{I}_{a2}$.

6.2   Zero Sequence Impedance of Y-Y Transformers

Figure 6.2.1 shows a wye-wye transformer with an impedance grounded neutral on both the primary and secondary windings. Application of a zero sequence voltage on the primary means that each phase voltage must have the same magnitude and phase angle. For that reason, they may as well be connected together as shown. If we assume complete symmetry for the transformer

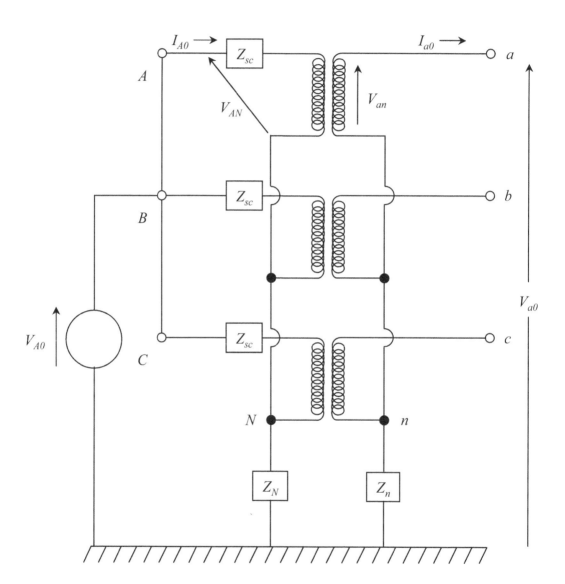

Figure 6.2.1: Wye-Wye Transformer with Zero Sequence Voltage Applied

$\left(\tilde{Z}_{aa} = \tilde{Z}_{bb} = \tilde{Z}_{cc} \text{ and } \tilde{Z}_{ab} = \tilde{Z}_{bc} = \tilde{Z}_{ca}\right)$ then it is apparent that the secondary voltages will also be identical in each phase.

If the internal equations for the "a" phase windings alone are considered, the following equation can be written:

$$\tilde{V}_{an} = \tilde{V}_{AN} - \tilde{Z}_{SC}\tilde{I}_{A0} \tag{6.2.1}$$

But note that:

$$\tilde{V}_{AN} = \tilde{V}_{A0} - 3\tilde{Z}_N\tilde{I}_{A0}$$
$$\tilde{V}_{an} = \tilde{V}_{a0} - 3\tilde{Z}_n\tilde{I}_{a0}$$
$$\tilde{I}_{a0} = \tilde{I}_{A0} \tag{6.2.2}$$

Combining Equations 6.2.1 and 6.2.2 shows that:

$$\tilde{V}_{A0} - \tilde{V}_{a0} = \left(3\tilde{Z}_N + \tilde{Z}_{SC} + 3\tilde{Z}_n\right)\tilde{I}_{A0} \tag{6.2.3}$$

Equation 6.2.3 can be represented in the per unit equivalent circuit of Figure 6.2.2 where the neutral impedances are multiplied by three and shown as line quantities. This does not change the drop caused by the neutral impedances because:

$$3\tilde{Z}_n\tilde{I}_{a0} = \left(3\tilde{Z}_n\right)\tilde{I}_{a0} \tag{6.2.4}$$

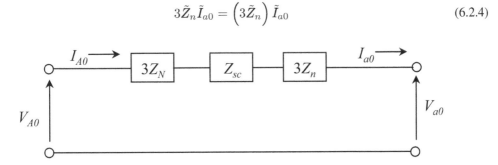

Figure 6.2.2: Equivalent Circuit for the Representation of the Transformer of Figure 6.2.1 in the Zero Sequence Network

Although it is not immediately apparent what base impedance applies to a neutral impedance, the manner in which it is used in Equations 6.2.2 and 6.2.3 requires that, like other system impedances, it be represented on a single-phase MVA base and line-to-neutral voltage base.

Note that if either the primary or secondary winding is ungrounded, then the corresponding grounding impedance $\left(\tilde{Z}_N \text{ or } \tilde{Z}_n\right)$ would be infinite and would be replaced by an open circuit in the equivalent circuit of the transformer in the zero sequence network of Figure 6.2.2.

6.3   Zero Sequence Impedance of Δ-Δ Transformer

Figure 6.3.1 shows a delta-delta transformer connected for a zero sequence test. It is immediately apparent that no zero sequence current can flow in the line on either side of the transformer, since zero sequence currents, being the same in each phase, must have a return path in order to flow. It might be argued that zero sequence currents could flow around the delta, but the fact remains that there in no driving voltage to produce such currents. Figure 6.3.2 shows the per unit equivalent of this transformer as it would appear in the zero sequence network.

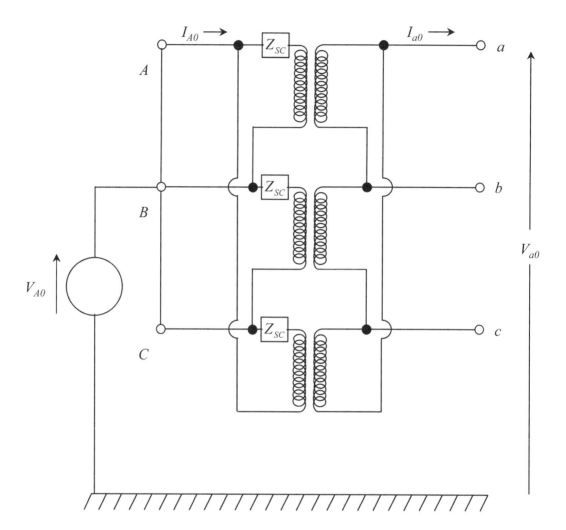

Figure 6.3.1: Delta-Delta Transformer with Zero Sequence Voltage Applied

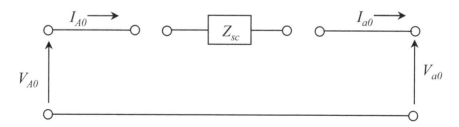

Figure 6.3.2: Equivalent Circuit for the Representation of the Transformer of Figure 6.3.1 in the Zero Sequence Network

6.4 Zero Sequence Impedance of Grounded Y-Δ Transformer

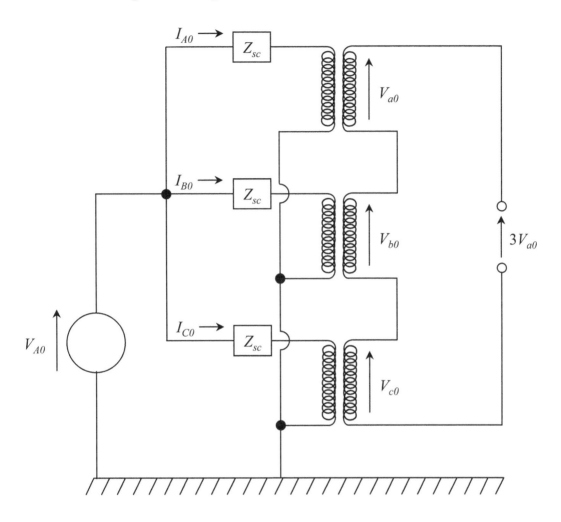

Figure 6.4.1: Wye-Delta Transformer with Zero Sequence Voltage Applied

Figure 6.4.1 shows a balanced wye-delta transformer with the wye neutral solidly grounded and with zero sequence voltage applied to the wye terminals. The delta is not yet closed. It will be useful to consider a Thevenin-Helmholtz equivalent of this transformer, as seen from the open delta terminals x-x. The open-circuit voltages, $\tilde{V}_{a0}$, $\tilde{V}_{b0}$, and $\tilde{V}_{c0}$ will be equal and additive so that the Thevenin-Helmholtz voltage is $3\tilde{V}_{a0}$. Then, shorting out the voltage source, $\tilde{V}_{A0} = \tilde{V}_{B0} = \tilde{V}_{C0}$, the short-circuit impedance seen looking into the terminals x-x is that of three (short circuited) transformers in a series, or $3\tilde{Z}_{SC}$. Noting that in per unit, the Thevenin-Helmholtz voltage, $3\tilde{V}_{a0} = 3\tilde{V}_{A0}$, the complete Thevenin-Helmholtz equivalent is shown in Figure 6.4.2.

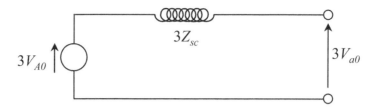

Figure 6.4.2: Thevenin-Helmholtz Equivalent Circuit for Figure 6.4.1 as Viewed from the Open Terminals

Since the terminals x-x in Figure 6.4.1 are actually closed, and since both the voltage and impedance of Figure 6.4.2 may be divided by three without changing the currents, $\tilde{I}_{a0}$, the zero sequence per unit equivalent of the wye-delta transformer with zero sequence voltage applied to the wye terminals is as shown in Figure 6.4.3.

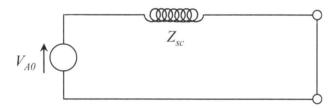

Figure 6.4.3: Thevenin-Helmholtz Equivalent Circuit with the Delta Closed

If a zero sequence voltage were applied to the delta terminals of a delta-wye (grounded or ungrounded) transformer, no zero sequence current could flow *regardless* of whether the wye side were opened or shorted. So, the complete equivalent of the wye-delta transformer is shown in Figure 6.4.4. The delta winding appears as a short-circuit to the zero sequence voltages applied to the wye winding. The resulting zero sequence currents flow through the wye windings and into the neutral. They also circulate in the delta winding but cannot flow in the line connected to the delta winding.

If the neutral of the wye winding were grounded through an impedance as shown in Figure 6.4.5, then, based on the foregoing argument, the following (internal) per unit equation can be written for one pair of windings:

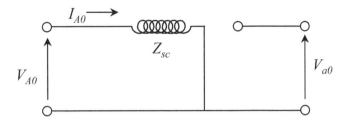

Figure 6.4.4: Equivalent Circuit Representation of the Transformer in Figure 6.4.1 in the Zero Sequence Network

$$\tilde{V}_{AN} - \tilde{I}_{A0}\tilde{Z}_{SC} = 0 \tag{6.4.1}$$

But note that:

$$\tilde{V}_{AN} = \tilde{V}_{A0} - 3\tilde{I}_{A0}\tilde{Z}_N \tag{6.4.2}$$

and combining Equations 6.4.1 and 6.4.2 shows that:

$$\tilde{V}_{A0} = \tilde{I}_{A0}\left(\tilde{Z}_{SC} + 3\tilde{Z}_N\right) \tag{6.4.3}$$

so that the complete per unit equivalent circuit for the transformer of Figure 6.4.5 is shown in Figure 6.4.6.

6.5   Zero Sequence Impedance of Three-Winding Transformers

Figure 6.5.1 shows a general per unit equivalent circuit of a three-phase, three-winding transformer. If each winding of the three-winding transformer is wye connected, then each branch of the transformer equivalent (Figure 6.5.1) is connected directly to the system by the Y-Y jumpers, and the resulting equivalent is a logical extension of the zero sequence equivalent for a two-winding wye-wye transformer shown in Figure 6.2.2 and can be derived in exactly the same manner. If any one of the three windings is delta connected, the corresponding branch of the transformer is shorted to the reference bus and the system connection is left open-circuited. If any of the three windings is connected as an ungrounded wye, we may simply state that $3\tilde{Z}_n = \infty$ and consider this a special case of the wye winding, which provides neither a through path for zero sequence currents nor a zero sequence "short-circuit."

If the three-phase transformer is made up of three single-phase units, the magnetizing impedance is usually high enough to be neglected, just as if it was in the positive (or negative) sequence case. Figure 6.5.2 shows a three-phase shell-type core construction in which a portion of the core is saturated under zero sequence excitation. The three-legged core design of Figure 6.5.3 provides no return path for zero sequence flux, except through the air. Then, since:

$$X = \omega L = \omega N^2 \wp = \frac{\omega N^2}{\Re} \tag{6.5.1}$$

the high reluctance path for zero sequence flux introduced by these two different constructions results in open circuit impedances of the order of 1.0 to 5.0 per unit for the shell design (with $\tilde{V}_{a0} = 1.0$) and the order of 0.50 to 1.0 for the three-legged core design. In these cases then, the shunt magnetizing branch of the transformer bank may warrant representation in the zero sequence equivalent circuit. In addition, due to zero sequence currents circulating in the tank walls, the zero

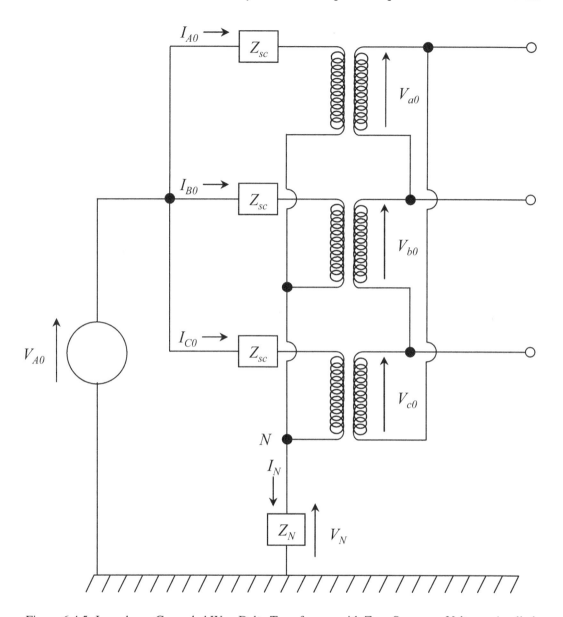

Figure 6.4.5: Impedance Grounded Wye-Delta Transformer with Zero Sequence Voltages Applied

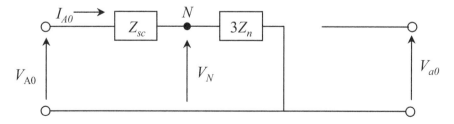

Figure 6.4.6: Equivalent Circuit Representation of the Transformer of Figure 6.4.5 in the Zero Sequence Network

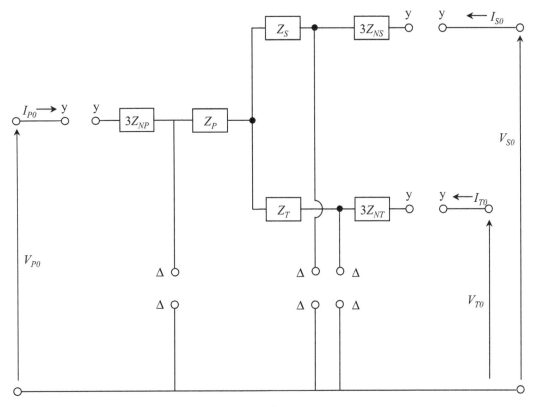

If winding is delta-connected, jumper Δ – Δ terminals.
If winding is wye-connected, jumper y – y terminals.

Figure 6.5.1: General Equivalent Circuit for a Three-Winding Transformer in the Zero Sequence Network

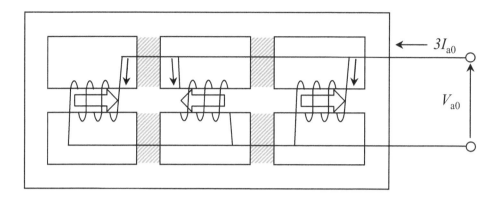

Figure 6.5.2: Schematic Diagram of a Three-Phase Shell-Type Transformer with Zero Sequence Voltages Applied (Shaded Areas are Saturated)

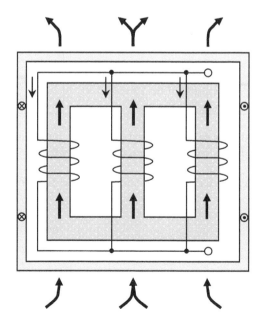

Figure 6.5.3: Schematic Diagram of a Three-Phase Core-Type Transformer with Zero Sequence Voltages Applied

sequence short-circuit impedance of the three-legged core design may be reduced on the order of 15% from its positive sequence value.

6.6 Grounding Transformers

If is often desirable to provide a zero sequence impedance to ground on otherwise ungrounded systems. A simple wye-grounded delta transformer would serve such a purpose, for example. Its shunt impedance would be very high in positive and negative sequence networks, but would be equal to $\tilde{Z}_{SC}$ in the zero sequence. It would be considerably cheaper than most transformers, since its continuous duty would be quite meager and its zero sequence rating can be based on emergency conditions of short duration. It would carry current only during faults.

Zig-Zag connected transformers are an even more practical means of effectively grounding the neutral of a system where no such ground can be made from existing transformer banks. Zig-zag transformers are similar to grounded wye/delta banks in that they provide a high shunt impedance in the positive and negative sequence networks, and a very low shunt impedance in the zero sequence network.

Figure 6.6.2 shows the internal equivalent circuit for the pair of windings on leg A of the zig-zag transformer illustrated in Figure 6.6.1, both windings having the same number of turns ($N_p = N_s$). On the basis of Figure 6.6.2, we have:

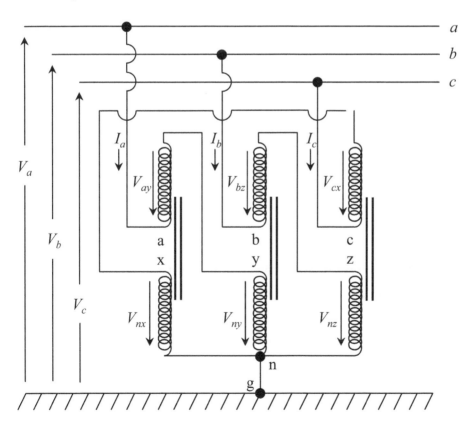

Figure 6.6.1: Connection Diagram for a Zig-Zag Grounding Transformer

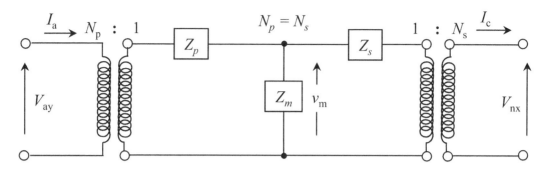

Figure 6.6.2: Internal Equivalent Circuit of the Windings on Core Leg A of the Zig-Zag Transformer in Figure 6.6.1

$$\tilde{V}_{ay} = N_p^2 \left[\tilde{Z}_p \tilde{I}_a + \tilde{Z}_m \left(\tilde{I}_a - \tilde{I}_c \right) \right]$$
$$\tilde{V}_{bz} = N_p^2 \left[\tilde{Z}_p \tilde{I}_b + \tilde{Z}_m \left(\tilde{I}_b - \tilde{I}_a \right) \right]$$
$$\tilde{V}_{cx} = N_p^2 \left[\tilde{Z}_p \tilde{I}_c + \tilde{Z}_m \left(\tilde{I}_c - \tilde{I}_b \right) \right] \tag{6.6.1}$$

$$\tilde{V}_{nx} = N_s^2 \left[\tilde{Z}_m \left(\tilde{I}_a - \tilde{I}_c \right) - \tilde{Z}_s \tilde{I}_c \right]$$
$$\tilde{V}_{ny} = N_s^2 \left[\tilde{Z}_m \left(\tilde{I}_b - \tilde{I}_a \right) - \tilde{Z}_s \tilde{I}_a \right]$$
$$\tilde{V}_{nz} = N_s^2 \left[\tilde{Z}_m \left(\tilde{I}_c - \tilde{I}_b \right) - \tilde{Z}_s \tilde{I}_b \right] \tag{6.6.2}$$

and on the basis of Figure 6.6.1 we have, with $N = N_p = N_s$:

$$\tilde{V}_{an} = \tilde{V}_{ay} - \tilde{V}_{ny} = N^2 \left[\left(\tilde{Z}_p + \tilde{Z}_s \right) \tilde{I}_a + \tilde{Z}_m \left(2\tilde{I}_a - \tilde{I}_b - \tilde{I}_c \right) \right]$$
$$\tilde{V}_{bn} = \tilde{V}_{bz} - \tilde{V}_{nz} = N^2 \left[\left(\tilde{Z}_p + \tilde{Z}_s \right) \tilde{I}_b + \tilde{Z}_m \left(2\tilde{I}_b - \tilde{I}_c - \tilde{I}_a \right) \right]$$
$$\tilde{V}_{cn} = \tilde{V}_{cx} - \tilde{V}_{nx} = N^2 \left[\left(\tilde{Z}_p + \tilde{Z}_s \right) \tilde{I}_c + \tilde{Z}_m \left(2\tilde{I}_c - \tilde{I}_a - \tilde{I}_b \right) \right] \tag{6.6.3}$$

6.6.1 Positive Sequence Equivalent Circuit for a Zig-Zag Transformer

For a positive sequence set of currents and voltages we have, on the basis of Equation 6.6.3:

$$\frac{\tilde{V}_{an}}{\tilde{I}_{a1}} = N^2 \left(\tilde{Z}_p + \tilde{Z}_s + 3\tilde{Z}_m \right)$$
$$\frac{\tilde{V}_{bn}}{\tilde{I}_{b1}} = N^2 \left(\tilde{Z}_p + \tilde{Z}_s + 3\tilde{Z}_m \right)$$
$$\frac{\tilde{V}_{cn}}{\tilde{I}_{c1}} = N^2 \left(\tilde{Z}_p + \tilde{Z}_s + 3\tilde{Z}_m \right) \tag{6.6.4}$$

The same results are obtained for a negative sequence set of currents and voltages. Figure 6.6.3 shows the appropriate positive and negative sequence equivalent networks.

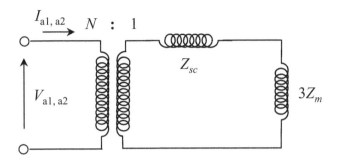

Figure 6.6.3: Positive and Negative Sequence Equivalent Circuit for a Zig-Zag Transformer

6.6.2 Zero Sequence Equivalent Circuit for a Zig-Zag Transformer

For a zero sequence set of currents and voltages, we have, on the basis of Equation 6.6.3:

$$\frac{\tilde{V}_{an}}{\tilde{I}_{a0}} = N^2 \left(\tilde{Z}_p + \tilde{Z}_s \right)$$

$$\frac{\tilde{V}_{bn}}{\tilde{I}_{b0}} = N^2 \left(\tilde{Z}_p + \tilde{Z}_s \right)$$

$$\frac{\tilde{V}_{cn}}{\tilde{I}_{c0}} = N^2 \left(\tilde{Z}_p + \tilde{Z}_s \right) \tag{6.6.5}$$

If an impedance $\tilde{Z}_n$ is connected between the neutral of the transformer and ground, the reader may easily verify that:

$$\frac{\tilde{V}_{ag}}{\tilde{I}_{a0}} = N^2 \left(\tilde{Z}_p + \tilde{Z}_s + 3\tilde{Z}_n \right) \tag{6.6.6}$$

Figure 6.6.4 illustrates the appropriate zero sequence equivalent network.

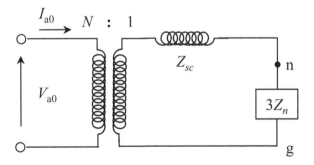

Figure 6.6.4: Zero Sequence Equivalent Network for a Zig-Zag Grounding Transformer

6.6.3 Rating of Zig-Zag Transformers

The nameplate of transformers specifically designed for zig-zag grounding service shows zero sequence impedance directly in ohms. Conversion to per unit impedance for the equivalent circuit of

Figure 6.6.4 requires division by the impedance base for the location in the power system at which the grounding transformer is connected. Since $\tilde{Z}_m$ is relatively large, the presence of the bank in the positive and negative sequence networks is generally ignored.

A transformer bank installed specifically for grounding duty has no appreciable current duty except under short-circuit conditions. For this reason, the rating of such banks is basically an overload rating and considers both current magnitude and time. Zig-zag ground transformer nameplates, for example, specify current rating for sixty seconds before exceeding an allowable temperature rise.

Even though the MVA rating of such a transformer requires this special interpretation, it can be argued that for a given current rating, a zig-zag grounding transformer requires less inherent MVA capability than a grounded wye-delta transformer to do the same job.

For example, according to Figure 6.6.5, the winding voltage for a zig-zag transformer is:

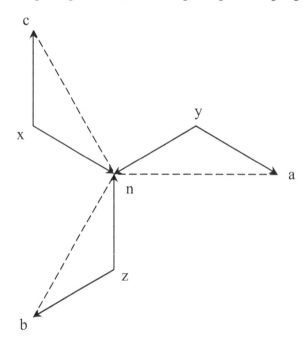

Figure 6.6.5: Voltage Phasor Diagram for a Zig-Zag Grounding Transformer

$$\tilde{V}_{ay} = \frac{\tilde{V}_{an}}{\sqrt{3}} \tag{6.6.7}$$

Thus, if the fault current requirement is $\tilde{I}_0$ amperes, connecting three single-phase transformers (having this short-time current rating) as a zig-zag transformer would require a total transformer kVA as follows:

$$kVA_{zig-zag} = \frac{3I_0kV_{an}}{\sqrt{3}} \tag{6.6.8}$$

The corresponding requirement for a grounded wye/delta bank would be:

$$kVA_{g-wye/delta} = 3I_0kV_{an} \tag{6.6.9}$$

Thus, zig-zag transformers require only 57.7% of the kVA of the corresponding wye/delta bank and are therefore often preferred for this application.

6.7 Three-Phase Autotransformers

In recent years there has been an increase in the application of three-phase autotransformers. From a positive and negative sequence point of view, since three-phase symmetry is assumed, a single-phase three-winding transformer equivalent circuit is applicable. Thus, for the autotransformer, with delta tertiary, the positive and negative sequence networks shown in Figure 6.7.1 are applicable, where:

$$\tilde{\tilde{Z}}_N = \frac{1}{2}\left(\tilde{\tilde{Z}}_{sc-c} + \tilde{\tilde{Z}}_{sc-t} - \tilde{\tilde{Z}}_{c-t}\right) \tag{6.7.1}$$

$$\tilde{\tilde{Z}}_L = \frac{1}{2}\left(\tilde{\tilde{Z}}_{sc-c} + \tilde{\tilde{Z}}_{c-t} - \tilde{\tilde{Z}}_{sc-t}\right) \tag{6.7.2}$$

$$\tilde{\tilde{Z}}_T = \frac{1}{2}\left(\tilde{\tilde{Z}}_{sc-t} + \tilde{\tilde{Z}}_{c-t} - \tilde{\tilde{Z}}_{sc-t}\right) \tag{6.7.3}$$

The autotransformer analysis is given in Reference [1], pgs. 135-143.

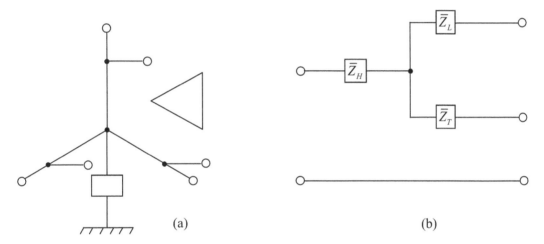

Figure 6.7.1: (a) Wye-Wye Autotransformer with Delta Tertiary; (b) Appropriate Positive and Negative Sequence Networks

The analysis for equivalent zero sequence models is somewhat more elusive. In particular, when a single neutral grounding impedance is applied to a neutral that is common to systems with different voltage bases, special precautions are required in deriving an equivalent per unit network. This topic is first introduced in Section 6.7.1, for a wye-wye transformer bank. The transition to the case of autotransformers then follows relatively easily.

6.7.1 Zero Sequence Impedance of Y-Y Transformer Bank

Figure 6.7.2 illustrates a wye-wye transformer bank with neutral grounding in both primary and secondary windings. We assume complete symmetry for the transformer ($\tilde{Z}_{aa} = \tilde{Z}_{bb} = \tilde{Z}_{cc}$) so that all phase currents and voltages are equal in magnitude and phase angle for the zero sequence test shown in Figure 6.7.2. Because of this symmetrical response, an equivalent single-phase representation as shown in Figure 6.7.3 may be used.

If the usual procedure for per unitizing system quantities is followed, then $\tilde{\tilde{Z}}_N$ is expressed in terms of the source system base and $\tilde{\tilde{Z}}_n$ upon the load system base, and the indicated transformer is

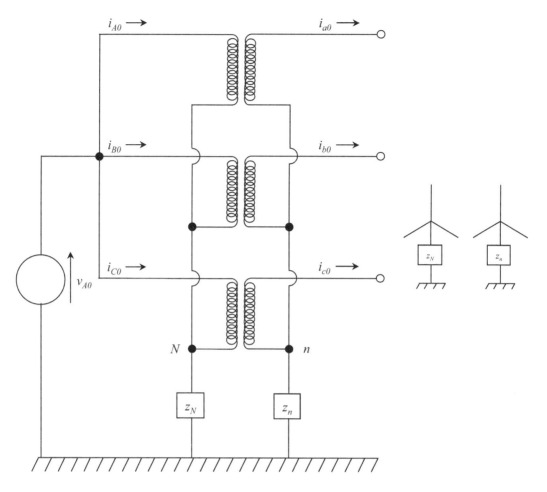

Figure 6.7.2: Wye-Wye Transformer with Zero Sequence Voltage Applied

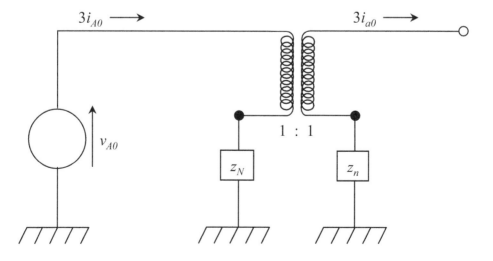

Figure 6.7.3: Equivalent Circuit for Figure 6.7.2

a 1:1 ideal transformer. For the sake of illustration, let the system be terminated in a balanced three-phase load, $\tilde{\tilde{Z}}_L$, wye connected with a solidly grounded neutral. The following voltage equation then pertains:

$$\tilde{V}_{a0} = \left[\tilde{\tilde{Z}}_N + \tilde{\tilde{Z}}_n + \frac{1}{3} \left(\tilde{\tilde{Z}}_p + \tilde{\tilde{Z}}_s + \tilde{\tilde{Z}}_L \right) \right] \left(3 \tilde{I}_{A0} \right) \tag{6.7.4}$$

$$\tilde{V}_{a0} = \left[\tilde{\tilde{Z}}_N + \tilde{\tilde{Z}}_n + \frac{1}{3} \left(\tilde{\tilde{Z}}_{s-c} + \tilde{\tilde{Z}}_L \right) \right] \left(3 \tilde{I}_{A0} \right) \tag{6.7.5}$$

or:

$$\frac{\tilde{V}_{a0}}{\tilde{I}_{a0}} = 3 \left(\tilde{\tilde{Z}}_N + \tilde{\tilde{Z}}_n \right) + \left(\tilde{\tilde{Z}}_{s-c} + \tilde{\tilde{Z}}_L \right) \tag{6.7.6}$$

whereby $\tilde{\tilde{Z}}_p$ and $\tilde{\tilde{Z}}_s$ represent the portions of transformer leakage reactance, which might (fictitiously) be associated with the primary and secondary windings, respectively. In Equation 6.7.6, the total short-circuit reactance of the transformer $\tilde{\tilde{Z}}_{s-c}$ replaces the $\tilde{\tilde{Z}}_p$ and $\tilde{\tilde{Z}}_s$ components. Figure 6.7.4 represents an equivalent zero sequence network for a wye-wye transformer bank.

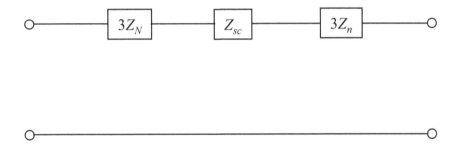

Figure 6.7.4: Equivalent Zero Sequence Network for Wye-Wye Transformer Bank

Figures 6.7.3 and 6.7.4 are based on the premise that the magnetizing impedance branch in the equivalent circuit is infinitely large. Therefore, the ampere-turns in the primary and secondary of the ideal transformer shown in Figure 6.7.3 must be equal. Consequently, if either primary or secondary wye is ungrounded, no zero sequence current can flow through either side of the wye-wye transformer bank. This can be reflected in the equivalent network (Figure 6.7.4) by allowing $\tilde{\tilde{Z}}_N$ and/or $\tilde{\tilde{Z}}_n$ to be replaced by an open circuit.

We consider now a somewhat more complex grounding arrangement as shown in Figure 6.7.5. In establishing system-base voltages for the purpose of per unitizing the system, an incompatibility develops in the region between N and n in the figure. This can be overcome, in a manner analogous to that discussed in Chapter 2, Section 2.5.2, by introducing ideal transformers that will relate the two different base voltages. Figure 6.7.4(b) demonstrates one manner in which this may be accomplished. The turns ratios of the ideal transformers are determined by the selected voltage bases V_N, V_n, and V_m. Then:

$$T = \frac{V_N}{V_m} \qquad t = \frac{V_n}{V_m} \tag{6.7.7}$$

and from the primary side, we see $\tilde{\tilde{Z}}_m / T^2$. From the secondary side, we see $\tilde{\tilde{Z}}_m / t^2$.

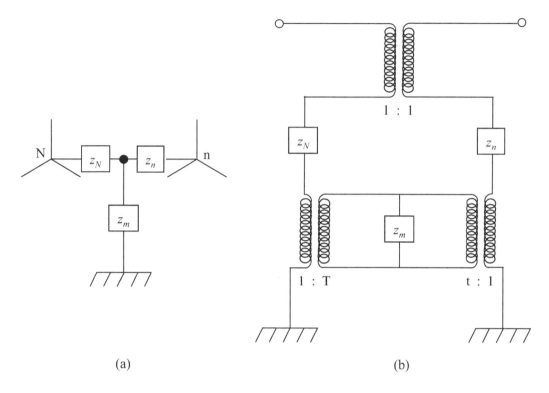

Figure 6.7.5: Generalized Wye-Wye Grounding Arrangement

For example, if $\tilde{\tilde{Z}}_m$ is expressed on the primary voltage base $\left[Z_{base} = (kV)^2/MVA\right]$, then $T = 1$, $t = V_n/V_N$. Then, from the secondary side of the system we see $\tilde{\tilde{Z}}_m(V_N/V_n^2$, which is consistent with the fact that a given ohmic impedance will be smaller in per unit on a system voltage base, which is larger, and vice versa. In selecting V_m equal to V_N or V_n, one or the other ideal transformers can be eliminated.

Figure 6.7.6 shows the appropriate zero sequence equivalent network for the wye-wye arrangement shown in Figure 6.7.5. If $\tilde{\tilde{Z}}_m$ is set equal to zero, Figure 6.7.6 reduces to that in Figure 6.7.4 as is to be expected. If, instead, $\tilde{\tilde{Z}}_N$ and $\tilde{\tilde{Z}}_n$ are set equal to zero, then we have the case for the two neutrals commonly grounded through a single impedance, $\tilde{\tilde{Z}}_m$.

Consider now the case whereby $\tilde{\tilde{Z}}_m \to \infty$. Due to the requirement of ampere-turn balance in ideal transformers, it becomes apparent that no zero sequence current can flow, unless $V_N = V_n$.

The ideal transformers used in Figure 6.7.6 can be eliminated in the following manner. The current through $3\tilde{\tilde{Z}}_m$ is equal to:

$$\tilde{\tilde{I}}_m = \tilde{\tilde{I}}_0 \left(\frac{V_m}{V_n} - \frac{V_m}{V_N}\right) \tag{6.7.8}$$

and the voltage across $3\tilde{\tilde{Z}}_m$ is:

$$\Delta\tilde{\tilde{V}}_m = 3\tilde{\tilde{I}}_0\tilde{\tilde{Z}}_m \left(\frac{V_m}{V_n} - \frac{V_m}{V_N}\right) \tag{6.7.9}$$

Consequently, we have:

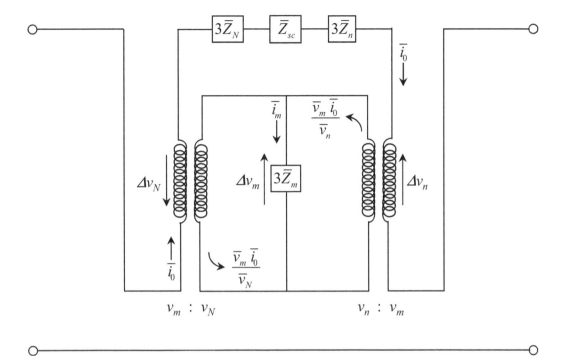

Figure 6.7.6: Zero Sequence Network for Figure 6.7.5

$$\Delta\tilde{\bar{V}}_N = -\left(\frac{V_m}{V_N}\right)\Delta\tilde{\bar{V}}_m = -3\tilde{\bar{I}}_0\tilde{\bar{Z}}_m\left(\frac{V_m}{V_n} - \frac{V_m}{V_N}\right)\left(\frac{V_m}{V_N}\right) \tag{6.7.10}$$

and:

$$\Delta\tilde{\bar{V}}_n = \left(\frac{V_m}{V_n}\right)\Delta\tilde{\bar{V}}_m = 3\tilde{\bar{I}}_0\tilde{\bar{Z}}_m\left(\frac{V_m}{V_n} - \frac{V_m}{V_N}\right)\left(\frac{V_m}{V_n}\right) \tag{6.7.11}$$

so that:

$$\Delta\tilde{\bar{Z}}_{eq} = \left(\frac{\Delta\tilde{\bar{V}}_N + \Delta\tilde{\bar{V}}_n}{\tilde{\bar{I}}_0}\right) = 3\Delta\tilde{\bar{Z}}_m\left(\frac{V_m}{V_n} - \frac{V_m}{V_n}\right)^2 \tag{6.7.12}$$

Thus, Figure 6.7.7 represents an applicable equivalent zero sequence network for the transformer grounding arrangement shown in Figure 6.7.5.

6.7.2 Zero Sequence Impedance of Y Autotransformer Banks

Figure 6.7.8(a) shows a neutral grounded, wye-connected autotransformer bank. Under zero sequence test conditions, the autotransformer may be considered as a two-winding transformer with a common neutral grounding. In this form, Figure 6.7.8(b) is a special case of Figure 6.7.5(a) so that the equivalent zero sequence network in Figure 6.7.7 may be readily adapted.

If the grounding impedance is expressed in per unit of the low voltage side, then in terms of the nomenclature in Figure 6.7.7, we would have:

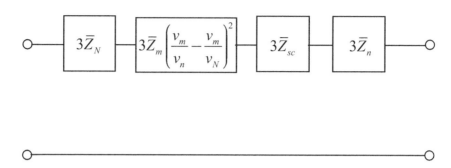

Figure 6.7.7: Zero Sequence Network for Figure 6.7.5

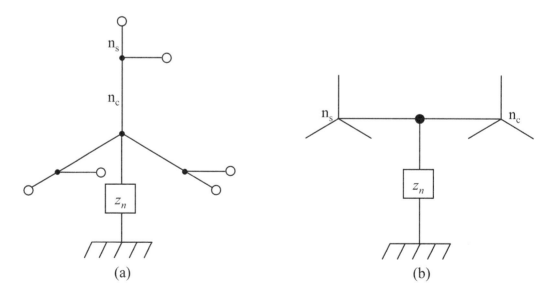

Figure 6.7.8: Wye-Connected Autotransformer Banks

$$3\tilde{\bar{Z}}_m \left(\frac{V_m}{V_n} - \frac{V_m}{V_N} \right)^2 = 3\tilde{\bar{Z}}_n \left(\frac{V_c}{V_c} - \frac{V_c}{V_{s-c}} \right)^2 \tag{6.7.13}$$

$$= 3\tilde{\bar{Z}}_n \left(1 - \frac{n_c}{n_s + n_c} \right)^2 \tag{6.7.14}$$

$$= 3\tilde{\bar{Z}}_n \left(\frac{n_s}{n_s + n_c} \right)^2 \tag{6.7.15}$$

$$= 3\tilde{\bar{Z}}_n \left(C.R. \right)^2 \tag{6.7.16}$$

where the coratio (CR) was defined in Chapter 2, Section 2.6. This is shown in Figure 6.7.9.

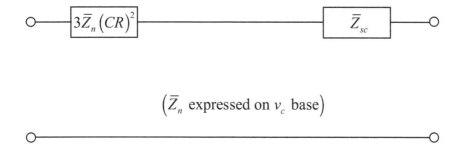

$$\left(\bar{Z}_n \text{ expressed on } v_c \text{ base} \right)$$

Figure 6.7.9: Zero Sequence Equivalent Network for Figure 6.7.8

If $\tilde{\bar{Z}}_n$ had been expressed in per unit on the high voltage side, then Figure 6.7.10 would be applicable.

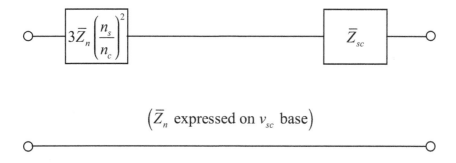

$$\left(\bar{Z}_n \text{ expressed on } v_{sc} \text{ base} \right)$$

Figure 6.7.10: Zero Sequence Equivalent Network for Figure 6.7.8

6.8 Zero Sequence Network for Autotransformer with Δ Tertiary

If a zero sequence test is performed along the lines indicated in Section 6.2, the concepts developed in Chapter 2 for multiwinding, single-phase transformers may be applied directly. Let $\tilde{\bar{Z}}_n$ (see Figure 6.8.1) be expressed in per unit on the low voltage side base. The resultant equivalent network is shown in Figure 6.8.2. Employing the results developed in the previous section, we have:

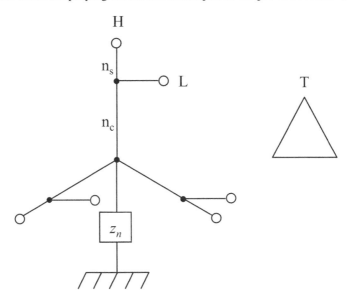

Figure 6.8.1: Autotransformer with Delta Tertiary

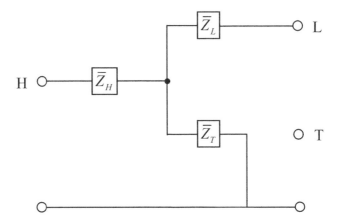

Figure 6.8.2: Zero Sequence Equivalent Network for Figure 6.8.1

$$\tilde{\bar{Z}}_H + \tilde{\bar{Z}}_L = \tilde{\bar{Z}}_{sc-c} + 3\tilde{\bar{Z}}_n \left(\frac{n_s}{n_s + n_c} \right)^2 \qquad (6.8.1)$$

which one obtains when the tertiary is open. When the high voltage winding is open, then:

$$\tilde{\bar{Z}}_L + \tilde{\bar{Z}}_T = \tilde{\bar{Z}}_{c-t} + 3\tilde{\bar{Z}}_n \tag{6.8.2}$$

which represents simply the equivalent zero sequence impedance of a grounded wye-delta transformer. In a similar manner, with the low voltage winding open-circuited, we have:

$$\tilde{\bar{Z}}_H + \tilde{\bar{Z}}_T = \tilde{\bar{Z}}_{sc-t} + 3\tilde{\bar{Z}}_n \left(\frac{n_c}{n_s + n_c}\right)^2 \tag{6.8.3}$$

where the $\tilde{\bar{Z}}_n$ (which is in terms of the LV base quantities) is referred to the HV base by means of $[n_c/(n_c + n_s)]^2$. Equations 6.8.1, 6.8.2, and 6.8.3 may be solved simultaneously to obtain:

$$\tilde{\bar{Z}}_H = \frac{1}{2}\left(\tilde{\bar{Z}}_{sc-c} + \tilde{\bar{Z}}_{sc-t} - \tilde{\bar{Z}}_{c-t}\right) - 3\tilde{\bar{Z}}_n \frac{n_s n_c}{(n_s + n_c)^2} \tag{6.8.4}$$

$$\tilde{\bar{Z}}_L = \frac{1}{2}\left(\tilde{\bar{Z}}_{sc-c} + \tilde{\bar{Z}}_{c-t} - \tilde{\bar{Z}}_{sc-t}\right) + 3\tilde{\bar{Z}}_n \frac{n_s}{(n_s + n_c)} \tag{6.8.5}$$

$$\tilde{\bar{Z}}_T = \frac{1}{2}\left(\tilde{\bar{Z}}_{sc-t} + \tilde{\bar{Z}}_{c-t} - \tilde{\bar{Z}}_{sc-c}\right) + 3\tilde{\bar{Z}}_n \frac{n_c}{(n_s + n_c)} \tag{6.8.6}$$

6.9 Autotransformer with Ungrounded Neutral and △ Tertiary

If, in the arrangement described in the previous section, the neutral impedance is allowed to become infinitely large ($\tilde{\bar{Z}}_n \to \infty$), the Equations 6.8.4, 6.8.5, and 6.8.6, together with Figure 6.8.2 may, on the surface, be somewhat misleading; one might expect that no zero sequence current can flow. The pitfall lies in the fact that the neutral impedance term is subtractive in the Equation 6.8.4 and additive in Equations 6.8.5 and 6.8.6.

In place of the wye network shown in Figure 6.8.2, an equivalent delta network may be used as shown in Figure 6.9.1. The conversion from wye to delta requires some algebraic manipulation, but with the help of the following identity (see [1], pg. 146):

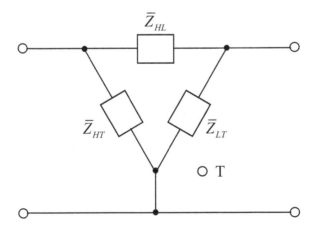

Figure 6.9.1: Alternate Network for Figure 6.8.2

$$\tilde{\bar{Z}}_{s-t} \equiv \tilde{\bar{Z}}_{sc-t}\left(\frac{n_c + n_s}{n_s}\right) + \tilde{\bar{Z}}_{sc-c}\left[\frac{n_c(n_c + n_s)}{n_s^2}\right] - \tilde{\bar{Z}}_{c-t}\left(\frac{n_c}{n_s}\right) \tag{6.9.1}$$

we obtain:

$$\tilde{\tilde{N}} = \frac{1}{2}\left(\tilde{\tilde{Z}}_{sc-t}\tilde{\tilde{Z}}_{c-t} + \tilde{\tilde{Z}}_{sc-c}\tilde{\tilde{Z}}_{sc-t} + \tilde{\tilde{Z}}_{c-t}\tilde{\tilde{Z}}_{sc-c}\right)$$
$$-\frac{1}{4}\left(\tilde{\tilde{Z}}_{s-c}^2 + \tilde{\tilde{Z}}_{c-t}^2 + \tilde{\tilde{Z}}_{sc-t}^2\right) \tag{6.9.2}$$

for compactness, let us define:

$$A = \left(\frac{n_s}{n_c+n_s}\right)^2 \qquad B = \frac{3\tilde{\tilde{Z}}_n}{n_c+n_s} \qquad C = \frac{3\tilde{\tilde{Z}}_n}{(n_c+n_s)^2}$$

$$\tilde{\tilde{Z}}_{NL} = \frac{\tilde{\tilde{N}} + 3A\tilde{\tilde{Z}}_n\tilde{\tilde{Z}}_{s-t}}{\frac{1}{2}\left(\tilde{\tilde{Z}}_{sc-t} + \tilde{\tilde{Z}}_{c-t} - \tilde{\tilde{Z}}_{sc-c}\right) + n_c B} \tag{6.9.3}$$

$$\tilde{\tilde{Z}}_{NL} = \frac{\tilde{\tilde{N}} + 3A\tilde{\tilde{Z}}_n\tilde{\tilde{Z}}_{s-t}}{\frac{1}{2}\left(\tilde{\tilde{Z}}_{sc-c} + \tilde{\tilde{Z}}_{sc-t} - \tilde{\tilde{Z}}_{c-t}\right) + n_s n_c C} \tag{6.9.4}$$

$$\tilde{\tilde{Z}}_{NL} = \frac{\tilde{\tilde{N}} + 3A\tilde{\tilde{Z}}_n\tilde{\tilde{Z}}_{s-t}}{\frac{1}{2}\left(\tilde{\tilde{Z}}_{sc-c} + \tilde{\tilde{Z}}_{c-t} - \tilde{\tilde{Z}}_{sc-t}\right) + n_s B} \tag{6.9.5}$$

For the special case where $\tilde{\tilde{Z}}_n \to \infty$, we obtain:

$$\tilde{\tilde{Z}}_{NL} = \frac{n_s^2}{n_c(n_c+n_s)}\tilde{\tilde{Z}}_{s-t} \tag{6.9.6}$$

$$\tilde{\tilde{Z}}_{LT} = \frac{n_s}{n_c}\tilde{\tilde{Z}}_{s-t} \tag{6.9.7}$$

$$\tilde{\tilde{Z}}_{HT} = \frac{n_s}{(n_c+n_s)}\tilde{\tilde{Z}}_{s-t} \tag{6.9.8}$$

Equations 6.9.6, 6.9.7, and 6.9.8 clearly indicate the nature of the problem. With $\tilde{\tilde{Z}} \to \infty$, no zero sequence current can flow in the common winding of the autotransformer. However, zero sequence current from H to L is coupled to the delta winding through Z_{s-t} with the series winding. Thus, ampere-turn balance is attained by the zero sequence current circulating in the delta winding.

If either the HV or the LV winding is open-circuited with respect to zero sequence currents, then the equivalent impedance seen by this current is:

$$\frac{\tilde{\tilde{Z}}_{HT}\left(\tilde{\tilde{Z}}_{LT} + \tilde{\tilde{Z}}_{HL}\right)}{\tilde{\tilde{Z}}_{HT} + \tilde{\tilde{Z}}_{LT} + \tilde{\tilde{Z}}_{HL}} \tag{6.9.9}$$

or:

$$\frac{\tilde{\tilde{Z}}_{LT}\left(\tilde{\tilde{Z}}_{HT} + \tilde{\tilde{Z}}_{HL}\right)}{\tilde{\tilde{Z}}_{HT} + \tilde{\tilde{Z}}_{LT} + \tilde{\tilde{Z}}_{HL}} \tag{6.9.10}$$

Now since:

$$\tilde{\tilde{Z}}_{HT} + \tilde{\tilde{Z}}_{LT} + \tilde{\tilde{Z}}_{HL} = \frac{n_s\tilde{\tilde{Z}}_{s-t}}{n_c(n_c+n_s)}\left[n_s - (n_s+n_c) + n_c\right] = 0 \tag{6.9.11}$$

we find that zero sequence current can flow.

On the other hand, with zero sequence voltage applied to the HV terminals and with the LV terminals short-circuited to ground, we have:

$$\frac{\tilde{\tilde{Z}}_{HL}\tilde{\tilde{Z}}_{LT}}{\tilde{\tilde{Z}}_{HL} + \tilde{\tilde{Z}}_{HT}} = \left(\frac{n_s}{n_c}\right)^2 \tilde{\tilde{Z}}_{s-t} \tag{6.9.12}$$

and vice versa with the H terminals shorted to ground and zero sequence voltage applied to the L terminals, we have:

$$\frac{\tilde{\tilde{Z}}_{HL}\tilde{\tilde{Z}}_{HT}}{\tilde{\tilde{Z}}_{HL} + \tilde{\tilde{Z}}_{HT}} = \left(\frac{n_s}{n_c + n_s}\right)^2 \tilde{\tilde{Z}}_{s-t} \tag{6.9.13}$$

Thus, despite the fact that an autotransformer with delta tertiary is ungrounded, zero sequence current *can* flow through the autotransformer due to the ampere-turn balance that can be obtained with the delta through the series portion of the winding.

6.10 References

[1] Edith Clarke. *Circuit Analysis of AC Power Systems - Volume II*. John Wiley and Sons, Inc., New York, New York, 1943.

6.11 Exercises

6.1 Three single-phase transformers are connected delta-wye (solidly grounded) 13.8 kV/69 kV in accordance with Figure 3.6.1(a) (on page 100). Each single-phase transformer is rated 20 MVA and has a short-circuit reactance of $j10\%$. Determine the reactance of the three-phase bank in ohms:

(a) as viewed from the 69 kV side.

(b) as viewed from the 13.8 kV side.

Determine also $\bar{X}_0$, $bar X_1$, and $\bar{X}_2$:

(c) as viewed from the high voltage side.

(d) as viewed from the low voltage side.

6.2 The transformer bank in Exercise 6.1 is supplied on the low voltage side from an infinite bus with a solidly grounded neutral through a transmission line having $\tilde{\tilde{Z}}_1 = \tilde{\tilde{Z}}_2 = j5\%$, and $\tilde{\tilde{Z}}_0 = j8\%$. If a SLGF occurs at the terminals of the transformer bank on the high voltage side, determine:

(a) $\tilde{I}_0$, $\tilde{I}_1$, and $\tilde{I}_2$ on the high and low voltage sides.

(b) $\tilde{V}_0$, $\tilde{V}_1$, and $\tilde{V}_2$ on the high and low voltage sides.

(c) Line currents on the high and low voltage sides.

(d) Line to neutral voltages on the high and low voltage sides.

(e) Line to line voltages on the high and low voltage sides.

6.3 Determine the equivalent network for a bank of three-winding, single-phase transformers connected wye(grounded)/wye(grounded)/delta, 66 kV/11 kV/2.2 kV in per unit on a 30 MVA base, given the following per unit short-circuit reactances taken at rated winding voltages:

$$H = 66 \text{ kV} \qquad L=11 \text{ kV} \qquad T=2.2 \text{ kV}$$
$$\bar{X}_{H-L} = 0.10 \text{ on } 30 \text{ MVA} \qquad \bar{X}_{H-T} = 0.06 \text{ on } 10 \text{ MVA}$$
$$\bar{X}_{L-T} = 0.14 \text{ on } 15 \text{ MVA}$$

Also determine the network reactances in ohms on:

(a) 66 kV base

(b) 11 kV base

(c) 2.2 kV base

6.4 If the transformer in Exercise 6.3 is supplied to the 11 kV side by a generator having a per unit zero sequence reactance of $j0.10$ on a 30 MVA base, what is the zero sequence reactance of the combination as measured from the 66 kV side? What is the corresponding reactance when the generator is disconnected?

6.5 Assume an autotransformer for connection between 220 kV and 110 kV lines, with an 11 kV delta connected tertiary. If all reactances are reduced to a 20 MVA base, they may be expressed as:

$$\bar{X}_{H-T} = 0.10 \qquad \bar{X}_{H-L} = 0.50 \qquad \bar{X}_{L-T} = 0.20$$

Determine the per unit reactances for the equivalent network. Find the zero sequence reactances as viewed from:

(a) the high voltage winding.

(b) the low voltage winding.

Assume that the transformer neutral is solidly grounded.

Chapter 7

Symmetrical Component Fault Analysis

In Chapter 4 we considered some examples of unbalanced operating conditions in a three-phase power system. We showed that some simple types of unbalances could be evaluated directly in terms of three-phase quantities. However, we also recognized that it was desirable to develop a more formal approach for dealing with more complex system unbalances. We therefore introduced the method of symmetrical components.

In Chapters 5 and 6, the symmetrical component representations of transmission lines and transformers were analyzed in sufficient detail to enable the reader to select a model appropriate to the objectives at hand.

In this chapter, we shall focus our attention upon the interconnections among symmetrical component sequence networks to represent the common types of unbalanced faults, which can occur in three-phase power systems. We shall therefore assume that in the entire power system there exist only transposed transmission line sections and three-phase transformers, which are symmetrical among the three-phases and that the *only three-phase system dissymmetry will be at the point of fault*. We shall conclude by considering how to deal with several such faults occurring simultaneously.

7.1 Symmetrical Three-Phase System

In a geometrically *symmetrical* three-phase network we have the form:

$$\begin{bmatrix} \tilde{V}_a \\ \tilde{V}_b \\ \tilde{V}_c \end{bmatrix} = \begin{bmatrix} \tilde{Z}_{aa} & \tilde{Z}_{ab} & \tilde{Z}_{ac} \\ \tilde{Z}_{ba} & \tilde{Z}_{bb} & \tilde{Z}_{bc} \\ \tilde{Z}_{ca} & \tilde{Z}_{cb} & \tilde{Z}_{cc} \end{bmatrix} \begin{bmatrix} \tilde{I}_a \\ \tilde{I}_b \\ \tilde{I}_c \end{bmatrix} \tag{7.1.1}$$

which, in terms of symmetrical components, appears as follows:

$$\begin{bmatrix} \tilde{V}_0 \\ \tilde{V}_1 \\ \tilde{V}_2 \end{bmatrix} = \begin{bmatrix} (\tilde{Z}_{aa} + 2\tilde{Z}_{ab}) & 0 & 0 \\ 0 & (\tilde{Z}_{bb} - \tilde{Z}_{ab}) & 0 \\ 0 & 0 & (\tilde{Z}_{cc} - \tilde{Z}_{ab}) \end{bmatrix} \begin{bmatrix} \tilde{I}_0 \\ \tilde{I}_1 \\ \tilde{I}_2 \end{bmatrix} \tag{7.1.2}$$

Thus, there is *no* mutual coupling among the symmetrical component networks for a geometrically symmetrical three-phase network.

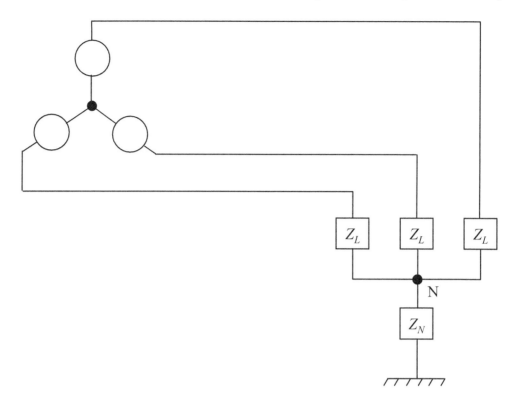

Figure 7.1.1: General Balanced Load Representation

7.1.1   Balanced Wye-Connected Load

Static loads are conventionally considered as lumped parameter elements with no mutual impedance between phases. For a balanced wye-connected three-phase load we have, per Figure 7.1.1:

$$\left[\begin{array}{c} \tilde{V}_a \\ \tilde{V}_b \\ \tilde{V}_c \end{array} \right] = \left[\begin{array}{ccc} (\tilde{Z}_L + \tilde{Z}_N) & \tilde{Z}_N & \tilde{Z}_N \\ \tilde{Z}_N & (\tilde{Z}_L + \tilde{Z}_N) & \tilde{Z}_N \\ \tilde{Z}_N & \tilde{Z}_N & (\tilde{Z}_L + \tilde{Z}_N) \end{array} \right] \left[\begin{array}{c} \tilde{I}_a \\ \tilde{I}_b \\ \tilde{I}_c \end{array} \right] \tag{7.1.3}$$

so that in terms of symmetrical components we have:

$$\left[\begin{array}{c} \tilde{V}_0 \\ \tilde{V}_1 \\ \tilde{V}_2 \end{array} \right] = \left[\begin{array}{ccc} \tilde{Z}_L + 3\tilde{Z}_N & 0 & 0 \\ 0 & \tilde{Z}_L & 0 \\ 0 & 0 & \tilde{Z}_L \end{array} \right] \left[\begin{array}{c} \tilde{I}_0 \\ \tilde{I}_1 \\ \tilde{I}_2 \end{array} \right] \tag{7.1.4}$$

The appropriate symmetrical component sequence network representations are shown in Figure 7.1.2. Observe that in the positive and negative sequence networks:

$$\tilde{V}_1 = \tilde{V}_{1N} = \tilde{V}_{1G} \tag{7.1.5}$$

and:

$$\tilde{V}_2 = \tilde{V}_{2N} = \tilde{V}_{2G} \tag{7.1.6}$$

so that the positive and negative sequence network top reference buses are to be labeled 1 and 2 respectively, while the positive and negative sequence network bottom reference buses are simul-

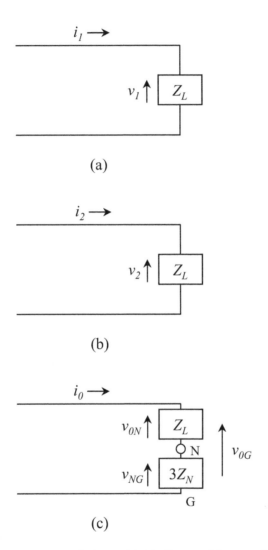

(a)

(b)

(c)

Figure 7.1.2: Representation of Figure 7.1.1 in Terms of Symmetrical Components

taneously N (load neutral bus) and G (ground, earth). In the zero sequence network, however, the load neutral and ground are separately identified such that:

$$\tilde{V}_{0G} = \tilde{V}_{0N} + \tilde{V}_{NG} \tag{7.1.7}$$

We shall determine the physical interpretation on the basis of the following example.

Example 7.1.1

   In Figure 7.1.2, let $\tilde{Z}_L = 1.0$ and $\tilde{Z}_N = j0.1$ and:

$$\tilde{I}_0 = 0.9848 - j0.1736 \quad \tilde{I}_1 = 2.0752 - j0.1684 \quad \tilde{I}_2 = 0.9397 + j0.3420$$

Then:

$$\tilde{V}_{0N} = 0.9848 - j0.1736 \quad \tilde{V}_{1N} = 2.0752 - j0.1684 \quad \tilde{V}_{2N} = 0.9397 + j0.3420$$

$$\tilde{V}_{NG} = 0.0521 + j0.2954$$

$$\tilde{V}_{aN} = 4.0000 + j0.0000 \quad \tilde{V}_{bN} = -0.9646 - j1.2438 \quad \tilde{V}_{cN} = -0.0806 + j0.7229$$

$$\tilde{V}_{aG} = 4.0521 + j0.2954 \quad \tilde{V}_{bG} = -0.9125 - j0.9484 \quad \tilde{V}_{cG} = -0.0285 + j1.0183$$

Figure 7.1.3(a) shows the appropriate phasor voltage diagram. In Figure 7.1.3(b), we have shown the phasor voltage diagram based on the associated line-to-line voltages, and also identified the geometric neutral G_n for the voltage triangle. We observe that $\tilde{V}_{0N}$ and $\tilde{V}_{0G}$ represent voltage displacements of N and G respectively from the geometric neutral of the voltage triangle.

   If the neutral is ungrounded in Figure 7.1.1, then $\tilde{I}_0 = 0$ and $\tilde{V}_{0N} = 0$. The basic voltage triangle remains unaffected, but since $\tilde{V}_{0N} = 0$, then N would lie at the geometric neutral G_n of the voltage triangle. The voltage $\tilde{V}_{NG}$ at the load is indeterminate, but is established by a neutral shift possibly occurring somewhere else in the system.

7.1.2   Balanced Delta-Connected Load

A balanced delta-connected load can always be converted to an equivalent balanced wye-connected load with an ungrounded neutral. Therefore, the results of Example 7.1.1 are also valid here except, of course, all line-to-neutral quantities at the load are for effective and not actually neutral.

7.2   Generator Representation

Figure 7.2.1(a) illustrates the equivalent network for a synchronous generator operating under balanced (positive sequence) steady-state conditions where:

$\bar{E}_i$ = open-circuit voltage of the generator at a given field current i_{fd}. (7.2.1)

$\bar{X}_{ad}$ = magnetizing reactance (direct axis) of the synchronous generator. (7.2.2)

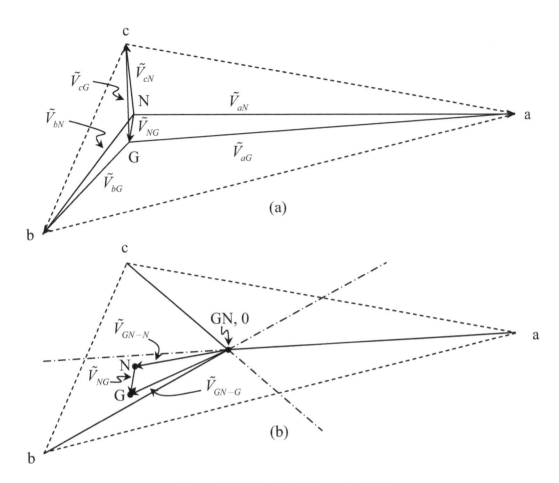

Figure 7.1.3: Results for Example 7.1.1

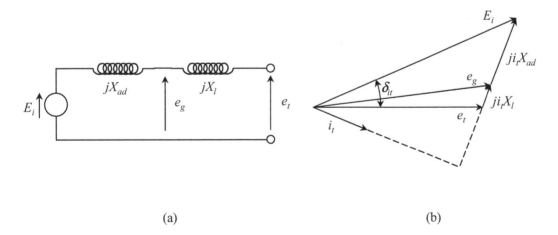

Figure 7.2.1: Positive Sequence Representation for a Synchronous Machine

$\bar{X}_\ell$ = stator winding leakage reactance. (7.2.3)

$\bar{V}_a$ = generator terminal voltage. (7.2.4)

$\bar{E}_g$ = voltage behind stator leakage reactance (measure of flux in the air gap). (7.2.5)

whereby:

$$\bar{X}_d = \bar{X}_\ell + \bar{X}_{ad} \qquad (7.2.6)$$

is commonly referred to as the *synchronous reactance* of the generator. Figure 7.2.1(b) depicts these quantities on a phasor diagram.

We shall investigate later the transient performance of synchronous machines. At that time we will develop the distinction between the synchronous reactance $\bar{X}_d$, the transient reactance $\bar{X}'_d$, and the subtransient reactance $\bar{X}''_d$. We shall merely accept here that $\bar{X}''_d$ is to be used in place of $\bar{X}_d$ when we are interested in knowing the fault current at the instant at which the fault occurs, and $\bar{X}'_d$ in place of $\bar{X}_d$ when we are interested in knowing the maximum fault current after the damper winding currents have decayed. $\bar{X}_d$ will be used in the positive sequence network when we wish to know the steady-state fault current after all transients have decayed.

In order to determine the nature of the negative sequence representation for a synchronous machine, we ascertain the performance of the machine when only negative sequence currents flow in the stator windings. Negative sequence stator currents will produce a traveling MMF wave in the air gap of the machine (just like positive sequence currents) that moves at synchronous speed, but in a direction opposite to that of the rotor. This is similar in behavior to the case of an induction motor under dynamic braking action.

For most practical hydroelectric generators with damper windings, and steam turbine generators with solid iron rotors, the effective impedance of the generator to negative sequence currents is due to the leakage reactance of the stator winding plus the effects of rotor iron or damper windings. Numerically we define:

$$\bar{X}_2 = \left(\frac{\bar{X}''_d + \bar{X}''_q}{2} \right) \qquad (7.2.7)$$

where $\bar{X}''_d$ and $\bar{X}''_q$ are the subtransient reactances of the machine in the direct axis and quadrature axis, respectively. The definition in Equation 7.2.7 is not the only possible definition; there are others but we shall postpone a discussion of this until later.

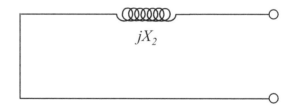

Figure 7.2.2: Negative Sequence Representation for a Synchronous Machine

With zero sequence currents flowing in the stator windings, all of the stator MMFs are acting into the rotor so that there is no significant net flux crossing the air gap. The zero sequence reactance of the synchronous machine is very nearly equal to the stator winding leakage reactance.

Figures 7.2.2 and 7.2.3 provide the negative and zero sequence representations for a synchronous machine.

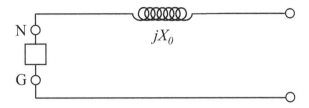

Figure 7.2.3: Zero Sequence Representation for a Synchronous Machine

7.3 Single-Line-to-Ground Fault (SLGF)

For the single-line-to-ground fault, with zero fault impedance, we have the following constraints *at the point of fault:*

$$\Delta I_b = \Delta I_c = 0 \quad V_{aG} = 0 \tag{7.3.1}$$

In terms of symmetrical components we have:

$$\Delta \tilde{I}_0 = \Delta \tilde{I}_1 = \Delta \tilde{I}_2 = \frac{\Delta \tilde{I}_a}{3} \tag{7.3.2}$$

$$\tilde{V}_{0G} + \tilde{V}_1 + \tilde{V}_2 = 0 \tag{7.3.3}$$

The appropriate interconnection amongst the sequence networks at the point of fault is shown in Figure 7.3.1.

In anticipation of more complex situations, we have shown the interconnections employing ideal 1:1 transformers in Figure 7.3.1(a). When, and only when, the ideal transformers can be replaced by direct interconnections without affecting proper current distribution, the interconnections shown in Figure 7.3.1(b) are appropriate.

For a SLGF *with* fault impedance $\tilde{Z}_F$ we have the constraints:

$$\Delta I_b = \Delta I_c = 0 \quad V_{aG} = \Delta \tilde{I}_a \tilde{Z}_F \tag{7.3.4}$$

so that:

$$\Delta \tilde{I}_0 = \Delta \tilde{I}_1 = \Delta \tilde{I}_2 = \frac{\Delta \tilde{I}_a}{3} \tag{7.3.5}$$

and:

$$\tilde{V}_{0G} + \tilde{V}_1 + \tilde{V}_2 = \Delta \tilde{I}_a \tilde{Z}_F = \Delta \tilde{I}_0 (3\tilde{Z}_F) \tag{7.3.6}$$

Figure 7.3.1(c) shows the appropriate interconnection.

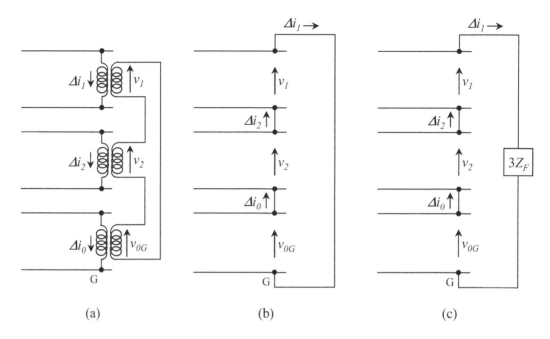

Figure 7.3.1: Symmetrical Component Interconnections for SLGF

7.4   Single-Line-to-Neutral Fault (SLNF)

It is possible for the fault to occur not only from line to ground, but also from line to neutral.
Consider the configuration in Figure 7.1.1 and assume a short circuit from line to neutral at the load.
We then have the constraints:

$$\Delta I_b = \Delta I_c = 0 \quad V_{aN} = 0 \tag{7.4.1}$$

which leads to:

$$\Delta \tilde{I}_0 = \Delta \tilde{I}_1 = \Delta \tilde{I}_2 = \frac{\Delta \tilde{I}_a}{3} \tag{7.4.2}$$

and:

$$\tilde{V}_{0N} + \tilde{V}_1 + \tilde{V}_2 = 0 \tag{7.4.3}$$

whereby we have assumed zero fault impedance.

The appropriate interconnections are shown in Figure 7.4.1. Observe that zero sequence current
will flow through the load even when $\tilde{Z}_N \to \infty$.

7.5   Line-to-Line Fault (LLF)

For a line-to-line fault, with *no fault impedance*, we have the constraints:

$$\Delta I_a = 0 \quad V_{bc} = 0 \tag{7.5.1}$$

Now:

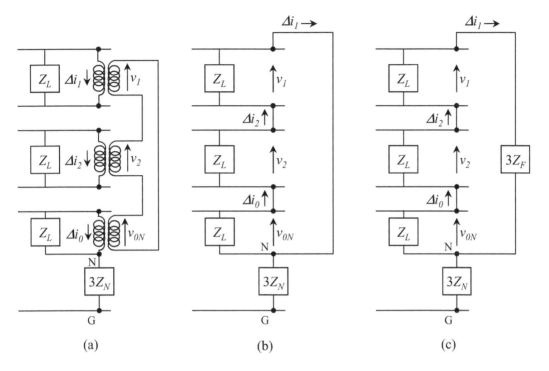

Figure 7.4.1: Symmetrical Component Interconnections for SLNF

$$\tilde{V}_b = \tilde{V}_{0G} + \tilde{a}^2 \tilde{V}_1 + \tilde{a} \tilde{V}_2 \tag{7.5.2}$$
$$\tilde{V}_c = \tilde{V}_{0G} + \tilde{a} \tilde{V}_1 + \tilde{a}^2 \tilde{V}_2 \tag{7.5.3}$$

so that:

$$\tilde{V}_{bc} = j\sqrt{3}(-\tilde{V}_1 + \tilde{V}_2) = 0 \tag{7.5.4}$$

and:

$$\tilde{V}_2 = \tilde{V}_1 \tag{7.5.5}$$

Also:

$$\Delta \tilde{I}_0 + \Delta \tilde{I}_1 + \Delta \tilde{I}_2 = 0 \tag{7.5.6}$$

However:

$$\Delta \tilde{I}_b + \Delta \tilde{I}_c = 0 \tag{7.5.7}$$
$$\Delta \tilde{I}_0 + \tilde{a}^2 \Delta \tilde{I}_1 + \tilde{a} \Delta \tilde{I}_2 + \Delta \tilde{I}_0 + \tilde{a} \Delta \tilde{I}_1 + \tilde{a}^2 \Delta \tilde{I}_2 = 0 \tag{7.5.8}$$
$$2\Delta \tilde{I}_0 - \Delta \tilde{I}_1 - \Delta \tilde{I}_2 = 0 \tag{7.5.9}$$

On the basis of Equation 7.5.6:

$$3\Delta \tilde{I}_0 = 0 \tag{7.5.10}$$

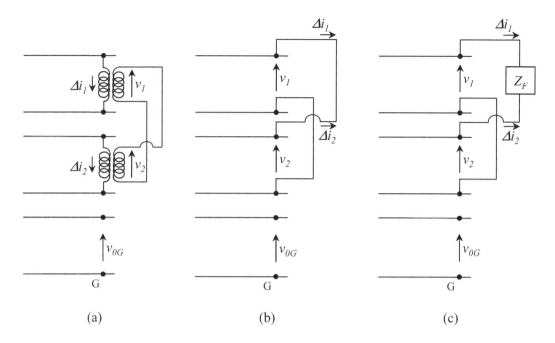

Figure 7.5.1: Symmetrical Component Interconnections for LLF

so that:

$$\Delta \tilde{I}_0 = 0 \tag{7.5.11}$$

and:

$$\Delta \tilde{I}_1 + \Delta \tilde{I}_2 = 0 \tag{7.5.12}$$

The appropriate interconnections are shown in Figures 7.5.1(a) and (b). Observe that the zero sequence network is not involved.

If we include fault impedance, then:

$$\tilde{V}_{bc} = \Delta \tilde{I}_b \tilde{Z}_F \tag{7.5.13}$$

so that per Equation 7.5.4:

$$j\sqrt{3}(-\tilde{V}_1 + \tilde{V}_2) = (\tilde{a}^2 \Delta \tilde{I}_1 + \tilde{a}\Delta \tilde{I}_2)\tilde{Z}_F \tag{7.5.14}$$

The current relationships remain as before so that:

$$\Delta \tilde{I}_0 = 0 \quad \Delta \tilde{I}_1 + \Delta \tilde{I}_2 = 0 \tag{7.5.15}$$

Therefore:

$$-\tilde{V}_1 + \tilde{V}_2 = -\Delta \tilde{I}_1 \tilde{Z}_F \tag{7.5.16}$$

The appropriate interconnections are shown in Figure 7.5.1(c).

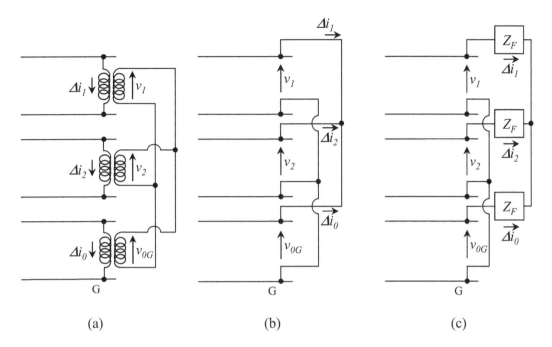

Figure 7.6.1: Symmetrical Component Interconnections for LLGF

7.6   Line-to-Line-to-Ground Fault (LLGF)

For a LLGF, with *no fault impedance*, we have the constraints:

$$\Delta \tilde{I}_a = 0 \quad V_{bG} = V_{cG} = 0 \tag{7.6.1}$$

so that:

$$\Delta \tilde{I}_0 + \Delta \tilde{I}_1 + \Delta \tilde{I}_2 = 0 \tag{7.6.2}$$

and:

$$\tilde{V}_{0G} = \tilde{V}_1 = \tilde{V}_2 = \frac{\tilde{V}_{aG}}{3} \tag{7.6.3}$$

Figures 7.6.1(a) and (b) show the proper interconnections among the sequence networks.

If we wish to allow for fault impedance, we must necessarily take the fault impedance from phases "b" and "c" to ground to be equal. Recall that simple interconnections among symmetrical component sequence networks are attained only when there is symmetry with respect to phase "a."

For a LLGF *with fault impedance* we have the constraints:

$$\Delta I_a = 0 \tag{7.6.4}$$
$$\tilde{V}_{bG} = \Delta \tilde{I}_b \tilde{Z}_F \tag{7.6.5}$$
$$\tilde{V}_{cG} = \Delta \tilde{I}_c \tilde{Z}_F \tag{7.6.6}$$

Then:

$$\begin{bmatrix} \tilde{V}_{aG} \\ \tilde{V}_{bG} \\ \tilde{V}_{cG} \end{bmatrix} = (\tilde{Z}_F) \begin{bmatrix} \Delta\tilde{I}_a \\ \Delta\tilde{I}_b \\ \Delta\tilde{I}_c \end{bmatrix} + \begin{bmatrix} \tilde{V}_{aG} \\ 0 \\ 0 \end{bmatrix} \tag{7.6.7}$$

and in terms of symmetrical components:

$$\begin{bmatrix} \tilde{V}_{0G} \\ \tilde{V}_1 \\ \tilde{V}_2 \end{bmatrix} = (\tilde{Z}_F) \begin{bmatrix} \Delta\tilde{I}_0 \\ \Delta\tilde{I}_1 \\ \Delta\tilde{I}_2 \end{bmatrix} + (\frac{\tilde{V}_{aG}}{3}) \begin{bmatrix} 1 \\ 1 \\ 1 \end{bmatrix} \tag{7.6.8}$$

or:

$$\begin{bmatrix} \tilde{V}_{0G} - \tilde{Z}_F\Delta\tilde{I}_0 \\ \tilde{V}_1 - \tilde{Z}_F\Delta\tilde{I}_1 \\ \tilde{V}_2 - \tilde{Z}_F\Delta\tilde{I}_2 \end{bmatrix} = (\frac{\tilde{V}_{aG}}{3}) \begin{bmatrix} 1 \\ 1 \\ 1 \end{bmatrix} \tag{7.6.9}$$

Figure 7.6.1(c) shows the appropriate interconnection among the sequence networks.

7.7   Line-to-Line-to Neutral Fault (LLNF)

Consider a line-to-line-to-neutral fault on phases "b" and "c," with *no* fault impedance. Then:

$$\Delta I_a = 0 \quad V_{bN} = V_{cN} = 0 \tag{7.7.1}$$

so that:

$$\Delta\tilde{I}_0 + \Delta\tilde{I}_1 + \Delta\tilde{I}_2 = 0 \tag{7.7.2}$$

and:

$$\tilde{V}_{0N} = \tilde{V}_1 = \tilde{V}_2 = \frac{\tilde{V}_{aN}}{3} \tag{7.7.3}$$

Figure 7.7.1 illustrates the appropriate interconnection among the sequence networks.

For a LLNF *with* fault impedance we proceed in the same manner as in the previous section and arrive at Figure 7.7.1(c) for the appropriate interconnections.

7.8   Single Open Conductor (SOC)

For a single open conductor on phase "a" we have the constraints:

$$I_a = 0 \quad \Delta V_b = \Delta V_c = 0 \tag{7.8.1}$$

so that:

$$\tilde{I}_0 + \tilde{I}_1 + \tilde{I}_2 = 0 \tag{7.8.2}$$

and:

$$\Delta\tilde{V}_0 = \Delta\tilde{V}_1 = \Delta\tilde{V}_2 = \frac{\Delta\tilde{V}_{aN}}{3} \tag{7.8.3}$$

Figure 7.8.1 illustrates the appropriate interconnections.

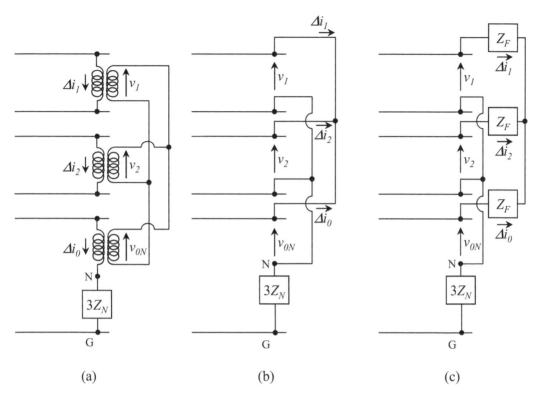

Figure 7.7.1: Symmetrical Component Interconnections for LLNF

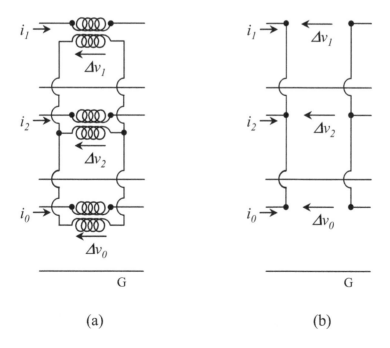

Figure 7.8.1: Symmetrical Component Interconnections for SOC

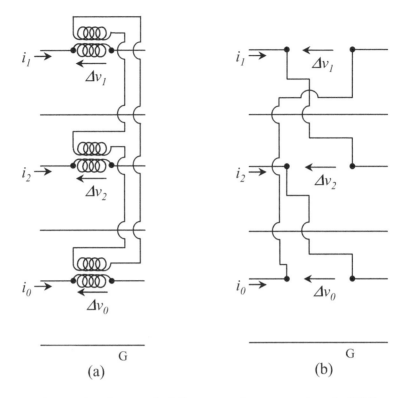

Figure 7.9.1: Symmetrical Component Interconnections for TOC

7.9   Two Open Conductors (TOC)

For open conductors in phases "b" and "c" we have the constraints:

$$\Delta V_a = 0 \quad I_b = I_c = 0 \tag{7.9.1}$$

so that:

$$\Delta \tilde{V}_0 + \Delta \tilde{V}_1 + \Delta \tilde{V}_2 = 0 \tag{7.9.2}$$

and:

$$\tilde{I}_0 = \tilde{I}_1 = \tilde{I}_2 = \frac{\tilde{I}_a}{3} \tag{7.9.3}$$

Figure 7.9.1 illustrates the appropriate interconnections.

7.10   Generalized Series Impedances

While the fault cases described in the previous sections are generally adequate for most practical fault studies, occasionally there arises a need to analyze cases of impedance unbalance that do not correspond exactly to these classic types of faults. We consider first the series unbalance, which is limited to the form:

$$\Delta \tilde{Z}_{aa} \neq \Delta \tilde{Z}_{bb} = \Delta \tilde{Z}_{cc} \tag{7.10.1}$$

per Figure 7.10.1. We write:

$$
\begin{bmatrix} \Delta\tilde{V}_a - \Delta\tilde{V}_n \\ \Delta\tilde{V}_b - \Delta\tilde{V}_n \\ \Delta\tilde{V}_c - \Delta\tilde{V}_n \end{bmatrix} = \begin{bmatrix} \Delta\tilde{Z}_{aa} & 0 & 0 \\ 0 & \Delta\tilde{Z}_{bb} & 0 \\ 0 & 0 & \Delta\tilde{Z}_{bb} \end{bmatrix} \begin{bmatrix} \tilde{I}_a \\ \tilde{I}_b \\ \tilde{I}_c \end{bmatrix}
\tag{7.10.2}
$$

with:

$$
\Delta\tilde{V}_n = \Delta\tilde{Z}_{nn}(\tilde{I}_a + \tilde{I}_b + \tilde{I}_c)
\tag{7.10.3}
$$

which transforms into:

$$
\begin{bmatrix} \Delta\tilde{V}_0 - \Delta\tilde{V}_n \\ \Delta\tilde{V}_1 \\ \Delta\tilde{V}_2 \end{bmatrix} = \left\{ \left[\Delta\tilde{Z}\right] + \left(\frac{\Delta\tilde{Z}_{aa} - \Delta\tilde{Z}_{bb}}{3}\right) \begin{bmatrix} 1 & 1 & 1 \\ 1 & 1 & 1 \\ 1 & 1 & 1 \end{bmatrix} \right\} \begin{bmatrix} \tilde{I}_0 \\ \tilde{I}_1 \\ \tilde{I}_2 \end{bmatrix}
$$

$$
= \left\{ \Delta\tilde{Z}_{bb}[U] + \left(\frac{\Delta\tilde{Z}_{aa} - \Delta\tilde{Z}_{bb}}{3}\right)[\xi] \right\} \begin{bmatrix} \tilde{I}_0 \\ \tilde{I}_1 \\ \tilde{I}_2 \end{bmatrix}
\tag{7.10.4}
$$

where for compactness we define

$$
\left[\Delta\tilde{Z}\right] = \begin{bmatrix} \Delta\tilde{Z}_{bb} & 0 & 0 \\ 0 & \Delta\tilde{Z}_{bb} & 0 \\ 0 & 0 & \Delta\tilde{Z}_{bb} \end{bmatrix}
$$

with:

$$
\Delta\tilde{V}_n = 3\Delta Z_{nn}\tilde{I}_0
\tag{7.10.5}
$$

Figure 7.10.1(b) shows the appropriate symmetrical component network interconnections employing ideal 1:1 transformers. We observe that when $\tilde{Z}_{aa} < \tilde{Z}_{bb}$ we have a negative impedance, which at times, may be inconvenient. We may alternatively write:

$$
\begin{bmatrix} \tilde{I}_a \\ \tilde{I}_b \\ \tilde{I}_c \end{bmatrix} = \begin{bmatrix} \Delta\tilde{Y}_{aa} & 0 & 0 \\ 0 & \Delta\tilde{Y}_{bb} & 0 \\ 0 & 0 & \Delta\tilde{Y}_{bb} \end{bmatrix} \begin{bmatrix} \Delta\tilde{V}_a - \Delta\tilde{V}_n \\ \Delta\tilde{V}_b - \Delta\tilde{V}_n \\ \Delta\tilde{V}_c - \Delta\tilde{V}_n \end{bmatrix}
\tag{7.10.6}
$$

with:

$$
\Delta\tilde{y}_{aa} = \frac{1}{\Delta\tilde{Z}_{aa}} \qquad \Delta\tilde{y}_{bb} = \frac{1}{\Delta\tilde{Z}_{bb}}
\tag{7.10.7}
$$

which transforms into:

$$
\begin{bmatrix} \tilde{I}_0 \\ \tilde{I}_1 \\ \tilde{I}_2 \end{bmatrix} = \left\{ \left[\Delta\tilde{Y}\right] + \left(\frac{\Delta\tilde{y}_{aa} - \Delta\tilde{y}_{bb}}{3}\right) \begin{bmatrix} 1 & 1 & 1 \\ 1 & 1 & 1 \\ 1 & 1 & 1 \end{bmatrix} \right\} \begin{bmatrix} \Delta\tilde{v}_0 - \Delta\tilde{v}_n \\ \Delta\tilde{v}_1 \\ \Delta\tilde{v}_2 \end{bmatrix}
$$

$$
= \left\{ \Delta\tilde{y}_{bb}[U] + \left(\frac{\Delta\tilde{y}_{aa} - \Delta\tilde{y}_{bb}}{3}\right)[\xi] \right\} \begin{bmatrix} \Delta\tilde{v}_0 - \Delta\tilde{v}_n \\ \Delta\tilde{v}_1 \\ \Delta\tilde{v}_2 \end{bmatrix}
\tag{7.10.8}
$$

where for compactness we define:

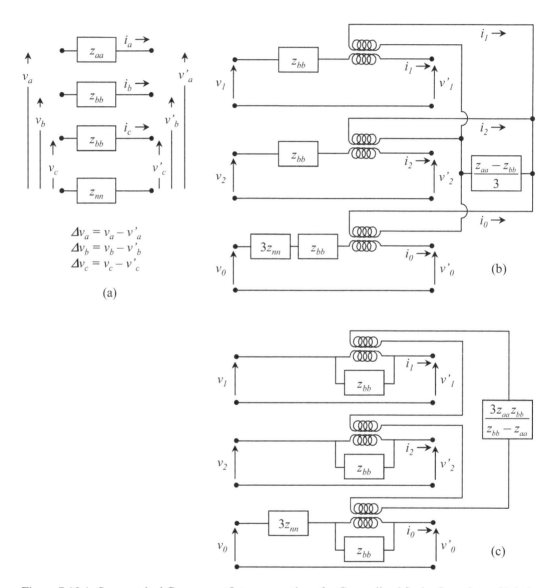

$$\Delta v_a = v_a - v'_a$$
$$\Delta v_b = v_b - v'_b$$
$$\Delta v_c = v_c - v'_c$$

(a)

Figure 7.10.1: Symmetrical Component Interconnections for Generalized Series Impedance Unbalance

$$\left[\Delta\tilde{Y}\right] = \begin{bmatrix} \Delta\tilde{y}_{bb} & 0 & 0 \\ 0 & \Delta\tilde{y}_{bb} & 0 \\ 0 & 0 & \Delta\tilde{y}_{bb} \end{bmatrix}$$

and with:

$$\left(\frac{\Delta\tilde{y}_{aa} - \Delta\tilde{y}_{bb}}{3}\right) = \left(\frac{\Delta\tilde{z}_{bb} - \Delta\tilde{z}_{aa}}{3\Delta\tilde{z}_{aa}\Delta\tilde{z}_{bb}}\right) \tag{7.10.9}$$

Figure 7.10.1(c) shows the appropriate symmetrical component network interconnections employing ideal 1:1 transformers and which is more convenient when $\tilde{z}_{aa} < \tilde{z}_{bb}$.

7.11 Generalized Shunt Impedance Unbalances

Figure 7.11.1(a) illustrates the constraints on an acceptable shunt impedance unbalance. The mathematical analysis is identical to that in Section 7.10, the only difference being in the nomenclature. For shunt impedances we employ $\tilde{V}_{\Phi N}$ or $\tilde{V}_{\Phi G}$ in place of $\Delta\tilde{V}_{\Phi}$, and $\Delta\tilde{I}_{Phi}$ in place of $\tilde{I}_{\Phi}$.

While the change in nomenclature is somewhat trivial, it is introduced to emphasize the difference between line-to-neutral voltages and line currents, and differential voltages and currents along the length of the line.

7.12 Simultaneous Faults

Often the occurrence of an unbalanced fault will lead to the inception of another unbalanced fault. When we speak of simultaneous faults, we do not imply that several faults occur at exactly the same instant in time, but rather that eventually they will coexist during a given period of time.

For all the fault conditions described in this chapter, we have shown an appropriate symmetrical component interconnection diagram employing ideal 1:1 transformers. These same interconnections apply directly for simultaneous faults in the power systems provided that:

(a) All faults are symmetrical with respect to phase "a."

(b) The faults appear at different locations in a power system.

In the event that several such faults occur at the same location in the power system, each fault is represented by its own interconnection, each interconnection being separated from one another by an incremental, zero impedance, line section.

Whether any of the 1:1 ideal transformers may be replaced by direct interconnections depends entirely upon the given system configuration and must be evaluated from situation to situation. An ideal transformer may be eliminated only, if by doing so, the current distribution in the network remains unchanged.

7.13 Faults Not Symmetrical with Respect to Phase "a"

For isolated faults in a power system, especially under the assumption that all parts of the power system are symmetrical except at the point of fault, it is immaterial as to which phase is faulted; selection of the fault with symmetry about phase "a" is acceptable and preferable.

However, for simultaneous faults this may not be possible. For example, the first fault might be a SLGF on the phase "a," followed by SLGF on phase "b." Considering the representation of the latter we have:

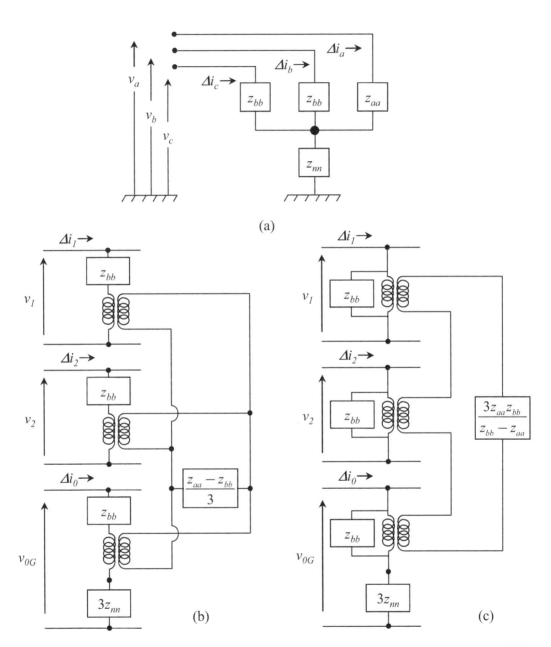

Figure 7.11.1: Symmetrical Component Interconnections for Generalized Shunt Impedance Unbalance

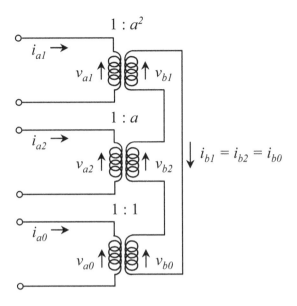

Figure 7.13.1: SLGF on Phase "b"

$$\tilde{V}_{bG} = 0 = \tilde{V}_{0G} + \tilde{a}^2\tilde{V}_1 + \tilde{a}\tilde{V}_2 \tag{7.13.1}$$

and since $\Delta I_a = \Delta I_c = 0$,

$$\Delta\tilde{I}_0 = \frac{\Delta\tilde{I}_b}{3} \tag{7.13.2}$$

$$\Delta\tilde{I}_1 = \frac{\tilde{a}\Delta\tilde{I}_b}{3} \tag{7.13.3}$$

$$\Delta\tilde{I}_2 = \frac{\tilde{a}^2\Delta\tilde{I}_b}{3} \tag{7.13.4}$$

so that:

$$\Delta\tilde{I}_0 = \tilde{a}^2\Delta\tilde{I}_1 = \tilde{a}\Delta\tilde{I}_2 \tag{7.13.5}$$

Figure 7.13.1 shows the modification to the symmetrical component interconnections when the fault is symmetrical to phase "b." Thus, all of the interconnections that have been described so far still remain valid when the original 1:1 ideal transformers are replaced by ideal phase-shifting transformers. Figure 7.13.2 shows the appropriate interconnections when the fault is symmetrical with respect to phase "c" which the reader may easily verify.

7.14 Exercises

7.1 A 13.2 kV delta-connected distribution system is grounded on one phase. What is the magnitude of the zero sequence voltage?

7.2 A fault occurs at one location on a power system. At this location, the open-circuit voltage is a purely positive sequence at 1.0 per unit. The Thevenin-Helmholtz equivalent impedances

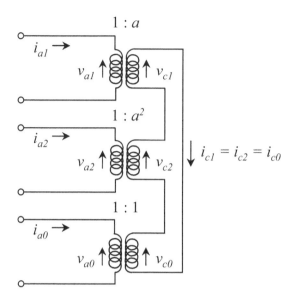

Figure 7.13.2: SLGF on Phase "c"

are:

$$\tilde{\tilde{Z}}_1 = j1.0 \qquad \tilde{\tilde{Z}}_2 = 0 \qquad \tilde{\tilde{Z}}_0 = 0.10 + j0.12$$

Determine the currents in each phase for the following types of faults (with zero fault impedance):

(a) SLFG

(b) LLFG

(c) LLF

(d) 3ϕ Fault

7.3 A synchronous generator, represented entirely by reactances, has an open-circuit voltage of 1.0 per unit. Assume that the absolute values of fault current (in per unit) for the different types of fault are:

(a) 0.5 for a 3ϕ Fault

(b) 0.7 for a LLF

(c) 1.1 for a SLGF

What are the per unit positive, negative, and zero sequence impedances?

7.4 In Figure 7.14.1, let: Generator: $j\bar{X}_1 = j1.35 \qquad j\bar{X}_2 = j0.186 \qquad j\bar{X}_0 = j0.05$

with balanced generator internal voltages and $\bar{E}_{a1} = 2.0$

Line : $j\bar{X}_1 = j\bar{X}_2 = j0.10 \qquad j\bar{X}_0 = j0.30$

Load: $\tilde{\tilde{Z}}_L = j1.0$

For a SLNF at "x," with no fault impedance, and with $Z_{nS} = 0, Z_{nL} \to \infty$

(a) Sketch the appropriate symmetrical component network interconnections.

(b) Evaluate $\tilde{V}_{0n}$, $\tilde{V}_{1n}$, and $\tilde{V}_{2n}$ at the fault.

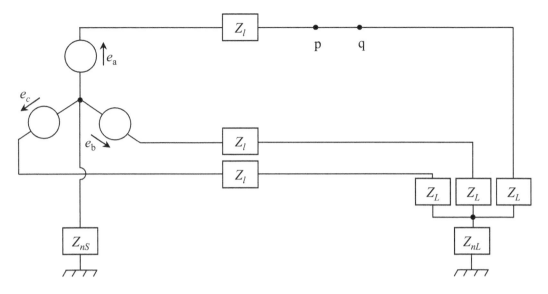

Figure 7.14.1: Diagram for Exercise 7.4

(c) Evaluate $\tilde{V}_{an}$, $\tilde{V}_{bn}$, and $\tilde{V}_{cn}$ at the fault.

(d) Evaluate $\tilde{V}_{ab}$, $\tilde{V}_{bc}$, and $\tilde{V}_{ca}$ at the fault.

(e) Sketch the voltage triangle for (d) and locate and identify the points:

$$
\begin{array}{ll}
\textit{Geometric neutral} & (ng) \\
\textit{Load neutral} & (nL) \\
\textit{Ground} & (G)
\end{array}
$$

7.5 In Figure 7.14.2, let $\tilde{\tilde{E}}_{a1} = 1.0$ and:

$$\tilde{\tilde{Z}}_L = j500 \qquad \tilde{\tilde{Z}}_C = -j745 \qquad \bar{Z}_{nS}$$

$$Z_{nS} \to \infty \qquad Z_{nC} = Z_{nL} = 0$$

For an open conductor at r-s:

(a) Sketch the appropriate symmetrical component network interconnections.

(b) Evaluate the voltage $\Delta\tilde{V}_{IS}$ across the open conductor.

(c) Suggest at least three practical methods for avoiding such high voltages. Since the solutions should be realistic, give a range of values for your solutions.

7.6 While the open-conductor fault exists in Exercise 7.5, a SLGF also occurs at point "y" in Figure 7.14.2. Sketch the symmetrical component network interconnections for this simultaneous fault condition.

7.7 Consider the configuration in Figure 7.14.3 of an open-delta load, a balanced set of three-phase source voltages, and ignore all other impedances:

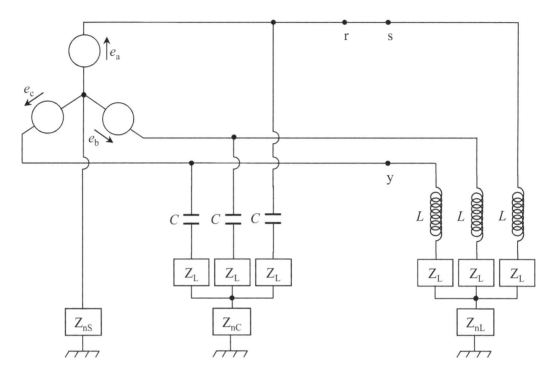

Figure 7.14.2: Diagram for Exercises 7.5 and 7.6

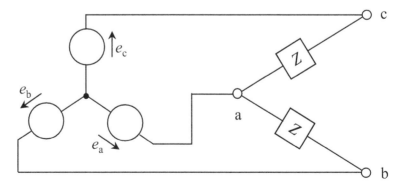

Figure 7.14.3: Diagram for Exercise 7.7

(a) Calculate the line currents in terms of e_a and Z in terms of physical (phase) quantities directly.

(b) Solve the problem in terms of symmetrical components from the three points of view shown in Figure 7.14.4.

(c) Observe that this problem may be interpreted in a number of different ways from a symmetrical component point of view. Given a valid interpretation, the solution itself often becomes rather mechanical.

(d) Observe that the additional consideration of line and source impedances does not materially affect the difficulty of the solution from a symmetrical component point of view. However, a physical (phase quantity) approach is considerably less direct.

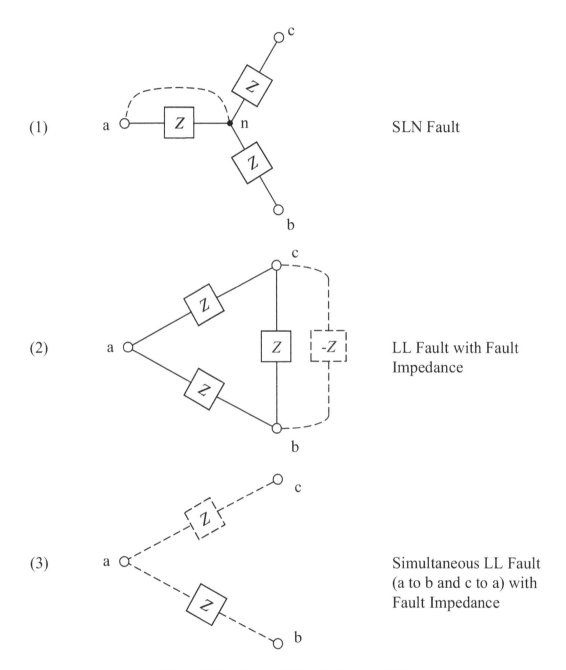

Figure 7.14.4: Points of View for Exercise 7.7b

Chapter 8

Design of Untransposed Transmission Lines

In Chapter 5, the basic principles underlying electromagnetic and electrostatic unbalance of untransposed transmission lines were presented in a manner appropriate for a general background in electric power engineering analysis. Even though the principles underlying these problems are relatively easy to understand, the actual analysis, especially for multicircuit lines can be a rather tedious task. For the electric power engineer who may be more deeply concerned with one or the other aspects of these problems, the references listed in Chapter 5 should be helpful. However, while the cited references generally include Appendices in which the analytical formulations are presented, page limitations often prevent more than a very concise presentation. The material in this chapter is somewhat more organized and more thorough than that in the cited references and reflects the author's experience over several years in various aspects of the analysis and design of untransposed transmission lines.

Matrix notation has been found indispensable in bringing to light the salient consequences of the phasing arrangement upon the performance of untransposed transmission lines. For the sake of clarity and conciseness, symbolic matrix notation will generally be employed. The expanded definitions for the various matrices are given in Appendix E.

8.1 Symmetrical Phase Impedance Matrix

Given a single-circuit three-phase transmission line section, with no ground wires and a single conductor per phase, the voltage drop along the line section is given, in terms of phase quantities, by:

$$
\begin{bmatrix} \Delta \tilde{V}_a \\ \Delta \tilde{V}_b \\ \Delta \tilde{V}_c \end{bmatrix} = \begin{bmatrix} \tilde{Z}_{aa} & \tilde{Z}_{ab} & \tilde{Z}_{ac} \\ \tilde{Z}_{ba} & \tilde{Z}_{bb} & \tilde{Z}_{ba} \\ \tilde{Z}_{ca} & \tilde{Z}_{cb} & \tilde{Z}_{cc} \end{bmatrix} \begin{bmatrix} \tilde{I}_a \\ \tilde{I}_b \\ \tilde{I}_c \end{bmatrix}
\tag{8.1.1}
$$

with:

$$
\tilde{Z}_{ii-e} = R_i + \frac{R_e}{3} + j \left(\frac{X_e}{3} + \frac{\mu_0}{2\pi} \ln \frac{1}{GMR_i} \right)
\tag{8.1.2}
$$

$$
\tilde{Z}_{ij-e} = \frac{R_e}{3} + j \left(\frac{X_e}{3} + \frac{\mu_0}{2\pi} \ln \frac{1}{d_{ij}} \right)
\tag{8.1.3}
$$

and in symbolic notation:

$$[\Delta \tilde{V}_\Phi] = [\tilde{Z}_\Phi][\tilde{I}_\Phi] \tag{8.1.4}$$

In view of the fact that:

$$\tilde{Z}_{ij} = \tilde{Z}_{ji} \tag{8.1.5}$$

$[\tilde{Z}_\Phi]$ is always a *symmetrical* matrix:

$$[\tilde{Z}_\Phi]^T = [\tilde{Z}_\Phi] \tag{8.1.6}$$

It is left for the reader to verify that the expanded impedance matrix including ground wires and subconductors is also always symmetrical, as well as the equivalent impedance matrix after eliminating the ground wire and subconductor equations.

8.2 Unsymmetrical Symmetrical Component Impedance Matrix

In terms of *symmetrical components* we have:

$$[\tilde{S}][\Delta \tilde{V}_\Phi] = [\tilde{S}][\tilde{Z}_\Phi][\tilde{S}]^{-1}[\tilde{S}][\tilde{I}_\phi] \tag{8.2.1}$$

and:

$$[\Delta \tilde{V}_c] = [\tilde{Z}_c][\tilde{I}_c] \tag{8.2.2}$$

with

$$\begin{bmatrix} \Delta \tilde{V}_0 \\ \Delta \tilde{V}_1 \\ \Delta \tilde{V}_2 \end{bmatrix} = \begin{bmatrix} \tilde{Z}_{00} & \tilde{Z}_{01} & \tilde{Z}_{02} \\ \tilde{Z}_{10} & \tilde{Z}_{11} & \tilde{Z}_{12} \\ \tilde{Z}_{20} & \tilde{Z}_{21} & \tilde{Z}_{22} \end{bmatrix} \begin{bmatrix} \tilde{I}_0 \\ \tilde{I}_1 \\ \tilde{I}_2 \end{bmatrix} \tag{8.2.3}$$

where:

$$[\tilde{S}] = \frac{1}{3} \begin{bmatrix} 1 & 1 & 1 \\ 1 & \tilde{a} & \tilde{a}^2 \\ 1 & \tilde{a}^2 & \tilde{a} \end{bmatrix} \qquad [\tilde{S}]^{-1} = \begin{bmatrix} 1 & 1 & 1 \\ 1 & \tilde{a}^2 & \tilde{a} \\ 1 & \tilde{a} & \tilde{a}^2 \end{bmatrix} \tag{8.2.4}$$

and:

Table 8.2.1			
$3\tilde{Z}_{00}$		$=$	$(\tilde{Z}_{aa} + \tilde{Z}_{bb} + \tilde{Z}_{cc}) + 2(\tilde{Z}_{bc} + \tilde{Z}_{ca} + \tilde{Z}_{ab})$
$3\tilde{Z}_{11}$	$= 3\tilde{Z}_{22}$	$=$	$(\tilde{Z}_{aa} + \tilde{Z}_{bb} + \tilde{Z}_{cc}) - 2(\tilde{Z}_{bc} + \tilde{Z}_{ca} + \tilde{Z}_{ab})$
$3\tilde{Z}_{01}$	$= 3\tilde{Z}_{20}$	$=$	$(\tilde{Z}_{aa} + a^2\tilde{Z}_{bb} + a\tilde{Z}_{cc}) - (\tilde{Z}_{bc} + a^2\tilde{Z}_{ca} + a\tilde{Z}_{ab})$
$3\tilde{Z}_{10}$	$= 3\tilde{Z}_{02}$	$=$	$(\tilde{Z}_{aa} + a\tilde{Z}_{bb} + a^2\tilde{Z}_{cc}) - (\tilde{Z}_{bc} + a\tilde{Z}_{ca} + a^2\tilde{Z}_{ab})$
$3\tilde{Z}_{12}$		$=$	$(\tilde{Z}_{aa} + a^2\tilde{Z}_{bb} + a\tilde{Z}_{cc}) + 2(\tilde{Z}_{bc} + a^2\tilde{Z}_{ca} + a\tilde{Z}_{ab})$
$3\tilde{Z}_{00}$		$=$	$(\tilde{Z}_{aa} + a\tilde{Z}_{bb} + a^2\tilde{Z}_{cc}) + 2(\tilde{Z}_{bc} + a\tilde{Z}_{ca} + a^2\tilde{Z}_{ab})$

In view of the fact that $\tilde{Z}_{01}$ is not necessarily equal to $\tilde{Z}_{10}$, $\tilde{Z}_{02}$ is not necessarily equal to $\tilde{Z}_{20}$, and $\tilde{Z}_{12}$ is not necessarily equal to $\tilde{Z}_{21}$, we find that, in general, $[\tilde{Z}_c]$ is *not* a *symmetrical* matrix.

We can prove this in a more generalized fashion as follows:

$$[\tilde{Z}_c] = [\tilde{S}][\tilde{Z}_\Phi][\tilde{S}]^{-1} \tag{8.2.5}$$

$$[\tilde{Z}_c]^T = \left([\tilde{S}]^T\right)^{-1}[\tilde{Z}_\Phi]^T[\tilde{S}]^T = [\tilde{S}]^{-1}[\tilde{Z}_\Phi][\tilde{S}] \neq [\tilde{Z}_c] \tag{8.2.6}$$

8.3 Selecting Phase "a" Central to Phases "b" and "c"

In those cases where one phase conductor is equidistant from the other two phase conductors, there is some advantage in assigning phase "a" to the central conductor. When this is possible, and if additionally, $\tilde{Z}_{bb} = \tilde{Z}_{cc}$, we have, per equations in Table 8.2.1:

$$\tilde{Z}_{01} = \tilde{Z}_{02} = \tilde{Z}_{10} = \tilde{Z}_{20} \tag{8.3.1}$$

$$\tilde{Z}_{12} = \tilde{Z}_{21} \tag{8.3.2}$$

so that:

$$[Z_c] = \begin{bmatrix} \tilde{Z}_{00} & \tilde{Z}_{01} & \tilde{Z}_{02} \\ \tilde{Z}_{10} & \tilde{Z}_{11} & \tilde{Z}_{12} \\ \tilde{Z}_{20} & \tilde{Z}_{21} & \tilde{Z}_{22} \end{bmatrix} \tag{8.3.3}$$

is a symmetrical matrix with only four significantly different matrix elements.

The condition $\tilde{Z}_{bb} = \tilde{Z}_{cc}$ will be met if the phase conductors are of identical material, and if the ground wires are symmetrically placed with respect to the central phase conductor.

8.4 Symmetrical Form for the Symmetrical Component Impedance Matrix

According to equations Table 8.2.1 we have:

$$\tilde{Z}_{11} = \tilde{Z}_{22} \quad \tilde{Z}_{01} = \tilde{Z}_{20} \quad \tilde{Z}_{10} = \tilde{Z}_{02} \tag{8.4.1}$$

If, in Equation 8.2.3, we interchange the second and third columns of the $[\tilde{Z}_c]$ matrix we obtain:

$$[Z'_c] = \begin{bmatrix} \tilde{Z}_{00} & \tilde{Z}_{02} & \tilde{Z}_{01} \\ \tilde{Z}_{10} & \tilde{Z}_{12} & \tilde{Z}_{11} \\ \tilde{Z}_{20} & \tilde{Z}_{22} & \tilde{Z}_{21} \end{bmatrix} \tag{8.4.2}$$

which is a symmetrical matrix.

The generally unsymmetrical matrix $[\tilde{Z}_c]$ can always be converted into the symmetrical matrix $[\tilde{Z}'_c]$ by merely interchanging the second and third columns of $[\tilde{Z}_c]$.

Consider the transposition matrix:

$$[T] = \begin{bmatrix} 1 & 0 & 0 \\ 0 & 0 & 1 \\ 0 & 1 & 0 \end{bmatrix} = [T]^{-1} \tag{8.4.3}$$

The transposition matrix [T] interchanges the second and third rows of a matrix when used as a premultiplier, and interchanges the second and third columns when used as a postmultiplier. Therefore:

$$[\tilde{Z}'_c] = [\tilde{Z}_c][T] \tag{8.4.4}$$

We may interject $[T][T]^{-1}$ in Equation 8.2.2 without affecting it:

$$\begin{aligned} [\Delta \tilde{V}_c] &= [\tilde{Z}_c][T][T]^{-1}[\tilde{I}_c] \\ &= [\tilde{Z}_c][T][T][\tilde{I}_c] \\ &= [\tilde{Z}'_c][T][\tilde{I}_c] \\ &= [\tilde{Z}'_c][\tilde{I}'_c] \end{aligned} \tag{8.4.5}$$

and in expanded form we have:

$$\begin{bmatrix} \Delta \tilde{V}_0 \\ \Delta \tilde{V}_1 \\ \Delta \tilde{V}_2 \end{bmatrix} = \begin{bmatrix} \tilde{Z}_{00} & \tilde{Z}_{02} & \tilde{Z}_{01} \\ \tilde{Z}_{10} & \tilde{Z}_{12} & \tilde{Z}_{11} \\ \tilde{Z}_{20} & \tilde{Z}_{22} & \tilde{Z}_{21} \end{bmatrix} \begin{bmatrix} \tilde{I}_0 \\ \tilde{I}_2 \\ \tilde{I}_1 \end{bmatrix} \tag{8.4.6}$$

Thus, Equation 8.4.6 is a valid set of equations, which may be obtained directly from Equation 8.2.3 by interchanging the second and third *columns* of $[\tilde{Z}_c]$, and interchanging the second and third *rows* of $[\tilde{I}_c]$.

To generalize:

$$[\tilde{Z}'_c] = [\tilde{Z}_c][T] = [\tilde{S}][\tilde{Z}_\Phi][\tilde{S}]^{-1}[T] = (3)[\tilde{S}][\tilde{Z}_\Phi][\tilde{S}] \tag{8.4.7}$$

$$[\tilde{Z}'_c]^T = [T]^T[\tilde{Z}_c]^T = [T]^T \left([\tilde{S}]^{-1}\right)^T [\tilde{Z}_\Phi]^T[\tilde{S}]^T = (3)[\tilde{S}][\tilde{Z}_\Phi][\tilde{S}] = [\tilde{Z}'_c] \tag{8.4.8}$$

proving that $[\tilde{Z}'_c]$ is always a symmetrical matrix.

Since the upper triangle of matrix elements is the transpose of the lower triangle of matrix elements in a symmetrical matrix, only the main diagonal elements and the lower triangle matrix elements need be stored in computer memory. Complete matrix inversion may be performed on this triangular matrix, thereby saving considerable computation time. There is also no necessity for printing the entire square matrix.

If none of these advantages are deemed significant, then the introduction of $[\tilde{Z}'_c]$ is somewhat academic and perhaps superfluous.

8.5   Equivalent Circuit Configuration

When ground wires are present in a single-circuit transmission line, an additional voltage equation is required for each ground wire in order to properly reflect the actual physical system. Thus, we have in symbolic notation:

$$\begin{bmatrix} \Delta \tilde{V}_\Phi \\ \Delta \tilde{V}_X \end{bmatrix} = \begin{bmatrix} \tilde{Z}_{\Phi\Phi} & \tilde{Z}_{\Phi X} \\ \tilde{Z}_{X\Phi} & \tilde{Z}_{XX} \end{bmatrix} \begin{bmatrix} \tilde{I}_\Phi \\ \tilde{I}_X \end{bmatrix} \tag{8.5.1}$$

Transforming into symmetrical components we have:

$$\begin{bmatrix} \tilde{S} & 0 \\ 0 & I \end{bmatrix} \begin{bmatrix} \Delta\tilde{V}_\Phi \\ \Delta\tilde{V}_X \end{bmatrix}$$

$$= \begin{bmatrix} \tilde{S} & 0 \\ 0 & I \end{bmatrix} \begin{bmatrix} \tilde{Z}_{\Phi\Phi} & \tilde{Z}_{\Phi X} \\ \tilde{Z}_{X\Phi} & \tilde{Z}_{XX} \end{bmatrix} \begin{bmatrix} \tilde{S}^{-1} & 0 \\ 0 & I \end{bmatrix} \begin{bmatrix} \tilde{S} & 0 \\ 0 & I \end{bmatrix} \begin{bmatrix} \tilde{I}_\Phi \\ \tilde{I}_X \end{bmatrix} \tag{8.5.2}$$

where I is the identity matrix and:

$$\begin{bmatrix} \Delta\tilde{V}_\Phi \\ \Delta\tilde{V}_X \end{bmatrix} = \begin{bmatrix} \tilde{Z}_{cc} & \tilde{S}\tilde{Z}_{\Phi X} \\ \tilde{Z}_{X\Phi}\tilde{S}^{-1} & \tilde{Z}_{XX} \end{bmatrix} \begin{bmatrix} \tilde{I}_c \\ \tilde{I}_X \end{bmatrix} \tag{8.5.3}$$

If we may assume the ground wires to be continuously grounded, then:

$$\begin{bmatrix} \Delta\tilde{V}_\Phi \\ 0 \end{bmatrix} = \begin{bmatrix} \tilde{Z}_{cc} & \tilde{S}\tilde{Z}_{\Phi X} \\ \tilde{Z}_{X\Phi}\tilde{S}^{-1} & \tilde{Z}_{XX} \end{bmatrix} \begin{bmatrix} \tilde{I}_c \\ \tilde{I}_X \end{bmatrix} \tag{8.5.4}$$

If we are not interested in $[\tilde{I}_X]$, then we may eliminate the ground wire equation and write:

$$\begin{aligned} [\Delta\tilde{V}_c] &= \left\{ [\tilde{Z}_{cc}] - [\tilde{S}][\tilde{Z}_{\Phi X}][\tilde{Z}_{XX}]^{-1}[\tilde{Z}_{X\Phi}][\tilde{S}]^{-1} \right\}[\tilde{I}_c] \\ &= [\tilde{S}]\left\{ [Z_{\Phi\Phi} - [\tilde{Z}_{\Phi X}][\tilde{Z}_{XX}]^{-1}[\tilde{Z}_{X\Phi}] \right\}[\tilde{S}]^{-1}[\tilde{I}_c] \end{aligned} \tag{8.5.5}$$

Thus, if we are not interested in $[\tilde{I}_X]$, we would eliminate the ground wire equation before transforming into symmetrical components and write:

$$[\tilde{Z}_{\Phi\Phi}]_{eq} = [\tilde{Z}_{\Phi\Phi}] - [\tilde{Z}_{\Phi X}][\tilde{Z}_{XX}]^{-1}[\tilde{Z}_{X\Phi} \tag{8.5.6}$$

where represents the physical impedance matrix for the equivalent single circuit line consisting of only three single equivalent phase conductors. Then:

$$[\Delta\tilde{V}_\Phi] = [\tilde{Z}_{\Phi\Phi}]_{eq}[\tilde{I}_\Phi] \tag{8.5.7}$$

This procedure is also described in Chapter 1, Section 1.1.7.

When the phase conductors in a single-circuit transmission line consist of a number of subconductors, an additional voltage equation is required for each subconductor in order to properly reflect the actual physical system. Thus, we have in symbolic notation:

$$\begin{bmatrix} \Delta\tilde{V}_\Phi \\ \Delta\tilde{V}_{\Phi'} \end{bmatrix} = \begin{bmatrix} \tilde{Z}_{\Phi\Phi} & \tilde{Z}_{\Phi\Phi'} \\ \tilde{Z}_{\Phi'\Phi} & \tilde{Z}_{\Phi'\Phi'} \end{bmatrix} \begin{bmatrix} \tilde{I}_\Phi \\ \tilde{I}_{\Phi'} \end{bmatrix} \tag{8.5.8}$$

Since all subconductors are electrically in parallel per phase, we have:

$$[\Delta\tilde{V}_{\Phi'}] = [\Delta\tilde{V}_\Phi] \tag{8.5.9}$$

By an appropriate subtraction of corresponding rows we obtain:

$$\begin{bmatrix} \Delta\tilde{V}_\Phi \\ 0 \end{bmatrix} = \begin{bmatrix} \tilde{Z}_{\Phi\Phi} & \tilde{Z}_{\Phi\Phi'} \\ (\tilde{Z}_{\Phi'\Phi} - \tilde{Z}_{\Phi\Phi}) & (\tilde{Z}_{\Phi'\Phi'} - \tilde{Z}_{\Phi\Phi'}) \end{bmatrix} \begin{bmatrix} \tilde{I}_\Phi \\ \tilde{I}_{\Phi'} \end{bmatrix} \tag{8.5.10}$$

and by an appropriate subtraction of corresponding *columns* we obtain:

$$\begin{bmatrix} \Delta \tilde{V}_\Phi \\ 0 \end{bmatrix} = \begin{bmatrix} \tilde{Z}_{\Phi\Phi} & (\tilde{Z}_{\Phi\Phi'} - \tilde{Z}_{\Phi\Phi}) \\ (\tilde{Z}_{\Phi'\Phi} - \tilde{Z}_{\Phi\Phi}) & (\tilde{Z}_{\Phi'\Phi'} - \tilde{Z}_{\Phi\Phi'} - \tilde{Z}_{\Phi'\Phi} + \tilde{Z}_{\Phi\Phi}) \end{bmatrix} \begin{bmatrix} \tilde{I}_\Phi \\ \tilde{I}_{\Phi'} \end{bmatrix} \qquad (8.5.11)$$

This last step, subtraction of columns, is not necessary if we are only interested in an analysis based purely upon numerical computation. However, if the analysis is to be carried out in terms of analog or model representation, then the last step is necessary in order to preserve a symmetrical impedance matrix.

Then, by the same technique employed in the ground wire elimination method, we obtain:

$$[\tilde{Z}_{\Phi\Phi}]_{eq} = [\tilde{Z}_{\Phi\Phi}] - [\tilde{Z}_{\Phi\Phi'} - \tilde{Z}_{\Phi\Phi}][\tilde{Z}_{\Phi'\Phi'} - \tilde{Z}_{\Phi\phi'} - \tilde{Z}_{\Phi'\Phi} + \tilde{Z}_{\Phi\Phi}]^{-1}[\tilde{Z}_{\Phi'\Phi} - \tilde{Z}_{\Phi\Phi}] \qquad (8.5.12)$$

and:

$$[\Delta \tilde{V}_\Phi] = [\tilde{Z}_{\Phi\Phi}]_{eq}[\tilde{I}_\Phi + \tilde{I}_{\Phi'}] \qquad (8.5.13)$$

where $[\tilde{Z}_{\Phi\Phi}]_{eq}$ represents the physical impedance matrix for the equivalent single-circuit line consisting of only three single equivalent phase conductors. This procedure is also described in Chapter 1, Section 1.1.7.

8.6   Phase Rotation

Transmission lines may be transposed, at some point along the line, either by phase rotation of all three-phase conductors or transposition of only two-phase conductors. We shall demonstrate how phase rotation changes the symmetrical component impedance matrix.

Given a set of voltage equations in terms of arbitrary conductor number:

$$\begin{bmatrix} \Delta \tilde{V}_1 \\ \Delta \tilde{V}_2 \\ \Delta \tilde{V}_3 \end{bmatrix} = \begin{bmatrix} \tilde{Z}_{11} & \tilde{Z}_{12} & \tilde{Z}_{13} \\ \tilde{Z}_{21} & \tilde{Z}_{22} & \tilde{Z}_{23} \\ \tilde{Z}_{31} & \tilde{Z}_{32} & \tilde{Z}_{33} \end{bmatrix} \begin{bmatrix} \tilde{I}_1 \\ \tilde{I}_2 \\ \tilde{I}_3 \end{bmatrix} \qquad (8.6.1)$$

Now assign phasing as follows:

$$\begin{bmatrix} \Delta \tilde{V}_a \\ \Delta \tilde{V}_b \\ \Delta \tilde{V}_c \end{bmatrix} = \begin{bmatrix} \tilde{Z}_{11} & \tilde{Z}_{12} & \tilde{Z}_{13} \\ \tilde{Z}_{21} & \tilde{Z}_{22} & \tilde{Z}_{23} \\ \tilde{Z}_{31} & \tilde{Z}_{32} & \tilde{Z}_{33} \end{bmatrix} \begin{bmatrix} \tilde{I}_a \\ \tilde{I}_b \\ \tilde{I}_c \end{bmatrix} \qquad (8.6.2)$$

$$[\Delta \tilde{V}_{\Phi-old}] = [\tilde{Z}][\tilde{I}_{\Phi-old}] \qquad (8.6.3)$$

for which we determine:

$$\begin{bmatrix} \Delta \tilde{V}_0 \\ \Delta \tilde{V}_1 \\ \Delta \tilde{V}_2 \end{bmatrix} = \begin{bmatrix} \tilde{Z}_{00} & \tilde{Z}_{01} & \tilde{Z}_{02} \\ \tilde{Z}_{10} & \tilde{Z}_{11} & \tilde{Z}_{12} \\ \tilde{Z}_{20} & \tilde{Z}_{21} & \tilde{Z}_{22} \end{bmatrix} \begin{bmatrix} \tilde{I}_0 \\ \tilde{I}_1 \\ \tilde{I}_2 \end{bmatrix} \qquad (8.6.4)$$

$$[\tilde{S}][\Delta \tilde{V}_{\Phi-old}] = [\tilde{S}][\tilde{Z}][\tilde{S}]^{-1}[\tilde{S}][\tilde{I}_{\Phi-old}] \qquad (8.6.5)$$

$$[\Delta \tilde{V}_{c-old}] = [\tilde{Z}_{c-old}][\tilde{I}_{c-old}] \qquad (8.6.6)$$

Now reassign the phasing such that, for the same set of conductors, we have:

$$\begin{bmatrix} \Delta \tilde{V}_c \\ \Delta \tilde{V}_a \\ \Delta \tilde{V}_b \end{bmatrix} = \begin{bmatrix} \tilde{Z}_{11} & \tilde{Z}_{12} & \tilde{Z}_{13} \\ \tilde{Z}_{21} & \tilde{Z}_{22} & \tilde{Z}_{23} \\ \tilde{Z}_{31} & \tilde{Z}_{32} & \tilde{Z}_{33} \end{bmatrix} \begin{bmatrix} \tilde{I}_c \\ \tilde{I}_a \\ \tilde{I}_b \end{bmatrix} \tag{8.6.7}$$

$$[\Delta \tilde{V}_{\Phi-new}] = [\tilde{Z}][\tilde{I}_{\Phi-new}] \tag{8.6.8}$$

Before we can convert into symmetrical components we must reorder the equations to follow in a normal a-b-c sequence. Thus, with:

$$[R] = \begin{bmatrix} 0 & 1 & 0 \\ 0 & 0 & 1 \\ 1 & 0 & 0 \end{bmatrix} \tag{8.6.9}$$

we obtain:

$$\begin{bmatrix} \Delta \tilde{V}_a \\ \Delta \tilde{V}_b \\ \Delta \tilde{V}_c \end{bmatrix} = \begin{bmatrix} 0 & 1 & 0 \\ 0 & 0 & 1 \\ 1 & 0 & 0 \end{bmatrix} \begin{bmatrix} \Delta \tilde{V}_c \\ \Delta \tilde{V}_a \\ \Delta \tilde{V}_b \end{bmatrix} \tag{8.6.10}$$

Thus:

$$[R][\Delta \tilde{V}_{\Phi-new}] = [R][\tilde{Z}][R]^{-1}[R][\tilde{I}_{\Phi-new}] \tag{8.6.11}$$

so that:

$$[\tilde{S}][R][\Delta \tilde{V}_{\Phi-new}] = [\tilde{S}][R][\tilde{Z}][R]^{-1}[\tilde{S}]^{-1}[\tilde{S}][R][\tilde{I}_{\Phi-new}] \tag{8.6.12}$$

$$[\Delta \tilde{V}_{c-new}] = [\tilde{Z}_{c-new}][\tilde{I}_{c-new}] \tag{8.6.13}$$

with:

$$\begin{aligned} [\tilde{Z}_{c-new}] &= \left\{ [\tilde{S}][R] \right\} [\tilde{Z}] \left\{ [\tilde{S}][R] \right\}^{-1} \\ &= \left\{ [\tilde{S}][R][\tilde{S}^{-1}] \right\} [\tilde{Z}_{c-old}] \left\{ [\tilde{S}][R][\tilde{S}]^{-1} \right\}^{-1} \end{aligned} \tag{8.6.14}$$

It is left for the reader to demonstrate that:

$$[\tilde{S}][R][\tilde{S}]^{-1} = [\tilde{R}_c] \tag{8.6.15}$$

with:

$$[\tilde{R}_c] = \begin{bmatrix} 1 & 0 & 0 \\ 0 & \tilde{a}^2 & 0 \\ 0 & 0 & \tilde{a} \end{bmatrix} \tag{8.6.16}$$

Therefore:

$$[\tilde{Z}_{c-new}] = [\tilde{R}_c][\tilde{Z}_{c-old}][\tilde{R}_c]^{-1} \tag{8.6.17}$$

and:

$$\begin{bmatrix} \Delta \tilde{V}_0 \\ \Delta \tilde{V}_1 \\ \Delta \tilde{V}_2 \end{bmatrix} = \begin{bmatrix} \tilde{Z}_{00} & \tilde{a}\tilde{Z}_{01} & \tilde{a}^2\tilde{Z}_{02} \\ \tilde{a}^2\tilde{Z}_{10} & \tilde{Z}_{11} & \tilde{a}\tilde{Z}_{12} \\ \tilde{a}\tilde{Z}_{20} & \tilde{a}^2\tilde{Z}_{21} & \tilde{Z}_{22} \end{bmatrix} \begin{bmatrix} \tilde{I}_0 \\ \tilde{I}_1 \\ \tilde{I}_2 \end{bmatrix} \tag{8.6.18}$$

Thus, the magnitudes and locations of the elements in $[\tilde{Z}_{c-new}]$ are identical to those in $[\tilde{Z}_{c-old}]$, but some of the elements are shifted in phase angle by 120 degrees or 240 degrees.

Similarly, for:

$$[\tilde{Z}_{c-new}] = [\tilde{R}_c]^2 [\tilde{Z}_{c-old}][\tilde{R}_c]^{-2} \qquad (8.6.19)$$

we have:

$$\begin{bmatrix} \Delta\tilde{V}_0 \\ \Delta\tilde{V}_1 \\ \Delta\tilde{V}_2 \end{bmatrix} = \begin{bmatrix} \tilde{Z}_{00} & \tilde{a}^2\tilde{Z}_{01} & \tilde{a}\tilde{Z}_{02} \\ \tilde{a}\tilde{Z}_{10} & \tilde{Z}_{11} & \tilde{a}^2\tilde{Z}_{12} \\ \tilde{a}^2\tilde{Z}_{20} & \tilde{a}\tilde{Z}_{21} & \tilde{Z}_{22} \end{bmatrix} \begin{bmatrix} \tilde{I}_0 \\ \tilde{I}_1 \\ \tilde{I}_2 \end{bmatrix} \qquad (8.6.20)$$

For a forward or a backward phase rotation, the magnitudes and locations of the elements in are identical to those in the original, but some of the elements are shifted in phase angle by 120 degrees or 240 degrees. Thus, given $[\tilde{Z}_{c-old}]$, then $[\tilde{Z}_{c-new}]$ can be established with a minimum of computational effort.

If a phase rotation is performed at the one-third points along a transmission line section, we add Equations 8.6.4, 8.6.18, and 8.6.20, and take one-third to obtain:

$$\begin{bmatrix} \Delta\tilde{V}_0 \\ \Delta\tilde{V}_1 \\ \Delta\tilde{V}_2 \end{bmatrix} = \begin{bmatrix} \tilde{Z}_{00} & 0 & 0 \\ 0 & \tilde{Z}_{11} & 0 \\ 0 & 0 & \tilde{Z}_{22} \end{bmatrix} \begin{bmatrix} \tilde{I}_0 \\ \tilde{I}_1 \\ \tilde{I}_2 \end{bmatrix} \qquad (8.6.21)$$

If phase rotations are performed at the one-third points along a transmission line section, we will always obtain Equation 8.6.21 for any line geometry.

8.7   Phase Transposition

Given a set of voltage equations in terms of conductor number.

$$\begin{bmatrix} \Delta\tilde{V}_1 \\ \Delta\tilde{V}_2 \\ \Delta\tilde{V}_3 \end{bmatrix} = \begin{bmatrix} \tilde{Z}_{11} & \tilde{Z}_{12} & \tilde{Z}_{13} \\ \tilde{Z}_{21} & \tilde{Z}_{22} & \tilde{Z}_{23} \\ \tilde{Z}_{31} & \tilde{Z}_{32} & \tilde{Z}_{33} \end{bmatrix} \begin{bmatrix} \tilde{I}_1 \\ \tilde{I}_2 \\ \tilde{I}_3 \end{bmatrix} \qquad (8.7.1)$$

Now assign phasing as follows:

$$\begin{bmatrix} \Delta\tilde{V}_a \\ \Delta\tilde{V}_b \\ \Delta\tilde{V}_c \end{bmatrix} = \begin{bmatrix} \tilde{Z}_{11} & \tilde{Z}_{12} & \tilde{Z}_{13} \\ \tilde{Z}_{21} & \tilde{Z}_{22} & \tilde{Z}_{23} \\ \tilde{Z}_{31} & \tilde{Z}_{32} & \tilde{Z}_{33} \end{bmatrix} \begin{bmatrix} \tilde{I}_a \\ \tilde{I}_b \\ \tilde{I}_c \end{bmatrix} \qquad (8.7.2)$$

$$[\Delta\tilde{V}_{\Phi-old}] = [\tilde{Z}][\tilde{I}_{\Phi-old}] \qquad (8.7.3)$$

for which we determine:

$$\begin{bmatrix} \Delta\tilde{V}_0 \\ \Delta\tilde{V}_1 \\ \Delta\tilde{V}_2 \end{bmatrix} = \begin{bmatrix} \tilde{Z}_{00} & \tilde{Z}_{01} & \tilde{Z}_{02} \\ \tilde{Z}_{10} & \tilde{Z}_{11} & \tilde{Z}_{12} \\ \tilde{Z}_{20} & \tilde{Z}_{21} & \tilde{Z}_{22} \end{bmatrix} \begin{bmatrix} \tilde{I}_0 \\ \tilde{I}_1 \\ \tilde{I}_2 \end{bmatrix} \qquad (8.7.4)$$

$$[\tilde{S}][\Delta\tilde{V}_{\Phi-old}] = [\tilde{S}][\tilde{Z}][\tilde{S}]^{-1}[\tilde{S}][\tilde{I}_{\Phi-old}] \qquad (8.7.5)$$

$$[\Delta\tilde{V}_{c-old}] = [\tilde{Z}_{c-old}][\tilde{I}_{c-old}] \qquad (8.7.6)$$

Now reassign the phasing such that, for the same set of conductors, we have:

$$\begin{bmatrix} \Delta\tilde{V}_a \\ \Delta\tilde{V}_c \\ \Delta\tilde{V}_b \end{bmatrix} = \begin{bmatrix} \tilde{Z}_{11} & \tilde{Z}_{12} & \tilde{Z}_{13} \\ \tilde{Z}_{21} & \tilde{Z}_{22} & \tilde{Z}_{23} \\ \tilde{Z}_{31} & \tilde{Z}_{32} & \tilde{Z}_{33} \end{bmatrix} \begin{bmatrix} \tilde{I}_a \\ \tilde{I}_c \\ \tilde{I}_b \end{bmatrix} \tag{8.7.7}$$

$$[\Delta\tilde{V}_{\Phi-new}] = [\tilde{Z}][\tilde{I}_{\Phi-new}] \tag{8.7.8}$$

Before we can convert into symmetrical components we must reorder the equations to follow in a normal a-b-c sequence. Thus, with:

$$[T] = \begin{bmatrix} 1 & 0 & 0 \\ 0 & 0 & 1 \\ 0 & 1 & 0 \end{bmatrix} \tag{8.7.9}$$

we obtain:

$$\begin{bmatrix} \Delta\tilde{V}_a \\ \Delta\tilde{V}_b \\ \Delta\tilde{V}_c \end{bmatrix} = \begin{bmatrix} 1 & 0 & 0 \\ 0 & 0 & 1 \\ 0 & 1 & 0 \end{bmatrix} \begin{bmatrix} \Delta\tilde{V}_a \\ \Delta\tilde{V}_c \\ \Delta\tilde{V}_b \end{bmatrix} \tag{8.7.10}$$

Thus:

$$[T][\Delta\tilde{V}_{\Phi-new}] = [T][\tilde{Z}][T]^{-1}[T][\tilde{I}_{\Phi-new}] \tag{8.7.11}$$

so that:

$$[\tilde{S}][T][\Delta\tilde{V}_{\Phi-new}] = [\tilde{S}][T][\tilde{Z}][T]^{-1}[\tilde{S}]^{-1}[\tilde{S}][R][\tilde{I}_{\Phi-new}] \tag{8.7.12}$$

$$[\Delta\tilde{V}_{c-new}] = [\tilde{Z}_{c-new}][\tilde{I}_{c-new}] \tag{8.7.13}$$

with:

$$\begin{aligned} [\tilde{Z}_{c-new}] &= \left\{ [\tilde{S}][T] \right\} [\tilde{Z}] \left\{ [\tilde{S}][T] \right\}^{-1} \\ &= \left\{ [\tilde{S}][T][\tilde{S}^{-1}] \right\} [\tilde{Z}_{c-old}] \left\{ [\tilde{S}][T][\tilde{S}]^{-1} \right\}^{-1} \end{aligned} \tag{8.7.14}$$

It is left for the reader to demonstrate that:

$$[\tilde{S}][T][\tilde{S}]^{-1} = [T] = [T]^{-1} \tag{8.7.15}$$

Therefore:

$$[\tilde{Z}_{c-new}] = [T][\tilde{Z}_{c-old}][T] \tag{8.7.16}$$

and:

$$\begin{bmatrix} \Delta\tilde{V}_0 \\ \Delta\tilde{V}_1 \\ \Delta\tilde{V}_2 \end{bmatrix} = \begin{bmatrix} \tilde{Z}_{00} & \tilde{Z}_{02} & \tilde{Z}_{02} \\ \tilde{Z}_{20} & \tilde{Z}_{22} & \tilde{Z}_{21} \\ \tilde{Z}_{10} & \tilde{Z}_{12} & \tilde{Z}_{11} \end{bmatrix} \begin{bmatrix} \tilde{I}_0 \\ \tilde{I}_1 \\ \tilde{I}_2 \end{bmatrix} \tag{8.7.17}$$

Thus, $[\tilde{Z}_{c-new}]$ contains the same elements as $[\tilde{Z}_{c-old}]$, but the second and third rows and columns are interchanged.

For a forward or a backward transposition, the magnitudes and locations of the elements in $[\tilde{Z}_{c-new}]$ are identical to those in the original $[\tilde{Z}_{c-old}]$, but the second rows and columns are interchanged. Thus, given $[\tilde{Z}_{c-old}]$, then $[\tilde{Z}_{c-new}]$ can be established with a minimum of computational effort.

The physical impact of transposing the phasing of the line conductors is not obvious from Equation 8.7.17. We might cite one example where there would be absolutely no change, namely the configuration symmetrical about phase "a." In this case, transposing phases "b" and "c" causes the second and third rows and columns in Equation 8.3.3 to be interchanged with no net change.

Since the transposition matrix [T] only interchanges phases "b" and "c," we need to employ the rotation matrix [R] together with [T] for other interchanges.

Consider the following equations whereby the left-hand columns represent the original phasing arrangement, and the right-hand columns represent the phasing after transposition.

$$
\begin{bmatrix} \Delta\tilde{V}_a \\ \Delta\tilde{V}_b \\ \Delta\tilde{V}_c \end{bmatrix} = [T] \begin{bmatrix} \Delta\tilde{V}_a \\ \Delta\tilde{V}_b \\ \Delta\tilde{V}_c \end{bmatrix} \qquad \text{(1) Transposing b and c}
$$

$$
\begin{bmatrix} \Delta\tilde{V}_a \\ \Delta\tilde{V}_b \\ \Delta\tilde{V}_c \end{bmatrix} = [R][T] \begin{bmatrix} \Delta\tilde{V}_c \\ \Delta\tilde{V}_b \\ \Delta\tilde{V}_a \end{bmatrix} \qquad \text{(2) Transposing c and a}
$$

$$
\begin{bmatrix} \Delta\tilde{V}_a \\ \Delta\tilde{V}_b \\ \Delta\tilde{V}_c \end{bmatrix} = [R]^2[T] \begin{bmatrix} \Delta\tilde{V}_b \\ \Delta\tilde{V}_a \\ \Delta\tilde{V}_c \end{bmatrix} \qquad \text{(3) Transposing a and b}
$$

$$(8.7.18)$$

For example, consider the transpositions in Figure 1.1.7(a) (page 17) for which we have:

$$
\begin{bmatrix} \Delta\tilde{V}_a \\ \Delta\tilde{V}_b \\ \Delta\tilde{V}_c \end{bmatrix} = [U] \begin{bmatrix} \Delta\tilde{V}_a \\ \Delta\tilde{V}_b \\ \Delta\tilde{V}_c \end{bmatrix}
$$

$$
\begin{bmatrix} \Delta\tilde{V}_a \\ \Delta\tilde{V}_b \\ \Delta\tilde{V}_c \end{bmatrix} = [R]^2[T] \begin{bmatrix} \Delta\tilde{V}_b \\ \Delta\tilde{V}_a \\ \Delta\tilde{V}_c \end{bmatrix}
$$

$$
\begin{bmatrix} \Delta\tilde{V}_a \\ \Delta\tilde{V}_b \\ \Delta\tilde{V}_c \end{bmatrix} = [R][T] \begin{bmatrix} \Delta\tilde{V}_c \\ \Delta\tilde{V}_a \\ \Delta\tilde{V}_b \end{bmatrix} \qquad (8.7.19)
$$

Then:

$$[\tilde{Z}_{c-net}] = (\frac{1}{3})\left\{ [\tilde{Z}_c] + [\tilde{R}_c]^2[T][\tilde{Z}_c][T][\tilde{R}_c]^{-2} + [\tilde{R}_c][\tilde{Z}_c][\tilde{R}_c]^{-1} \right\} \qquad (8.7.20)$$

But we have just established that for an initial configuration symmetrical about phase "a":

$$[T][\tilde{Z}_c][T] = [\tilde{Z}_c] \qquad (8.7.21)$$

Therefore:

$$[\tilde{Z}_{c-net}] = (\frac{1}{3})\left\{ [\tilde{Z}_c] + [\tilde{R}_c]^2[\tilde{Z}_c][\tilde{R}_c]^{-2} + [\tilde{R}_c][\tilde{Z}_c][\tilde{R}_c]^{-1} \right\} \qquad (8.7.22)$$

so that we have, in effect, complete phase rotations, and $[\tilde{Z}_{c-net}]$ takes the form of Equation 8.6.21.

Chapter 9

Other Component Systems

In Chapter 4, a system of symmetrical components was developed wherein the three-phase voltages were expressed in terms of the "a" phase components of each of three symmetrical sets of phasors. There are actually an infinite number of other possible component sets that could be used to advantage in the solution of power system problems, each with some more or less desirable characteristics. What are the characteristics that a desirable transformation should possess?

One would certainly like to represent portions of three-phase systems, which are mutually coupled amongst the three-phases, in terms of independent uncoupled single-phase component networks; e.g., the transformation should result in the diagonalization of a given three-phase impedance matrix. In Chapter 4, it was shown that the symmetrical component transformation satisfies this requirement if the physical system possesses three-phase symmetry (e.g., a transposed three-phase transmission line) but not otherwise.

If diagonalization cannot be obtained, then at least the interconnections between the component networks should be such that considerable simplification is obtained in this manner. The alpha, beta, and zero components (or Clarke components as they are commonly called) to be discussed in this chapter possess both of these characteristics. In addition, these latter components are better suited for transient studies than are symmetrical components as will also be discussed later.

The aforementioned transformations convert variables from one stationary reference frame to another stationary reference frame. This is in contrast to Park's transformation (d-q-0 components) in electric machine theory where the stationary reference frame (three-phase quantities) is converted to a rotating reference frame (dc quantities). In less formal language, all of these transformations are merely changes in variables introduced to simplify the analysis of the problem.

In a more general sense, it is often desirable (sometimes necessary) to find a transformation for a given problem that will definitely diagonalize a coefficient matrix. A specific transformation would be defined for each problem. How this is accomplished in detail is beyond the scope of this book. We might mention, however, that the transformation is obtained as the solution of what the mathematicians formally call the "eigenvalue problem." In simpler terms, a complex system is broken down into its natural or characteristic modes. Such modal analyses are used in studying travelling wave phenomena on multiconductor systems and in corona and radio noise prediction studies.

9.1 Clarke Components

The following two equations define the Clarke transformation:

$$\begin{bmatrix} \tilde{V}_0 \\ \tilde{V}_\alpha \\ \tilde{V}_\beta \end{bmatrix} = (\frac{1}{3}) \begin{bmatrix} 1 & 1 & 1 \\ 2 & -1 & -1 \\ 0 & \sqrt{3} & -\sqrt{3} \end{bmatrix} \begin{bmatrix} \tilde{V}_a \\ \tilde{V}_b \\ \tilde{V}_c \end{bmatrix} \qquad (9.1.1)$$

$$\begin{bmatrix} \tilde{V}_a \\ \tilde{V}_b \\ \tilde{V}_c \end{bmatrix} = \begin{bmatrix} 1 & 1 & 0 \\ 1 & -\frac{1}{2} & \frac{\sqrt{3}}{2} \\ 1 & -\frac{1}{2} & -\frac{\sqrt{3}}{2} \end{bmatrix} \begin{bmatrix} \tilde{V}_0 \\ \tilde{V}_\alpha \\ \tilde{V}_\beta \end{bmatrix} \qquad (9.1.2)$$

and in terms of symbolic matrix notation:

$$[\tilde{V}_\alpha] = [\alpha][\tilde{V}_\Phi] \qquad (9.1.3)$$

$$[\tilde{V}_\Phi] = [\alpha]^{-1}[\tilde{V}_\alpha] \qquad (9.1.4)$$

Comparing Equations 9.1.1 and 9.1.2 to the symmetrical component transformation defined by Equations 4.3.12 and 4.3.13, we observe that in the former, the elements of the transformation matrix are real numbers whereas in the latter, they are complex numbers. Therefore, where symmetrical components are suited for transforming steady-state phasor voltages and currents, Clarke components are equally well suited for transforming transient and nonsinusoidal voltages and currents.

Given the equation:

$$[\tilde{V}_\Phi] = [\tilde{Z}_\Phi][\tilde{I}_\Phi] \qquad (9.1.5)$$

Then:

$$\begin{aligned} [\tilde{V}_\alpha] &= [\alpha][\tilde{V}_\Phi] = [\alpha][\tilde{Z}_\Phi][\alpha]^{-1}[\alpha][\tilde{I}_\Phi] \\ &= [\alpha][\tilde{Z}_\Phi][\alpha]^{-1}[\tilde{I}_\alpha] \end{aligned} \qquad (9.1.6)$$

and:

$$[\tilde{Z}_\alpha] = [\alpha][\tilde{Z}_\Phi][\alpha]^{-1} = \begin{bmatrix} \tilde{Z}_{00} & \tilde{Z}_{0\alpha} & \tilde{Z}_{0\beta} \\ \tilde{Z}_{\alpha 0} & \tilde{Z}_{\alpha\alpha} & \tilde{Z}_{\alpha\beta} \\ \tilde{Z}_{\beta 0} & \tilde{Z}_{\beta\alpha} & \tilde{Z}_{\beta\beta} \end{bmatrix} \qquad (9.1.7)$$

Table 9.1.1 defines each $[\tilde{Z}_\alpha]$ element for a three-phase, three-wire transmission line.

Table 9.1.1
Three-phase, Three-wire Transmission Line

$3\tilde{Z}_{00}$	$=$		$(\tilde{Z}_{aa} + \tilde{Z}_{bb} + \tilde{Z}_{cc}) + 2(\tilde{Z}_{bc} + \tilde{Z}_{ca} + \tilde{Z}_{ab})$
$3\tilde{Z}_{\alpha\alpha}$	$=$		$(2\tilde{Z}_{aa} + \frac{\tilde{Z}_{bb}+\tilde{Z}_{cc}}{2}) + 2(\tilde{Z}_{bc} - 2\tilde{Z}_{ca} - 2\tilde{Z}_{ab})$
$3\tilde{Z}_{\beta\beta}$	$=$		$\frac{3}{2}(\tilde{Z}_{bb} + \tilde{Z}_{cc}) - 3(\tilde{Z}_{bc})$
$3\tilde{Z}_{\alpha 0}$	$=$	$6\tilde{Z}_{0\alpha}$	$= \quad (2\tilde{Z}_{aa} - \tilde{Z}_{bb} - \tilde{Z}_{cc}) - (2\tilde{Z}_{bc} - \tilde{Z}_{ca} - \tilde{Z}_{ab})$
$3\tilde{Z}_{\beta 0}$	$=$	$6\tilde{Z}_{0\beta}$	$= \quad \sqrt{3}(\tilde{Z}_{bb} - \tilde{Z}_{cc}) - \sqrt{3}(\tilde{Z}_{ca} - \tilde{Z}_{ab})$
$3\tilde{Z}_{\alpha\beta}$	$=$	$3\tilde{Z}_{\beta\alpha}$	$= \quad -\frac{\sqrt{3}}{2}(\tilde{Z}_{bb} - \tilde{Z}_{cc}) - \sqrt{3}(\tilde{Z}_{ca} - \tilde{Z}_{ab})$

For a symmetrical three-phase system in which:

$$\tilde{Z}_{aa} = \tilde{Z}_{bb} = \tilde{Z}_{cc} \qquad \tilde{Z}_{ab} = \tilde{Z}_{bc} = \tilde{Z}_{ca} \tag{9.1.8}$$

we obtain:

$$
\begin{aligned}
\tilde{Z}_{00} &= \tilde{Z}_{aa} + 2\tilde{Z}_{ab} & (9.1.9) \\
\tilde{Z}_{\alpha\alpha} &= \tilde{Z}_{\beta\beta} = \tilde{Z}_{aa} - \tilde{Z}_{ab} & (9.1.10) \\
\tilde{Z}_{\alpha 0} &= \tilde{Z}_{0\alpha} = \tilde{Z}_{\beta 0} = \tilde{Z}_{0\beta} = \tilde{Z}_{\alpha\beta} = \tilde{Z}_{\beta\alpha} = 0 & (9.1.11)
\end{aligned}
$$

9.2 Clarke Component Impedances in Terms of Symmetrical Component Impedances

Because of the general popularity of symmetrical components, power system component impedances are often known in terms of symmetrical components rather than in terms of phase quantities. Table 9.2.1 shows the appropriate relationships.

If the three-phase power system is symmetrical, then:

$$\tilde{Z}_{01} = \tilde{Z}_{10} = \tilde{Z}_{02} = \tilde{Z}_{20} = \tilde{Z}_{12} = \tilde{Z}_{21} = 0 \tag{9.2.1}$$

in which case:

$$
\begin{aligned}
\tilde{Z}_{00} &= \tilde{Z}_{00} & (9.2.2) \\
\tilde{Z}_{\alpha\alpha} &= \tilde{Z}_{\beta\beta} = \frac{1}{2}(\tilde{Z}_{11} + \tilde{Z}_{22}) & (9.2.3) \\
\tilde{Z}_{\alpha\beta} &= -\tilde{Z}_{\beta\alpha} = \frac{j}{2}(\tilde{Z}_{11} - \tilde{Z}_{22}) & (9.2.4) \\
\tilde{Z}_{0\alpha} &= \tilde{Z}_{\alpha 0} = \tilde{Z}_{0\beta} = \tilde{Z}_{\beta 0} = 0 & (9.2.5)
\end{aligned}
$$

Table 9.2.1.
Three-phase, Three-wire Transmission Line

$$
\begin{aligned}
\tilde{Z}_{00} &= \tilde{Z}_{00} \\
\tilde{Z}_{\alpha\alpha} &= \tfrac{1}{2}(\tilde{Z}_{11} + \tilde{Z}_{22} + \tilde{Z}_{12} + \tilde{Z}_{21}) \\
\tilde{Z}_{\beta\beta} &= \tfrac{1}{2}(\tilde{Z}_{11} + \tilde{Z}_{22} - \tilde{Z}_{12} - \tilde{Z}_{21}) \\
\tilde{Z}_{0\alpha} &= \tfrac{1}{2}(\tilde{Z}_{01} + \tilde{Z}_{02}) \\
\tilde{Z}_{\alpha 0} &= (\tilde{Z}_{20} + \tilde{Z}_{10}) \\
\tilde{Z}_{0\beta} &= \tfrac{j}{2}(\tilde{Z}_{01} - \tilde{Z}_{02}) \\
\tilde{Z}_{\beta 0} &= j(\tilde{Z}_{20} - \tilde{Z}_{10}) \\
\tilde{Z}_{\alpha\beta} &= \tfrac{j}{2}(\tilde{Z}_{11} - \tilde{Z}_{22} - \tilde{Z}_{12} + \tilde{Z}_{21}) \\
\tilde{Z}_{\beta\alpha} &= -\tfrac{j}{2}(\tilde{Z}_{11} - \tilde{Z}_{22} + \tilde{Z}_{12} - \tilde{Z}_{21})
\end{aligned}
$$

Observe that for synchronous machines, for which $\tilde{Z}_{22} \neq \tilde{Z}_{11}$, we also have $\tilde{Z}_{\alpha\beta} = -\tilde{Z}_{\beta\alpha}$. Thus, we not only have a mutual coupling between the alpha and beta networks, but the mutual impedance is nonreciprocal. While this presents no problem in an algebraic analysis, a representation in terms of an analog model is somewhat difficult.

9.3 Clarke Component Networks

For a symmetrical three-phase system, and for $\tilde{Z}_{22} = \tilde{Z}_{11}$, and where there are no Clarke component intersequence mutual impedances, we have:

$$
\begin{bmatrix} \Delta\tilde{V}_0 \\ \Delta\tilde{V}_\alpha \\ \Delta\tilde{V}_\beta \end{bmatrix} = \begin{bmatrix} \tilde{V}_0 - \tilde{V}_{0'} \\ \tilde{V}_\alpha - \tilde{V}_{\alpha'} \\ \tilde{V}_\beta - \tilde{V}_{\beta'} \end{bmatrix} = \begin{bmatrix} \tilde{Z}_{00} & 0 & 0 \\ 0 & \tilde{Z}_{\alpha\alpha} & 0 \\ 0 & 0 & \tilde{Z}_{\beta\beta} \end{bmatrix} \begin{bmatrix} \tilde{I}_0 \\ \tilde{I}_\alpha \\ \tilde{I}_\beta \end{bmatrix}
\tag{9.3.1}
$$

If the system source voltages are balanced three-phase, then:

$$
\begin{bmatrix} \tilde{V}_0 \\ \tilde{V}_\alpha \\ \tilde{V}_\beta \end{bmatrix} = (\frac{\tilde{V}_a}{3}) \begin{bmatrix} 1 & 1 & 1 \\ 2 & -1 & -1 \\ 0 & \sqrt{3} & -\sqrt{3} \end{bmatrix} \begin{bmatrix} 1 \\ \tilde{a}^2 \\ \tilde{a} \end{bmatrix}
\tag{9.3.2}
$$

and:

$$
\begin{bmatrix} \tilde{V}_0 \\ \tilde{V}_\alpha \\ \tilde{V}_\beta \end{bmatrix} = (\tilde{V}_a) \begin{bmatrix} 0 \\ 1 \\ -j \end{bmatrix}
\tag{9.3.3}
$$

Thus, a balanced three-phase voltage source has both alpha and beta components. The alpha source voltage is the same as the positive sequence source voltage. The beta source voltage is equal to the positive sequence source voltage but shifted in phase by -90 degrees. Thus, in terms of Clarke components, we now have two active single-phase networks instead of only one, as for symmetrical components. The appropriate Clarke component networks are shown in Figure 9.3.1.

If the beta voltages and currents are advanced by 90 degrees, the networks appear as shown in Figure 9.3.2. For three-phase systems in which $\tilde{Z}_{\alpha\alpha} = \tilde{Z}_{\beta\beta} = \tilde{Z}_{11} = \tilde{Z}_{22}$, the alpha and beta impedance networks are identical. This will lead to simplification of solutions for certain fault cases.

The networks shown in Figures 9.3.1 and 9.3.2 are to be regarded as Thevenin-Helmholtz equivalents of much more complex networks, as was the interpretation of similar networks for symmetrical components. The source voltages $\tilde{E}_\alpha$ and $\tilde{E}_\beta$ are the open-circuit voltages as viewed from some prefault point in the three-phase system. $\tilde{Z}_{00}$, $\tilde{Z}_{\alpha\alpha}$, and $\tilde{Z}_{\beta\beta}$ reflect the entire system Clarke component impedances as seen by their respective component voltages at the network terminals.

As with symmetrical components, it will be necessary to know how each system element appears to alpha, beta, and zero components. Generally, the alpha and beta impedances are the same as the positive sequence impedances, and the zero sequence characteristics are identical to those for symmetrical components. The phase shift in wye-delta transformers will, however, be treated differently for Clarke components than for symmetrical components.

9.4 Tests for Clarke Component Impedances

For a symmetrical three-phase system, the Clarke transformation converts the mutually coupled three-phase network into three uncoupled single-phase networks, each with its own voltages, currents, and impedances.

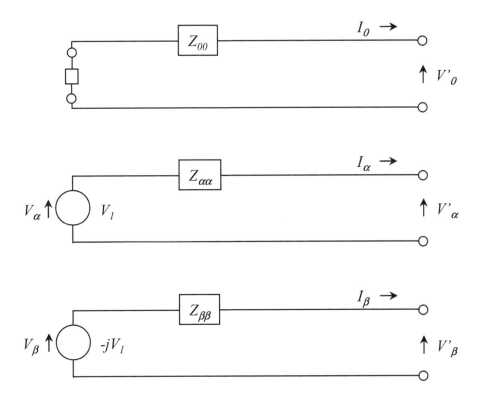

Figure 9.3.1: Clarke Component Networks

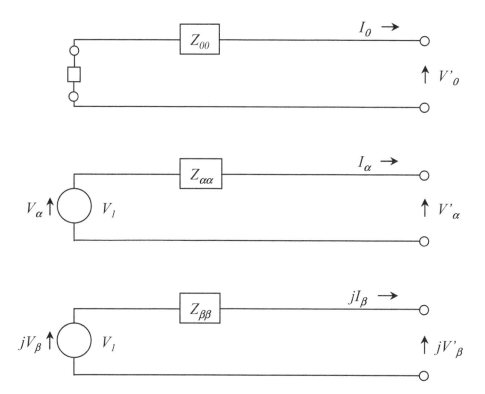

Figure 9.3.2: Clarke Component Networks with Modified Beta Network

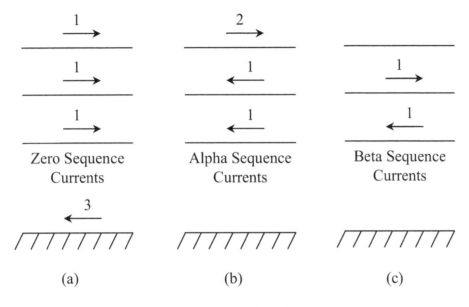

Figure 9.4.1: Clarke (α, β, 0) Components

The zero sequence currents are the same as for symmetrical components and may be visualized as shown in Figure 9.4.1(a). The alpha currents may be visualized as shown in Figure 9.4.1(b); this property is reflected in the second row of Equation 9.1.1. The beta currents may be visualized as shown in Figure 9.4.1(c); this property is reflected in the third row of Equation 9.1.1.

Figure 9.4.2 illustrates the manner in which a test might be performed on a three-phase system in order to determine the alpha impedance for that portion of the network. We have:

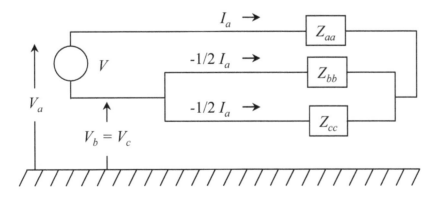

Figure 9.4.2: Loop Method for Determining α Impedances

$$\begin{bmatrix} \Delta \tilde{V}_a \\ \Delta \tilde{V}_b \\ \Delta \tilde{V}_c \end{bmatrix} = \begin{bmatrix} \tilde{Z}_{aa} & \tilde{Z}_{ab} & \tilde{Z}_{ac} \\ \tilde{Z}_{ba} & \tilde{Z}_{bb} & \tilde{Z}_{bc} \\ \tilde{Z}_{ca} & \tilde{Z}_{cb} & \tilde{Z}_{cc} \end{bmatrix} \begin{bmatrix} \tilde{I}_a \\ \tilde{I}_b \\ \tilde{I}_c \end{bmatrix}$$

$$= (\tilde{I}_a) \begin{bmatrix} \tilde{Z}_{aa} & \tilde{Z}_{ab} & \tilde{Z}_{ac} \\ \tilde{Z}_{ba} & \tilde{Z}_{bb} & \tilde{Z}_{bc} \\ \tilde{Z}_{ca} & \tilde{Z}_{cb} & \tilde{Z}_{cc} \end{bmatrix} \begin{bmatrix} 1 \\ -\frac{1}{2} \\ -\frac{1}{2} \end{bmatrix}$$

$$= (\tilde{I}_a) \begin{bmatrix} \tilde{Z}_{aa} - \frac{1}{2}(\tilde{Z}_{ab} + \tilde{Z}_{ac}) \\ \tilde{Z}_{ba} - \frac{1}{2}(\tilde{Z}_{bb} + \tilde{Z}_{bc}) \\ \tilde{Z}_{ca} - \frac{1}{2}(\tilde{Z}_{cb} + \tilde{Z}_{cc}) \end{bmatrix} \tag{9.4.1}$$

Now:

$$\Delta \tilde{V}_b = \Delta \tilde{V}_c \tag{9.4.2}$$

so that:

$$\begin{aligned} \Delta \tilde{V}_b &= \frac{1}{2}(\Delta \tilde{V}_b + \Delta \tilde{V}_c) \\ &= \frac{1}{2}[\tilde{Z}_{ba} + \tilde{Z}_{ca} - \tilde{Z}_{bc} - \frac{1}{2}(\tilde{Z}_{bb} + \tilde{Z}_{cc})](\tilde{I}_a) \end{aligned} \tag{9.4.3}$$

Then:

$$\begin{aligned} \Delta \tilde{V} &= \Delta \tilde{V}_a - \Delta \tilde{V}_b \\ &= \frac{1}{4}[(4\tilde{Z}_{aa} + \tilde{Z}_{bb} + \tilde{Z}_{cc}) - 2(2\tilde{Z}_{ab} + 2\tilde{Z}_{ca} - \tilde{Z}_{bc})](\tilde{I}_a) \end{aligned} \tag{9.4.4}$$

Thus:

$$\tilde{Z}_{\alpha\alpha} = \frac{2}{3} \frac{\Delta \tilde{V}}{\tilde{I}_a} \tag{9.4.5}$$

Figure 9.4.3 illustrates the manner in which a test might be performed on a three-phase system in order to determine the beta impedance for that portion of the network. We have:

$$\begin{bmatrix} \Delta \tilde{V}_a \\ \Delta \tilde{V}_b \\ \Delta \tilde{V}_c \end{bmatrix} = \begin{bmatrix} \tilde{Z}_{aa} & \tilde{Z}_{ab} & \tilde{Z}_{ac} \\ \tilde{Z}_{ba} & \tilde{Z}_{bb} & \tilde{Z}_{bc} \\ \tilde{Z}_{ca} & \tilde{Z}_{cb} & \tilde{Z}_{cc} \end{bmatrix} \begin{bmatrix} \tilde{I}_a \\ \tilde{I}_b \\ \tilde{I}_c \end{bmatrix}$$

$$= (\tilde{I}_b) \begin{bmatrix} \tilde{Z}_{aa} & \tilde{Z}_{ab} & \tilde{Z}_{ac} \\ \tilde{Z}_{ba} & \tilde{Z}_{bb} & \tilde{Z}_{bc} \\ \tilde{Z}_{ca} & \tilde{Z}_{cb} & \tilde{Z}_{cc} \end{bmatrix} \begin{bmatrix} 0 \\ 1 \\ -1 \end{bmatrix}$$

$$= (\tilde{I}_b) \begin{bmatrix} (\tilde{Z}_{ab} - \tilde{Z}_{ac}) \\ (\tilde{Z}_{bb} - \tilde{Z}_{bc}) \\ (\tilde{Z}_{cb} - \tilde{Z}_{cc}) \end{bmatrix} \tag{9.4.6}$$

Then:

$$\Delta \tilde{V} = \Delta \tilde{V}_b - \Delta \tilde{V}_c = [\Delta \tilde{Z}_{bb} + \Delta \tilde{Z}_{cc} - 2\Delta \tilde{Z}_{bc}](\tilde{I}_b) \tag{9.4.7}$$

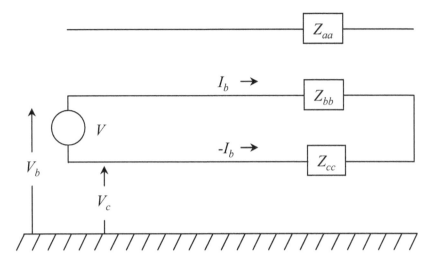

Figure 9.4.3: Loop Method for Determining β Impedance

Thus:

$$\tilde{Z}_{\beta\beta} = \frac{1}{2}\frac{\Delta\tilde{V}}{\tilde{I}_b} \qquad (9.4.8)$$

9.5 Three-Phase Fault

For a three-phase fault:

$$\tilde{V}_a = \tilde{V}_b = \tilde{V}_c = 0 \qquad (9.5.1)$$
$$\Delta\tilde{I}_a + \Delta\tilde{I}_b + \Delta\tilde{I}_c = 0 \qquad (9.5.2)$$

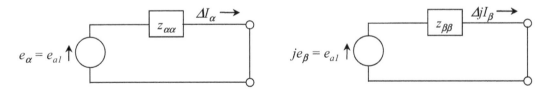

Figure 9.5.1: Clarke Component Network for a Three-Phase Fault

Then by Equation 9.1.1:

$$\Delta\tilde{I}_0 = 0 \qquad (9.5.3)$$
$$\tilde{V}_\alpha = 0 \qquad (9.5.4)$$
$$\tilde{V}_\beta = 0 \qquad (9.5.5)$$

Equations 9.5.3 to 9.5.5 are satisfied by the network interconnections shown in Figure 9.5.1. Observe that the alpha and beta network solutions are identical and that the solution for $\tilde{I}_\alpha$, for example, also gives the solution for $j\,\tilde{I}_\beta$.

9.6   Line-to-Ground Fault (SLGF)

A line-to-ground fault on phase "a" is shown in Figure 9.6.1. For this case:

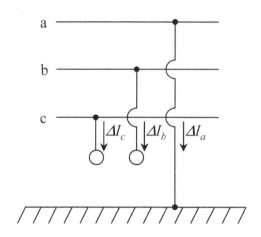

Figure 9.6.1: SLGF on Phase "a"

$$\tilde{V}_a \;=\; 0 \tag{9.6.1}$$
$$\Delta \tilde{I}_b \;=\; \Delta \tilde{I}_c = 0 \tag{9.6.2}$$

In terms of Clarke components:

$$\tilde{V}_\alpha + \tilde{V}_0 \;= 0 \tag{9.6.3}$$
$$\Delta \tilde{I}_\beta \;= 0 \tag{9.6.4}$$
$$\Delta \tilde{I}_\alpha \;= 2\Delta \tilde{I}_0 \tag{9.6.5}$$

Now, Equation 9.6.3 indicates a series interconnection between the alpha and zero component networks, but Equation 9.6.5 does not satisfy such an interconnection. The matter can be alleviated as follows:

$$\tilde{V}_0 = 0 - \Delta \tilde{I}_0 = 0 - (2\Delta \tilde{I}_0)(\frac{\tilde{Z}_{00}}{2}) \tag{9.6.6}$$

which implies that we modify the zero sequence network by taking one-half of all the zero sequence impedances while indicating $2\Delta \tilde{I}_0$ as the network current. This difficulty could have been avoided by employing the orthonormalized form of the Clarke's transformation. Figure 9.6.2 shows the appropriate Clarke component network interconnections. It should be noted that although the beta fault current component is zero, beta voltages are not zero.

9.7   Line-to-Line Faults (LLF)

Figure 9.7.1 shows a line-to-line fault from phase "b" to "c." For this case:

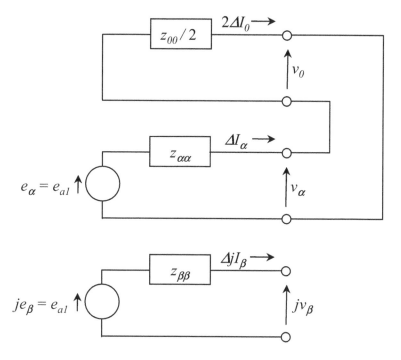

Figure 9.6.2: Clarke Component Network Interconnections for a SLGF

$$\Delta \tilde{I}_a = 0 \tag{9.7.1}$$
$$\Delta \tilde{I}_b + \Delta \tilde{I}_c = 0 \tag{9.7.2}$$
$$\tilde{V}_b = \tilde{V}_c \tag{9.7.3}$$

Then:

$$\Delta \tilde{I}_0 = \Delta \tilde{I}_\alpha = 0 \tag{9.7.4}$$
$$\tilde{V}_\beta = 0 \tag{9.7.5}$$

Figure 9.7.2 shows the appropriate Clarke component network interconnections. It should be noted that the zero sequence network is not at all involved. The fault *currents* are determined entirely by the beta network. Again note that the *voltages* must also take into account the alpha network.

9.8 Clarke Component Network Interconnections - Series Impedance Unbalance

Given a generally symmetrical three-phase system except for a series section with phase impedance unbalance as illustrated in Figure 9.8.1, then, according to Equations 9.1.9 through 9.1.11:

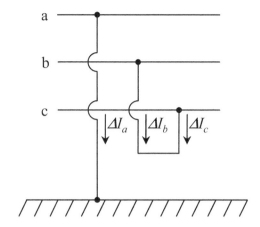

Figure 9.7.1: LLF Between Phases "b" and "c"

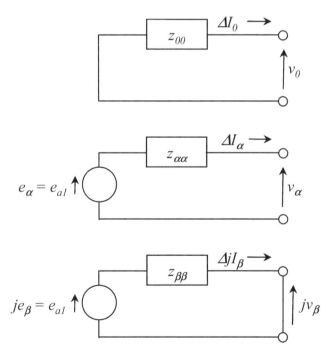

Figure 9.7.2: Clarke Component Network Interconnections for a LLF

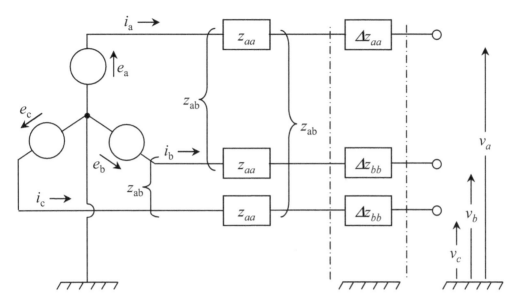

Figure 9.8.1: Series Impedance Unbalance

$$\Delta\tilde{Z}_{00} = \frac{1}{3}(\Delta\tilde{Z}_{aa} + 2\Delta\tilde{Z}_{bb}) \tag{9.8.1}$$

$$\Delta\tilde{Z}_{\alpha\alpha} = \frac{1}{3}(2\Delta\tilde{Z}_{aa} + \Delta\tilde{Z}_{bb}) \tag{9.8.2}$$

$$\Delta\tilde{Z}_{\beta\beta} = \Delta\tilde{Z}_{bb} \tag{9.8.3}$$

$$\Delta\tilde{Z}_{\alpha 0} = 2\Delta\tilde{Z}_{0\alpha} = \frac{2}{3}(\Delta\tilde{Z}_{aa} - \Delta\tilde{Z}_{bb}) \tag{9.8.4}$$

$$\Delta\tilde{Z}_{\beta 0} = 2\Delta\tilde{Z}_{0\beta} = 0 \tag{9.8.5}$$

$$\Delta\tilde{Z}_{\alpha\beta} = \Delta\tilde{Z}_{\beta\alpha} = 0 \tag{9.8.6}$$

and the Clarke component system of voltage equations become (for the unbalanced section):

$$\begin{bmatrix} \Delta\tilde{V}_0 \\ \Delta\tilde{V}_\alpha \\ \Delta\tilde{V}_\beta \end{bmatrix} = \frac{1}{3} \begin{bmatrix} (\Delta\tilde{Z}_{aa} + 2\Delta\tilde{Z}_{bb}) & (\Delta\tilde{Z}_{aa} - \Delta\tilde{Z}_{bb}) & 0 \\ 2(\Delta\tilde{Z}_{aa} - \Delta\tilde{Z}_{bb}) & (2\Delta\tilde{Z}_{aa} + \Delta\tilde{Z}_{bb}) & 0 \\ 0 & 0 & 3\Delta\tilde{Z}_{bb} \end{bmatrix} \begin{bmatrix} \tilde{I}_0 \\ \tilde{I}_\alpha \\ \tilde{I}_\beta \end{bmatrix} \tag{9.8.7}$$

Unfortunately, Equation 9.8.7 contains an impedance matrix that is not symmetrical and therefore cannot be represented physically as it stands. However, if the zero sequence network is modified, a symmetrical impedance matrix may be formulated that will allow for physical representation in terms of interconnections between component networks. Thus, we may rewrite Equation 9.8.7 in the following form:

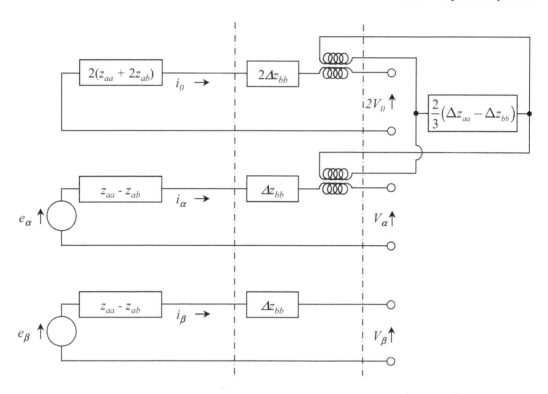

Figure 9.8.2: Clarke Component Networks for Figure 9.8.1 ($\Delta \tilde{Z}_{aa} > \Delta \tilde{Z}_{bb}$)

$$
\begin{bmatrix} 2\Delta \tilde{V}_0 \\ \Delta \tilde{V}_\alpha \\ \Delta \tilde{V}_\beta \end{bmatrix} = \frac{1}{3} \begin{bmatrix} 2(\Delta \tilde{Z}_{aa} + 2\Delta \tilde{Z}_{bb}) & 2(\Delta \tilde{Z}_{aa} - \Delta \tilde{Z}_{bb}) & 0 \\ 2(\Delta \tilde{Z}_{aa} - \Delta \tilde{Z}_{bb}) & (2\Delta \tilde{Z}_{aa} + \Delta \tilde{Z}_{bb}) & 0 \\ 0 & 0 & 3\Delta \tilde{Z}_{bb} \end{bmatrix} \begin{bmatrix} \tilde{I}_0 \\ \tilde{I}_\alpha \\ \tilde{I}_\beta \end{bmatrix}
$$

$$
= \left\{ (\Delta \tilde{Z}_{bb}) \begin{bmatrix} 2 & 0 & 0 \\ 0 & 1 & 0 \\ 0 & 0 & 1 \end{bmatrix} + \frac{2}{3}(\Delta \tilde{Z}_{aa} - \Delta \tilde{Z}_{bb}) \begin{bmatrix} 1 & 1 & 0 \\ 1 & 1 & 0 \\ 0 & 0 & 0 \end{bmatrix} \right\} \begin{bmatrix} \tilde{I}_0 \\ \tilde{I}_\alpha \\ \tilde{I}_\beta \end{bmatrix}
$$

$$(9.8.8)$$

When $\Delta \tilde{Z}_{aa} > \Delta \tilde{Z}_{bb}$, the system of component networks illustrated in Figure 9.8.2 will satisfy Equation 9.8.8. The mutual coupling transformers are ideal 1:1 transformers. If the remainder of the system is such that no ambiguity in current distribution results, the ideal transformers may be eliminated and replaced by direct interconnections. When $\Delta \tilde{Z}_{aa} = \Delta \tilde{Z}_{bb}$, no coupling exists between the networks, as is to be expected.

When $\tilde{E}_a$, $\tilde{E}_b$, and $\tilde{E}_c$ are a set of balanced voltages, then $\tilde{E}_0 = 0$ and only $\tilde{E}_\alpha$ and $\tilde{E}_\beta$ remain as source voltages.

When $\Delta \tilde{E}_{aa} \to \infty$, representing a single open conductor (SOC) case, Figure 9.8.2 clearly illustrates that power will continue to flow to the remainder of the system even though voltages and currents are unbalanced.

When $\Delta \tilde{Z}_{bb} > \Delta \tilde{Z}_{aa}$, a more appropriate equivalent network may be devised. We rewrite Equation 9.8.7 as follows:

$$
\begin{bmatrix} 2\Delta\tilde{I}_0 \\ \Delta\tilde{I}_\alpha \\ \Delta\tilde{I}_\beta \end{bmatrix} = \frac{1}{3\Delta\tilde{Z}_{aa}\Delta\tilde{Z}_{bb}} \left[\tilde{Z}\right] \begin{bmatrix} \Delta\tilde{V}_0 \\ \Delta\tilde{V}_\alpha \\ \Delta\tilde{V}_\beta \end{bmatrix}
$$

$$
= \left\{ \left(\frac{\Delta\tilde{Z}_{aa}}{\Delta\tilde{Z}_{aa}\Delta\tilde{Z}_{bb}}\right)[A] + \frac{2}{3}\left(\frac{\Delta\tilde{Z}_{bb}-\Delta\tilde{Z}_{aa}}{\Delta\tilde{Z}_{aa}\Delta\tilde{Z}_{bb}}\right)[B] \right\} \begin{bmatrix} \Delta\tilde{V}_0 \\ \Delta\tilde{V}_\alpha \\ \Delta\tilde{V}_\beta \end{bmatrix}
$$

$$(9.8.9)$$

where, for compactness, we define:

$$
\left[\tilde{Z}\right] = \begin{bmatrix} 2(\Delta\tilde{Z}_{bb}+2\Delta\tilde{Z}_{aa}) & 2(\Delta\tilde{Z}_{bb}-\Delta\tilde{Z}_{aa}) & 0 \\ 2(\Delta\tilde{Z}_{bb}-\Delta\tilde{Z}_{aa}) & (2\Delta\tilde{Z}_{bb}+\Delta\tilde{Z}_{aa}) & 0 \\ 0 & 0 & 3\Delta\tilde{Z}_{aa} \end{bmatrix}
$$

and

$$
[A] = \begin{bmatrix} 2 & 0 & 0 \\ 0 & 1 & 0 \\ 0 & 0 & 1 \end{bmatrix} \qquad [B] = \begin{bmatrix} 1 & 1 & 0 \\ 1 & 1 & 0 \\ 0 & 0 & 0 \end{bmatrix}
$$

which follows directly from a similar development used in Chapter 7, Section 7.10.

The system of Clarke component networks in Figure 9.8.3 will satisfy Equation 9.8.9. Again, when there is no ambiguity in current distribution, the ideal 1:1 transformers may be replaced by a direct interconnection.

When $\Delta\tilde{Z}_{bb} \to \infty$, representing the two open conductor (TOC) case, Figure 9.8.3 illustrates that power will continue to flow to the remainder of the system even though the voltages and currents become unbalanced.

9.9   Y-Δ Transformers

Figure 9.9.1 shows a wye-delta transformer with beta currents entering the wye-connected primary winding. Observe that alpha currents leave the delta-connected secondary winding. We have:

$$
\tilde{I}_b = -\left(\frac{N_p}{N_s}\right)\tilde{I}_C \tag{9.9.1}
$$

$$
\tilde{I}_c = \left(\frac{N_p}{N_s}\right)\tilde{I}_B \tag{9.9.2}
$$

$$
\tilde{I}_a = \left(\frac{N_p}{N_s}\right)(\tilde{I}_C - \tilde{I}_B) \tag{9.9.3}
$$

$$
\tilde{V}_{bc} = \left(\frac{N_s}{N_p}\right)\tilde{V}_A \tag{9.9.4}
$$

$$
\tilde{V}_{ca} = \left(\frac{N_s}{N_p}\right)\tilde{V}_B \tag{9.9.5}
$$

$$
\tilde{V}_{ab} = \left(\frac{N_s}{N_p}\right)\tilde{V}_C \tag{9.9.6}
$$

Then:

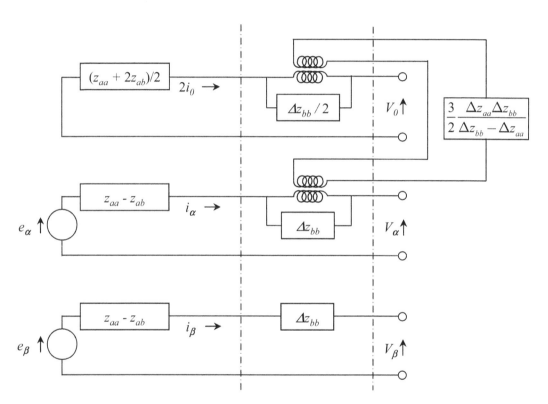

Figure 9.8.3: Clarke Component Networks for Figure 9.8.1 ($\Delta \tilde{Z}_{bb} > \Delta \tilde{Z}_{aa}$)

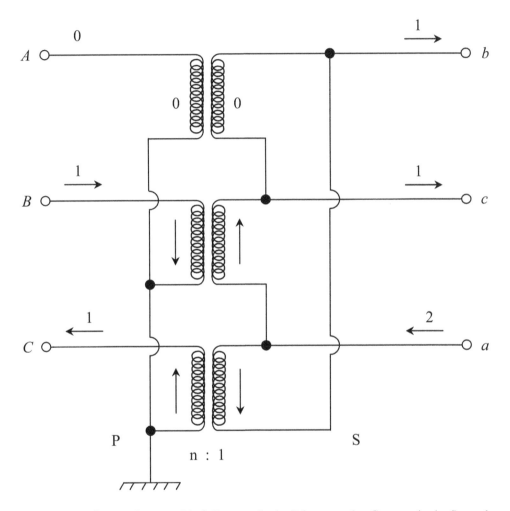

Figure 9.9.1: Y-Δ Transformer with β Currents in the Primary and α Currents in the Secondary

$$\tilde{I}_{\beta p} \;=\; \frac{1}{\sqrt{3}}(\tilde{I}_B - \tilde{I}_C) = -\frac{1}{\sqrt{3}}\Big(\frac{N_s}{N_p}\Big)\tilde{I}_t \tag{9.9.7}$$

$$\tilde{V}_{\beta p} \;=\; \frac{1}{\sqrt{3}}(\tilde{V}_B - \tilde{V}_C) = -\frac{1}{\sqrt{3}}\Big(\frac{N_p}{N_s}\Big)(\tilde{V}_{ab} - \tilde{V}_{ca}) \tag{9.9.8}$$

But:

$$\tilde{I}_{\alpha s} \;=\; \frac{1}{3}(2\tilde{I}_a - \tilde{I}_b - \tilde{I}_c) = \tilde{I}_a = -\sqrt{3}\Big(\frac{N_p}{N_s}\Big)\tilde{I}_{\beta p} \tag{9.9.9}$$

$$\tilde{V}_{\alpha s} \;=\; \frac{1}{3}(2\tilde{V}_a - \tilde{V}_b - \tilde{V}_c) = \frac{1}{3}(\tilde{V}_{ab} - \tilde{V}_{ca})$$

$$\;=\; -\frac{1}{\sqrt{3}}\Big(\frac{N_s}{N_p}\Big)\tilde{V}_{\beta p} \tag{9.9.10}$$

If we wish to express these quantities in per unit, we select the base quantities as follows:

Primary	Secondary	
$\overline{MVA_{base}}$	$\overline{MVA_{base}}$	(9.9.11)
kV_{p-base}	$kV_{s-base} = kV_{p-base}(Ns/Np)(\frac{1}{\sqrt{3}})$	(9.9.12)
kI_{p-base}	$kI_{s-base} = kI_{p-base}(Np/Ns)(\sqrt{3})$	(9.9.13)

Therefore:

$$\tilde{\tilde{I}}_{\alpha s} \;=\; -\tilde{\tilde{I}}_{\beta p} \tag{9.9.14}$$

$$\tilde{\tilde{V}}_{\alpha s} \;=\; -\tilde{\tilde{V}}_{\beta p} \tag{9.9.15}$$

By a similar analysis it can be shown that, for the wye-delta transformer shown in Figure 9.9.1, α currents on the primary side will produce β line currents on the secondary side according to the relationships:

$$\tilde{\tilde{I}}_{\beta s} \;=\; \tilde{\tilde{I}}_{\alpha p} \tag{9.9.16}$$

$$\tilde{\tilde{V}}_{\beta s} \;=\; \tilde{\tilde{V}}_{\alpha p} \tag{9.9.17}$$

For the alternate wye-delta connection shown in Figure 9.9.2, we have the following relationships:

$$\tilde{\tilde{I}}_{\alpha s} \;=\; \tilde{\tilde{I}}_{\beta p} \tag{9.9.18}$$

$$\tilde{\tilde{V}}_{\alpha s} \;=\; \tilde{\tilde{V}}_{\beta p} \tag{9.9.19}$$

$$\tilde{\tilde{I}}_{\beta s} \;=\; -\tilde{\tilde{I}}_{\alpha p} \tag{9.9.20}$$

$$\tilde{\tilde{V}}_{\beta s} \;=\; -\tilde{\tilde{V}}_{\alpha p} \tag{9.9.21}$$

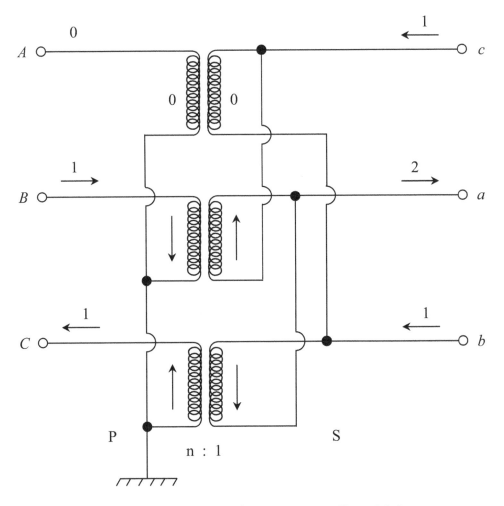

Figure 9.9.2: Alternate Y-Δ Connection (vs. Figure 9.9.1)

Example 9.9.1

Consider the sample three-phase system shown in Figure 9.9.3, with balanced source voltages and balanced load, supplied through a wye-delta transformer connected as shown in Figure 9.9.1. The appropriate Clarke component networks are shown in Figure 9.9.4.

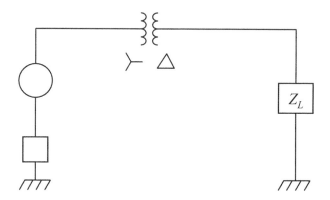

Figure 9.9.3: Sample Three-Phase System for Example 9.9.1

We determine the currents at the balanced voltage source, ignoring the phase shift through the transformers. Thus:

$$\tilde{\tilde{I}}_{0p} = 0$$

$$\tilde{\tilde{I}}_{\alpha p} = \frac{\tilde{\tilde{E}}_\alpha}{\tilde{\tilde{Z}}_L + \tilde{\tilde{Z}}_t} = \frac{\tilde{\tilde{E}}_a}{\tilde{\tilde{Z}}_L + \tilde{\tilde{Z}}_t}$$

$$\tilde{\tilde{I}}_{\beta p} = \frac{\tilde{\tilde{E}}_\beta}{\tilde{\tilde{Z}}_L + \tilde{\tilde{Z}}_t} = \frac{-j\tilde{\tilde{E}}_a}{\tilde{\tilde{Z}}_L + \tilde{\tilde{Z}}_t} = -j\tilde{\tilde{I}}_{\alpha p}$$

We now calculate the secondary currents and introduce the appropriate phase shifts.

$$\tilde{\tilde{I}}_{0s} = 0$$

$$\tilde{\tilde{I}}_{\alpha s} = -\tilde{\tilde{I}}_{\beta p} = j\tilde{\tilde{I}}_{\alpha p}$$

$$\tilde{\tilde{I}}_{\beta s} = \tilde{\tilde{I}}_{\alpha p}$$

Then, according to Equation 9.1.2, we have:

$$
\begin{aligned}
\tilde{\tilde{I}}_{ap} &= &&&&& \tilde{\tilde{I}}_{\alpha p} \\
\tilde{\tilde{I}}_{bp} &= -\tfrac{1}{2}\tilde{\tilde{I}}_{\alpha p} + \tfrac{\sqrt{3}}{2}\tilde{\tilde{I}}_{\beta p} &&= (-\tfrac{1}{2} - j\tfrac{\sqrt{3}}{2})\tilde{\tilde{I}}_{\alpha p} &&= \tilde{a}^2\tilde{\tilde{I}}_{\alpha p} \\
\tilde{\tilde{I}}_{cp} &= -\tfrac{1}{2}\tilde{\tilde{I}}_{\alpha p} - \tfrac{\sqrt{3}}{2}\tilde{\tilde{I}}_{\beta p} &&= (-\tfrac{1}{2} + j\tfrac{\sqrt{3}}{2})\tilde{\tilde{I}}_{\alpha p} &&= \tilde{a}\tilde{\tilde{I}}_{\alpha p} \\
\tilde{\tilde{I}}_{as} &= &&&&& \tilde{\tilde{I}}_{\alpha s} &&= j\tilde{\tilde{I}}_{ap} \\
\tilde{\tilde{I}}_{bs} &= -\tfrac{1}{2}\tilde{\tilde{I}}_{\alpha s} + \tfrac{\sqrt{3}}{2}\tilde{\tilde{I}}_{\beta s} &&= (-\tfrac{1}{2} - j\tfrac{\sqrt{3}}{2})\tilde{\tilde{I}}_{\alpha s} &&= \tilde{a}^2\tilde{\tilde{I}}_{\alpha s} &&= j\tilde{\tilde{I}}_{bp} \\
\tilde{\tilde{I}}_{cs} &= -\tfrac{1}{2}\tilde{\tilde{I}}_{\alpha s} - \tfrac{\sqrt{3}}{2}\tilde{\tilde{I}}_{\beta s} &&= (-\tfrac{1}{2} + j\tfrac{\sqrt{3}}{2})\tilde{\tilde{I}}_{\alpha s} &&= \tilde{a}\tilde{\tilde{I}}_{\alpha s} &&= j\tilde{\tilde{I}}_{cp}
\end{aligned}
$$

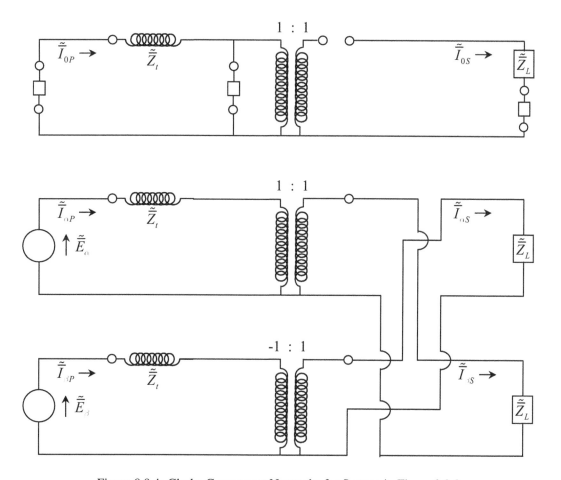

Figure 9.9.4: Clarke Component Networks for System in Figure 9.9.3

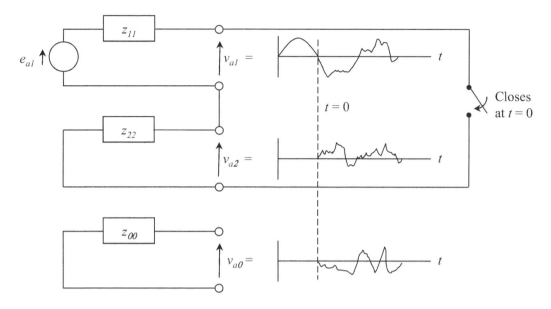

Figure 9.10.1: Transient Response for SLGF Per Symmetrical Components

9.10   Transient Solutions by Component Systems

Figure 9.10.1 shows symmetrical component networks interconnected to study transient behavior upon the sudden closing of switch S1; i.e., the sudden application of a SLGF.

Justifying such an approach, to begin with, requires some thought. Certainly the networks must be valid equivalents for all frequencies that will be produced by the transient. If the problem is to be solved analytically and a Laplace transform approach is to be used, the impedances will be expressed in operational form. Therefore, the Thevenin-Helmholtz reduction of a complex system to the form shown in Figure 9.10.1, for sinusoidal fundamental frequency excitation, is not valid.

Often an analog or model approach is preferred, and the solution might appear in the form of oscillographic recordings as shown in Figure 9.10.1. According to Equation 4.3.12, the solution for the phase "a" voltage is straightforward, even for the transient waveforms; i.e., the oscillograms for $\tilde{V}_{a0}$, $\tilde{V}_{a1}$, and $\tilde{V}_{a2}$ need only be added, point by point. Unless the frequency content of the transient waveforms can be identified, it is not possible to obtain solutions for the phase "b" and "c" voltages. Thus, symmetrical components are not a practical means for solving most transient problems.

Suppose that the SLGF problem were set up in terms of Clarke components, as shown in Figure 9.10.2. Now, according to Equation 9.1.2, we observe that all three-phase voltages may be determined by arithmetic scaling and addition. The fact that the elements in the Clarke transformation are real numbers permits the use of Clarke components for the solutions of transient problems as well as ordinary fundamental frequency problems.

9.11   References

[1] Edith Clarke. *Circuit Analysis of AC Power Systems - Volume I*. John Wiley and Sons, Inc., New York, New York, 1943.

[2] Tsungchao Hsiao. *Fault Analysis By Modified Alpha, Beta, and Zero Components*. AIEE Trans., PAS-81, 1962, pgs. 136-142.

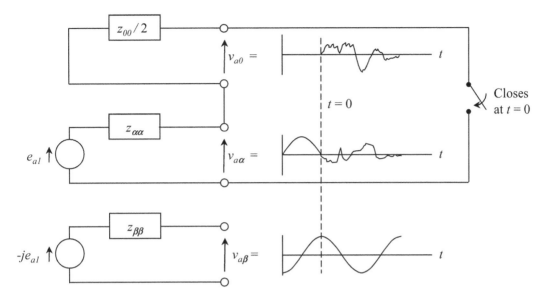

Figure 9.10.2: Transient Response for SLGF Per Clarke Components

9.12   Exercises

9.1 Three single-phase transformers, with per unit short-circuit impedances of $j0.10$, are connected wye-delta as shown in Figure 9.12.1. The per unit load impedances are $j0.90$. Assume a balanced source voltage of 1.0 per unit. For the LLF shown, determine the currents in each phase at the source, in the lines before the load, in the load, in the fault, and in the delta windings:

(a) By means of symmetrical components.

(b) By means of Clarke components.

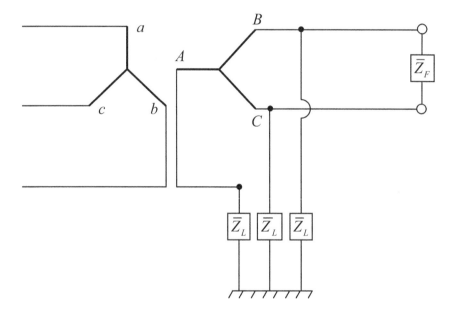

Figure 9.12.1: System for Exercise 9.1

Appendices

Appendix A

Principles of Electricity and Magnetism

This appendix contains a review of basic electromagnetic concepts and serves to place in both historical and physical perspective the principles underlying the concepts presented in the various chapters of this book. This field has been written about extensively. References [1] to [28] discuss this topic in greater depth than allowed here.

The attractive powers of lodestones and amber were recognized over two thousand years ago. However, no serious attempt was made to understand the nature of these powers or the laws governing these forces until the 17th century. In the course of subsequent historical development, magnetostatics was treated in a manner similar to electrostatics, the unit magnetic pole being defined analogous to the unit electric charge. There evolved a magnetic system of units and an electrostatic system of units, strongly parallel to one another but with no relationship between the two. Even as late as 1774, the Academy of Bavaria had offered a prize for the best dissertation in answer to the question "Is there a real and physical analogy between electric and magnetic forces?" to which there was no immediate response.

It was not until 1819 that Hans Christian Oersted observed that a permanent magnet (compass needle) moved to a position at a right angle to a straight current-carrying conductor, thus recognized that an electric current produces a magnetic field and that magnetism and an electric charge in motion were indeed related. That this relationship could not have been discovered much earlier is due to the fact that no source of conduction current existed until Alessandro Volta built the "Voltaic pile," formed from dozens of little plates about one inch in diameter, of different materials alternately stacked, each pair separated by cardboard soaked in a salt solution.

The interest sparked by this discovery produced a torrent of experimental and analytical results by subsequent investigators, each result being presented on its own merits and system of units. It was not until 1873 that James Clerk Maxwell was able to demonstrate that all of these various results could be predicted on the basis of a single electromagnetic model defined in mathematical form by a relatively few, brilliantly concise, equations.

We begin, therefore, with a presentation of Maxwell's equations since they will be used as the starting point for developing concepts in the various portions of this book.

A.1 Maxwell's Equations

In his famous treatise, published in 1873, James Clerk Maxwell (1831–1879) proposed a single model, in mathematical form, to reflect all of the electromagnetic phenomena as they were known to him. The remarkable model, structured on the basis of experimental evidence of others and ingenious deductions of his own, and expressed in terms of vector field theory, defines the electromagnetic field in terms of five vector quantities, $\vec{E}$, $\vec{H}$, $\vec{D}$, $\vec{B}$, $\vec{J}$, and three scalar quantities, μ, ϵ, and σ. Moon and Spencer [17] propose an additional vector that represents the total force per unit charge and includes the force upon the charge when it is in motion $\vec{v}$ with respect to the $\vec{B}$ field. This model, satisfactory for engineering purposes, consists of five partial differential equations supplemented by three auxiliary equations (Equations A.1.1 to A.1.3).

$$curl\,\vec{H} = \nabla \times \vec{H} = \vec{J} + \frac{\partial \vec{D}}{\partial t} \quad [A/m^2] \tag{A.1.1}$$

$$curl\,\vec{E} = \nabla \times \vec{E} = -\frac{\partial \vec{B}}{\partial t} \quad [V/m^2] \tag{A.1.2}$$

$$\vec{B} = \mu\vec{H} \quad [Wb/m^2] \tag{A.1.3}$$

$$\vec{D} = \epsilon\vec{E} \quad [C/m^2] \tag{A.1.4}$$

$$div\vec{B} = \nabla \cdot \vec{B} = 0 \quad [Wb/m^3] \tag{A.1.5}$$

$$div\vec{D} = \nabla \cdot \vec{D} = \rho \quad [C/m^3] \tag{A.1.6}$$

$$\vec{F} = \vec{E} + \left(\vec{v} + \vec{B}\right) \quad [V/m] = [N/C] \tag{A.1.7}$$

$$\vec{J} = \sigma\vec{E} \quad [A/m^2] \tag{A.1.8}$$

From a practical point of view, the following integral forms of the equations are often more useful.

$$\int\int \left(\nabla \times \vec{H}\right) \cdot d\vec{S} = \oint \vec{H} \cdot d\ell = \int\int \left(\vec{J} + \frac{\partial \vec{D}}{\partial t}\right) \cdot d\vec{S}$$

$$= I + \epsilon \left[\frac{\partial \left(\int\int \vec{E} \cdot d\vec{S}\right)}{\partial t}\right] \tag{A.1.9}$$

$$\int\int \left(\nabla \times \vec{E}\right) \cdot d\vec{S} = \oint \vec{E} \cdot d\ell = -\frac{\partial \left(\int\int \vec{B} \cdot d\vec{S}\right)}{\partial t} = -\frac{\partial \lambda}{\partial t} \tag{A.1.10}$$

$$\int\int\int \left(\nabla \cdot \vec{B}\right) dv = \int_S \int \vec{B} \cdot d\vec{S} = 0 \tag{A.1.11}$$

$$\int\int\int \left(\nabla \cdot \vec{D}\right) dv = \int_S \int \vec{D} \cdot d\vec{S} = \int\int\int \rho\,dv = Q \tag{A.1.12}$$

Often it may be difficult or impractical to apply these equations directly to practical engineering problems. However, the electric power engineer will be confronted from time to time with non-textbook situations of electromagnetic configurations in which the ability to derive an appropriate practical formulation, beginning with Maxwell's equations, will be desirable and often essential. Therefore, in the various chapters of this book, we begin with Maxwell's equations but derive a more practical formulation subject to applicable constraints appropriate to the associated operating conditions.

In order that the reader may achieve confidence in his interpretation of the various Maxwell equations, the following sections contain a review of the laws of Coulomb, Gauss, Ampere, Faraday, Lenz, etc., to show that the various individual theorems and laws, often discovered on the basis of purely experimental evidence, are but special cases of the unified theory of Maxwell and, at the same time, more tractable models for solving practical engineering problems.

These topics are presented in historical sequence of discovery as appropriate for our purpose and also to place in perspective the relatively short period of time over which these major developments have occurred.

A.2 Electrostatics

Amber, which is a fossilized resin from an extinct variety of pine tree, has the curious property of attracting small pieces of paper, chaff, and other light bodies when rubbed. The awareness of this attractive property can be traced back to the writings of Thales of Miletus (640–546 B.C.), although there is no record of electric repulsions having been observed by the Greek philosophers.

It is somewhat surprising that no apparent effort was made to analyze this phenomenon for some 2200 years until 1600 A.D. when Dr. William Gilbert (1540–1603) published his great work "De Magnete." For his investigations, Gilbert invented the *Versorium* (rotating metal needle). This instrument, somewhat like a compass needle but not made of iron, was used to detect electric attraction. He noted that not only amber, but many other materials, can be electrified by rubbing and that every material is attracted by an electrified body. He classified such materials as glass, sulfur, sealing wax, resins, rock crystals, etc., as *electrica* (Latin for amber), things that attract in the same manner as amber.

Otto von Guericke (1602–1686) constructed the first machine for the production of static electricity, and was also perhaps the first to discover electrostatic repulsion. Stephen Gray (1670–1736) in 1729, and John Theophilus Desaguliers (1683–1744) in 1739 recognized and determined that only a limited class of materials, notably the metals, could convey electricity easily. The terms *conductors* and *non-conductors* were established.

Charles Francois de Cisternay du Fay (1698–1793) observed in 1734 that:

> There are two distinct kinds of electricity, very different from one another; one of which
> I will call vitreous, and the other resinous electricity.

The former was associated with electrified glass after having been rubbed with silk, the latter associated with resins after having been rubbed with fur. Charles du Fay is also credited with the recognition of the fact that bodies with like kinds of electricity repel each other whereas bodies with unlike kinds of electricity attract one other. The designation in 1777 of these two forms of electricity as positive and negative electricity is attributed to Georg Christoph Lichtenberg (1744–1799), who arbitrarily assigned positive to electrified glass, and negative to electrified hard rubber. We now recognize that an electric charge is a fundamental property of matter even though this is not particularly obvious in a macroscopic sense due to the generally balanced distribution of positive protons and negative electrons. From a practical point of view we speak of a body as having a net charge when this balance has been disturbed. Since a charge can neither be created nor destroyed, based upon experimental evidence, then the total charge in a closed system cannot change. Consequently, the net charge in a closed system is conserved.

Although electrostatic electrification had fascinated people for thousands of years, it was not until the 18th century that the following was finally postulated:

1. Two point charges exert a force upon each other along the line joining them, inversely proportional to the square of the distance between them.

2. These forces are proportional to the product of the charge magnitudes, being attractive for unlike charges and repulsive for like charges.

While this had already been tested indirectly by Daniel Bernoulli (1700–1782), Henry Cavendish (1731–1810) and Joseph Priestley (1733–1804), it was not until 1785 that Charles Augustin de Coulomb (1736–1806) obtained direct verification of the law by means of a torsion balance. Coulomb's law may be expressed quantitatively by:

$$F = \frac{Q_1 Q_2}{4\pi\epsilon r^2} \quad [N] \tag{A.2.1}$$

where the charges are taken to be sufficiently small so that their separation distance r between their physical centers approaches that of their electrical centers. The term represents an electric property of the medium in which the force field resides. We write:

$$\epsilon = \epsilon_r \epsilon_0 \quad [F/m] \tag{A.2.2}$$

$$\epsilon_0 = \frac{10^{-9}}{36\pi} \quad [F/m] \tag{A.2.3}$$

$$\epsilon_r \equiv \text{relative permittivity} \tag{A.2.4}$$

and ϵ_0 is defined as the permittivity or absolute dielectric constant of free space (vacuum), and ϵ_r is the relative permittivity of the dielectric.

On this basis, the charge of an electron is:

$$e = 1.6008 \times 10^{-19} \quad [C] \tag{A.2.5}$$

A.3   Electric Field Intensity

The appropriate vector representation for Coulomb's law is written as:

$$\vec{F} = \frac{Q_1 Q_2}{4\pi\epsilon r^2} \vec{r} \quad [N] \tag{A.3.1}$$

which applies to a pair of point charges. For an electric field due to more than two point charges, and assuming that the dielectric material is homogeneous and isotropic such that it remains constant, then the principle of superposition may be invoked, in which case the total force acting upon a specific charged particle is the vector sum of the individual forces between it and all the other charges. In particular, for a distribution of n charges:

$$\vec{F}_i = Q_i \sum_{\substack{j=1 \\ j \neq i}}^{n} \frac{Q_j}{4\pi\epsilon r_{ij}^2} \vec{r}_{ij} \quad [N] \tag{A.3.2}$$

Thus, $\vec{F}$ is the net vector summation of forces on the ith point charge due to all other point charges in the field.

If we write:

$$\vec{F}_i = \vec{E}_i Q_i \quad [N] \tag{A.3.3}$$

or:

$$\vec{E}_i = \frac{\vec{F}_i}{Q_i} = \sum_{\substack{j=1 \\ j \neq i}}^{n} \frac{Q_j}{4\pi\epsilon r_{ij}^2} \, \vec{r}_{ij} \qquad [N/C] \qquad \text{(A.3.4)}$$

we have defined a convenient measure of the force per unit charge at the point i due to the action of all other charges in the field. On the basis of equal action and reaction, we therefore make the premise that the space about a charge possesses some peculiar condition that is present by virtue of that charge itself. Thus, we define the field strength $\vec{E}_i$ at a point i in the field, due to all other charges in the field, whether there exists a charge Q_i at point i or not. Whereas most authors introduce the concept of a vanishingly small unit test charge Q_t against which $\vec{E}_i$ can be measured in a given field, and then proceed with a rather unsatisfying limiting process such that the unit test charge is not supposed to alter the original field, we shall accept the above mathematical definition as a more suitable, legitimate alternative.

The vector $\vec{E}$ is called the *electric field intensity* with dimensions of [N/C] or, in practical units, [V/m]. It is a fundamental quantity in Maxwell's equations, as we saw in Equation A.1.2:

$$curl \, \vec{E} = \nabla \times \vec{E} = -\frac{\partial \vec{B}}{\partial t} \qquad [V/m^2]$$

and it is a concept that enables us to assign *vector* characteristics as a function of *single* point charges rather than by pairs of point charges as is the case in Coulomb's law.

Therefore, associated with every point charge Q we have:

$$\vec{E} = \frac{Q}{4\pi\epsilon r^2} \, \vec{r} \qquad [V/m] \qquad \text{(A.3.5)}$$

and by *vector* superposition, the $\vec{E}$ field at a point due to a distribution of point charges can be determined.

In a more general sense, given a known distribution of point charges, we have, according to Equation A.3.4):

$$\vec{E}_P = \sum_{j=1}^{n} \frac{Q_j}{4\pi\epsilon r_{Pj}^2} \, \vec{r}_{Pj} \qquad [N/C] \qquad \text{(A.3.6)}$$

where $\vec{E}_P$ is the electric field intensity at point P in space due to a distribution of n point charges Q_j, with no charge at point P.

Example A.3.1

Let a set of point charges per unit length q be distributed uniformly along the line filament of length L as illustrated in Figure A.3.1. Determine the field, $\vec{E}_P$ at P(x_1, y_1). The solution is as follows:

$$d\vec{E} = \frac{qe^{j\theta} \, dx}{4\pi\epsilon r^2}$$

$$x_1 - x = r \cos \theta \qquad y_1 = r \sin \theta$$

$$dx = \frac{y_1 \, d\theta}{\sin^2 \theta}$$

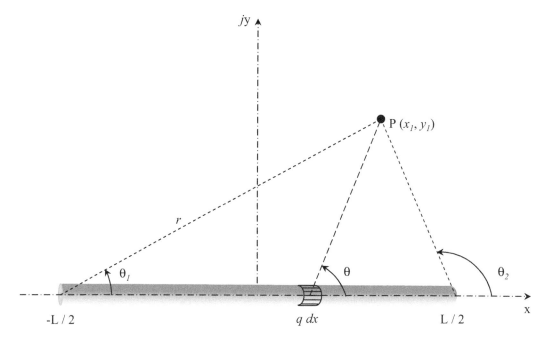

Figure A.3.1: Electric Field Intensity about Line Conductor

$$\theta_1 = \arctan \frac{y_1}{x_1 + \dfrac{L}{2}} \quad \theta_2 = \arctan \frac{y_1}{x_1 - \dfrac{L}{2}}$$

$$\vec{E} = \int_{\theta_1}^{\theta_2} d\vec{E} = -\frac{jq\left(e^{j\theta_1} - e^{j\theta_2}\right)}{4\pi\epsilon\, y_1} = \frac{q\,\sin\left(\dfrac{\theta_2 - \theta_1}{2}\right)\, e^{j\left(\frac{\theta_2 - \theta_1}{2}\right)}}{2\pi\epsilon\, y_1}$$

If the line filament is infinitely long, then:

$$\theta_1 = 0 \quad \theta_2 = \pi$$

and:

$$\vec{E} = \frac{jq}{2\pi\epsilon\, y_1}$$

According to the result in Example A.3.1, we find that the electric field intensity about an infinitely long, uniformly charged (q, [C/m]) line filament is radial to the filament and equal to:

$$E = \frac{q}{2\pi\epsilon\, r} \qquad [V/m] \tag{A.3.7}$$

A.4   Gauss's Law and Electric Flux Density

The result in Equation A.3.7 shows that the electric field intensity depends not only upon the charge density q, but also upon the permittivity of the medium in which the field resides. We define a new vector:

$$\vec{D} = \epsilon \vec{E} \qquad [C/m^3] \tag{A.4.1}$$

which is independent of the characteristics of the surrounding medium but is a direct measure of the charge density on the conductor.

For the point charge Q we have [per Equation A.3.5]:

$$D = \frac{Q}{4\pi r^2} \qquad [C/m^3] \tag{A.4.2}$$

and for the infinitely long charged line we have [per Equation A.3.7]:

$$D = \frac{q}{2\pi r} \qquad [C/m^3] \tag{A.4.3}$$

and it is observed that, for these geometrically symmetrical configurations, D is simply the total charge divided by the area of the surface surrounding the charge and therefore may be interpreted as *electric induction* or *electric flux density*. We may visualize this field in terms of lines of electric flux, each line emanating from, or terminating upon, a charge regardless of the permittivity of the medium surrounding the charge.

Maxwell's equation (Equation A.1.6) states that:

$$div\,\vec{D} = \nabla \cdot \vec{D} = \rho \qquad [C/m^3]$$

which is simply the mathematical way of saying that point charges act as sources or sinks for lines of electric induction. Then, by Gauss's theorem (Equation A.1.12):

$$\iiint \left(\nabla \cdot \vec{D} \right) dv = \int_S \int \vec{D} \cdot d\vec{S} = \iiint \rho\,dv = Q$$

which states that the integral of the flux density vector field D normal to and over the surface enclosing a region is exactly equal to the net quantity of electric charge contained within the enclosed region.

Thus, for charged electrode configurations that possess a suitable geometric symmetry, the electric flux density can be calculated very easily. Subsequently the electric field intensity can be determined directly on the basis of Equation A.4.1.

Maxwell's equation (Equation A.1.12 is the integral form of Equation A.1.6 and is simply the mathematical way of saying that point charges act as sources or sinks for lines of electric induction.

Example A.4.1
Let a set of point charges per unit length q be distributed uniformly along the line filament of infinite length. Determine the E field about the line filament.

Because of the geometric symmetry, the D field will be purely radial to the conductor. Then:

$$D = \frac{q}{2\pi r} \qquad [C/m^3]$$

$$E = \frac{D}{\epsilon} = \frac{q}{2\pi\epsilon r} \qquad [C/m^3]$$

Example A.4.1 illustrates that when, due to some geometric symmetry, the nature of the electric field is known a priori, the application of Gauss's law is considerably easier to employ than the distribution of point charges as illustrated in Example A.3.1. On the other hand, for a finite length of charged conductor, the latter method is easier to employ than Gauss's law.

A.5   Scalar Potential

Since work or energy is required to cause a separation of electric charges, we may also visualize the space about a charged conductor to have different levels of potential energy. The energy required to move a positive unit point charge in an existing E field is given by:

$$v_2 - v_1 = -\int \vec{E} \cdot d\vec{r} \qquad [J/C] = [V] \tag{A.5.1}$$

whereby the negative sign before the integral indicates a decrease in energy level as the integration proceeds in the direction of the E field. For a single positive point charge, for example, we have:

$$v_2 - v_1 = -\int_{r_1}^{r_2} \frac{Q}{4\pi\epsilon r^2} \, dr = \frac{Q}{4\pi\epsilon} \left(\frac{1}{r_2} - \frac{1}{r_1} \right) \qquad [V] \tag{A.5.2}$$

or:

$$v_2 = v_1 + \frac{Q}{4\pi\epsilon} \left(\frac{r_2 - r_1}{r_1 r_2} \right) \qquad [V] \tag{A.5.3}$$

For practical engineering purposes, only potential differences are of interest. Merely for the sake of convenience, occasionally an absolute zero potential is arbitrarily assigned to some electrode against which other relative potentials are defined. If and when it becomes necessary to determine the true absolute zero equipotential surface in an electric field, no such arbitrary assumption is required provided that the complete system of charged electrodes is taken into account and where it is recognized that:

$$\sum Q_i = 0 \qquad [C] \tag{A.5.4}$$

By virtue of the dot product in Equation A.5.1, the potential is a scalar quantity. Therefore, the scalar potential at any point in space can be determined easily as the arithmetic sum of the contributions due to any distribution of point charges.

Example A.5.1
Let a set of point charges per unit length q be distributed uniformly along the line filament of length L as illustrated in Figure A.3.1. Determine the potential at P(x_1, y_1).
Solution:

$$r^2 = (x_1 - x)^2 + y_1^2$$

$$dv = \frac{-q\,dx}{4\pi\epsilon_0 \sqrt{(x_1 - x)^2 + y_1^2}}$$

$$v(x_1, y_1) = -q \int_{-\frac{L}{2}}^{+\frac{L}{2}} \frac{dx}{4\pi\epsilon_0 \sqrt{(x_1 - x)^2 + y_1^2}}$$

$$= \frac{q}{4\pi\epsilon_0} \ln \left(\frac{x_1 + \dfrac{L}{2} + \sqrt{\left(x_1 + \dfrac{L}{2}\right)^2 + y_1^2}}{x_1 - \dfrac{L}{2} + \sqrt{\left(x_1 - \dfrac{L}{2}\right)^2 + y_1^2}} \right) \qquad [V]$$

A comparison of Examples A.3.1 and A.5.1 bring to light the advantages of dealing with scalar quantities rather than vector quantities. Several other important aspects of Example A.5.1 need to be mentioned. First, the constraint of uniform linear charge density along the line filament is somewhat academic. If the line filament is a metallic conductor, the line charges will, in fact, distribute themselves in a manner to give a constant potential over the surface of the conductor.

Second, we have assumed a set of positive charges and ignored the associated set of negative charges. This too is somewhat academic, and the result in Example A.5.1 represents an absolute potential only if it is assumed that the associated set of negative charges are located infinitely remote from the charged line filament.

Equation A.5.1 is strictly valid only in time-invariant situations; i.e., pure electrostatic fields with no time-varying magnetic flux-linkages. Under these constraints it follows that the integral of E over a closed path must be zero since the initial and final values of the scalar potential are identical. We may therefore formulate this mathematically as:

$$\oint \vec{E} \cdot d\ell = 0 \tag{A.5.5}$$

which is a special case of Maxwell's equation (Equation A.1.10).

The vector differential equivalent of Equation A.5.5 is:

$$curl\,\vec{E} = \nabla \times \vec{E} = 0 \tag{A.5.6}$$

which is a special case of Maxwell's equation (Equation A.1.2). From an extrapolated point of view, Equation A.5.5 is also a form of Kirchhoff's voltage law as applied in elementary circuit analysis. Also, since the curl of a gradient is always equal to zero according to the rules of vector calculus, we may write:

$$\vec{E} = -\nabla\phi \qquad [V/m] \tag{A.5.7}$$

thereby introducing ϕ [V] for the scalar potential field. Then, in any region devoid of free electric charges, such that $\nabla \cdot \vec{D} = 0$, we have:

$$\nabla^2 \phi = 0 \tag{A.5.8}$$

which is Laplace's equation. Thus, the scalar potential field is a solution of Laplace's equation, and this forms the basis of the various field mapping techniques.

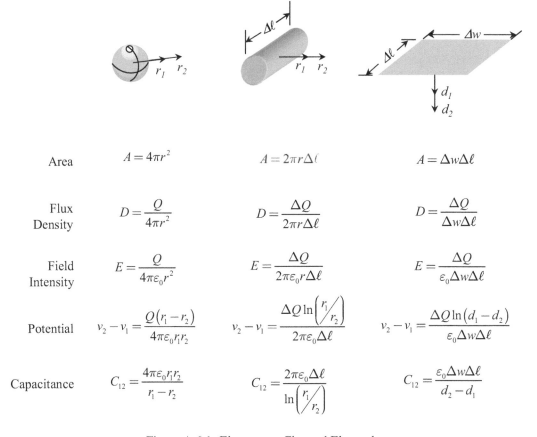

Area	$A = 4\pi r^2$	$A = 2\pi r \Delta \ell$	$A = \Delta w \Delta \ell$
Flux Density	$D = \dfrac{Q}{4\pi r^2}$	$D = \dfrac{\Delta Q}{2\pi r \Delta \ell}$	$D = \dfrac{\Delta Q}{\Delta w \Delta \ell}$
Field Intensity	$E = \dfrac{Q}{4\pi \varepsilon_0 r^2}$	$E = \dfrac{\Delta Q}{2\pi \varepsilon_0 r \Delta \ell}$	$E = \dfrac{\Delta Q}{\varepsilon_0 \Delta w \Delta \ell}$
Potential	$v_2 - v_1 = \dfrac{Q(r_1 - r_2)}{4\pi \varepsilon_0 r_1 r_2}$	$v_2 - v_1 = \dfrac{\Delta Q \ln\left(\dfrac{r_1}{r_2}\right)}{2\pi \varepsilon_0 \Delta \ell}$	$v_2 - v_1 = \dfrac{\Delta Q \ln(d_1 - d_2)}{\varepsilon_0 \Delta w \Delta \ell}$
Capacitance	$C_{12} = \dfrac{4\pi \varepsilon_0 r_1 r_2}{r_1 - r_2}$	$C_{12} = \dfrac{2\pi \varepsilon_0 \Delta \ell}{\ln\left(\dfrac{r_1}{r_2}\right)}$	$C_{12} = \dfrac{\varepsilon_0 \Delta w \Delta \ell}{d_2 - d_1}$

Figure A.6.1: Elementary Charged Electrodes

A.6   Potential and Capacitance Coefficients

Figure A.6.1 summarizes the electric field characteristics of some elementary forms of charged electrodes. In particular, the associated equipotential surfaces are of the form of concentric spheres, concentric cylinders, or parallel plane surfaces. It is well known that a thin perfectly conducting surface can replace an equipotential surface without disturbing the electric field between the charged electrode and the introduced conducting surface, so that in effect we can think of the charged electrodes shown in Figure A.6.1 in terms of two-electrode systems in each case. Figure A.6.1 includes the functional relationship between the scalar potential between the two electrodes and the equal and opposite charges residing on these electrodes. It is to be observed that in all these examples we have a form:

$$v_2 - v_1 = PQ \qquad [V] \tag{A.6.1}$$

whereby P is always a function of the electrode geometry and the permittivity of the medium in which the field resides. The *potential coefficient* (P) is a generalized term that defines the relationship between charges on a pair of electrodes and the resultant voltage difference between the electrodes. Conversely, the *capacitance coefficient* (C, Farads) is a generalized term that defines the relationship between the voltage difference between a pair of electrodes and the resultant charges on the electrodes:

$$Q = C\,(v_2 - v_1) \qquad [C] \qquad\qquad (A.6.2)$$

For a two-electrode system we have $C = 1/P$. For a multi-electrode system, the relationship between C and P is more complex.

A.7 Electric Current

In 1739, John Theophilus Desaguliers (1683–1744) determined that only a limited class of materials, notably the metals, could convey electricity easily, and he called these *conductors*. Up to this time there existed no means for sustaining a continuous flow of electricity; experiments dealt only with static electricity and, at best, only with discharges of finite quantities of electricity.

In 1780, Luigi Galvani (1737–1798) observed that the legs of a dissected frog twitched when a nearby electrostatic generator produced a spark while the point of a metal scalpel was in contact with a nerve of the frog's leg. Galvani attributed this, erroneously, to inherent animal electricity.

Alessandro Volta (1745–1827) repeated Galvani's experiments and attributed the source of electricity simply to the action of two dissimilar metals brought in contact with two points on the same nerve filament. In 1800, he built the" *Voltaic pile*," formed from dozens of little plates about one inch in diameter, of different metals alternately stacked, each pair separated by cardboard soaked in a salt solution. Even though Volta maintained an erroneous theory of contact potential for the explanation of the performance of his Voltaic pile, he had invented a source of continuous current, for a long time referred to as the *Galvanic current*.

Conduction current (I) is defined to be due to a flow or transport of electric charge. In an electrolyte, both positive and negative ions are in motion in opposite directions. In gas and vacuum tubes, and in metallic conductors, the current is due to the motion of electrons alone. Current density due to conduction is defined by:

$$J = \rho \frac{dx}{dt} \qquad [A/m^2] \qquad\qquad (A.7.1)$$

where ρ is charge density $[C/m^3]$. The total current passing through a given area is:

$$I = \int\int \vec{J} \cdot d\vec{S} \qquad [A] \qquad\qquad (A.7.2)$$

Displacement current is defined due to a relative displacement of bound electrons. A time-varying electric field will stress and distort the lattice structure of the atoms of nonconducting media causing a relative displacement of bound electrons. Displacement current density is defined by:

$$\frac{\partial \vec{D}}{\partial t} = \epsilon \frac{\partial \vec{E}}{\partial t} \qquad [A/m^2] \qquad\qquad (A.7.3)$$

Even the best of metallic conductors exhibit *resistance* to conduction current. We define resistance (R) by:

$$R = \frac{\rho \ell}{A} = \frac{\ell}{\sigma A} \qquad [\Omega] \qquad\qquad (A.7.4)$$

where ρ is specific resistivity of the medium in $[\Omega\text{-m}]$, ℓ is the length of the conductor in $[m]$, A is the cross-sectional area normal to the current in $[m^2]$, and σ is the specific conductivity of the medium in $[S/m]$. Occasionally specific resistivity is expressed in terms of $[\Omega/m^3]$, which may appear to be dimensionally incorrect, but is intended to mean the resistance of a cube of material whereby the sides of the cube are each one meter in length.

The relationship between current density and electric field intensity proposed by George Simon Ohm (1789–1849) is given by the elementary form of Ohm's law (1827):

$$J = \sigma E \qquad [A/m^2] \tag{A.7.5}$$

The similarity in form between Equations A.7.5 and A.4.1 suggests that the solutions obtained for electrostatic fields, e.g., Figure A.6.1 can be applied directly to problems of current distribution (dc current) in homogeneous media, such as earth, merely by substituting $G = 1/R$ and I for C and Q, respectively. Conversely, electrostatic field problems may be solved by means of resistance models (lumped resistance grids, electrolytic tanks, Teledeltos paper).

A.8   Magnetism

The magnetic properties of naturally occurring lodestones were first observed by the Greeks in the district of Magnesia in Asia Minor over 2000 years ago. The Chinese were perhaps the first to put magnetism to practical use by employing suspended lodestones for navigational purposes. The recognition that steel needles could be permanently magnetized by means of natural magnets led to the development of the magnetic compass.

In 1269, Pierre le Pelerin de Maricourt (Petrus Peregrinus de Maricourt) recognized that breaking a magnetized needle again resulted in a dipole magnet, and his "*Epistola Petri ad Sygerium*" is the first treatise on the magnet ever written. In 1600, William Gilbert (1544–1603) published his book "*De Magnete*" in which he describes the results of his experiments and proposes that the Earth is also a magnet.

The law that two magnetic poles exert a force on one another inversely proportional to the square of the distance between them was stated by John Michell (1724–1793) in 1750, and subsequently by Tobias Mayer (1723–1762) and Johann Heinrich Lambert (1728–1777). The law is usually attributed to Charles Augustin de Coulomb (1736–1806) who put it on a firm foundation by his experiments with the torsion balance in 1785.

In the course of historical development, magnetostatics were treated in a manner similar to electrostatics, the unit magnetic pole being taken analogous to the unit electric charge. There developed a magnetic system of units parallel to the electrostatic system of units, with no relationship between them. Even as late as 1774, the Academy of Bavaria had offered a prize for the best dissertation in answer to the question, "Is there a real and physical analogy between electric and magnetic forces?"

A.9   Electromagnetism

It was not until 1819 that Hans Christian Oersted (1777–1851) showed that a permanent magnet (compass needle) placed itself at right angles to a straight current-carrying conductor, thus establishing that an electric current produces a magnetic field and that magnetism and an electric charge in motion were related.

A.9.1   Ampere's Force Law

In 1820 Andre Marie Ampere (1775–1836) stated the force relationship between two parallel current elements:

$$dF = \frac{\mu_0 \, (i_1 d\ell_1) \, (i_2 d\ell 2)}{4\pi r^2} \left(1 - 1.5 \cos^2 \phi \right) \qquad [N] \tag{A.9.1}$$

where ϕ is the angle between the direction of the current elements and the line connecting the two current elements, and μ_0 is the magnetic permeability of free space. Also due to Ampere are the following two laws:

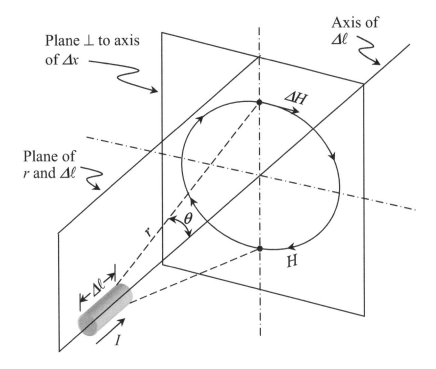

Figure A.9.1: Biot-Savart Law

1 Two parallel current elements, similarly directed, will attract each other; two parallel current elements, oppositely directed, will repel each other.

2 Two current-carrying conductors, free to move, will align themselves such as to be parallel to one another with currents similarly directed.

A.9.2 Biot-Savart Law

In 1820 Jean Baptiste Biot (1774–1862) and Felix Savart (1791–1841) determined by experiment that the relationship between magnetic field intensity and current, as illustrated in Figure A.9.1, is given by:

$$dH = \frac{I \sin \theta}{4\pi r^2} \, d\ell \qquad [A/m] \tag{A.9.2}$$

in other words:

> Each element of length of current-carrying conductor $d\ell$ causes a contribution dH to the magnetic field intensity at a point P(x, y) proportional to $d\ell$, I, and $\sin \theta$, and inversely proportional to the square of the distance r. dH is a contribution to the vector field, which lies in a plane perpendicular to the axis of $d\ell$.

Maxwell's equations (Equations A.1.1 and A.1.9) are simply generalizations of the Biot-Savart law. The Biot-Savart law often permits a quantitative determination of the H field for an irregular configuration of current-carrying conductors when other methods are not practical.

Example A.9.1

Consider the finite length of straight current-carrying conductor in free space, illustrated in Figure A.9.2. We have:

$$dH = \frac{I\,dx}{4\pi r^2}\,\sin\theta \qquad [A/m]$$

with

$$r\cos\theta = x_1 - x$$
$$r\sin\theta = y_1$$
$$dx = -dr\cos\theta + r\sin\theta\,d\theta$$
$$0 = dr\sin\theta + r\cos\theta\,d\theta$$

we have:

$$dx = \frac{r\cos^2\theta}{\sin\theta}\,d\theta + r\sin\theta\,d\theta = \frac{r\,d\theta}{\sin\theta}$$

$$dH = \frac{I\sin\theta}{4\pi y_1}\,d\theta$$

and:

$$H = \frac{I}{4\pi y_1}\,(\cos\theta_1 - \cos\theta_2)$$

$$= \frac{I}{4\pi y_1}\left[\frac{\left(x_1 + \dfrac{L}{2}\right)}{\sqrt{\left(x_1 + \dfrac{L}{2}\right)^2 + y_1^2}} - \frac{\left(x_1 - \dfrac{L}{2}\right)}{\sqrt{\left(x_1 - \dfrac{L}{2}\right)^2 + y_1^2}} \right] \qquad [A/m]$$

At $x_1 = 0$ we have:

$$H\big|_{x_1=0} = \frac{I}{2\pi y_1\,\sqrt{1 + \left(\dfrac{2y_1}{L}\right)^2}} \qquad [A/m]$$

Because of the continuity of current law, the results obtained for the finite length of conductor can, at best, represent only a portion of the complete result for a closed-current path configuration.

Example A.9.2

Consider the circular loop current-carrying conductor illustrated in Figure A.9.3. We wish to determine H on the axis of the loop current at some point z_1. Each conductor element $R\,d\alpha$ will

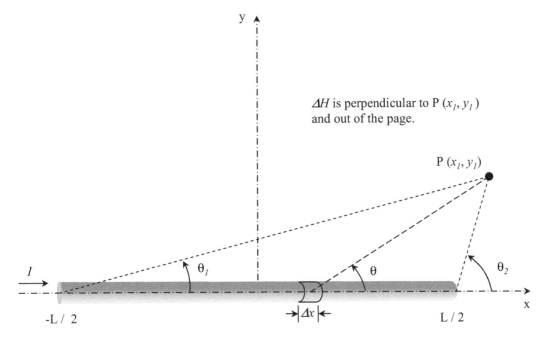

Figure A.9.2: Finite Length of Straight Current-Carrying Conductor in Free Space

contribute a dH vector at the point z_1 having a component along the z-axis and a component perpendicular to the z-axis. For the complete current loop, the latter will contribute a net of zero. Since $\theta = \pi/2$ in the Biot-Savart law for this case, we have:

$$dH = \frac{IR\,d\alpha}{4\pi r^2}\,\cos\phi$$

with:

$$r\sin\phi = z_1$$
$$r\cos\phi = R$$
$$r^2 = z_1^2 + R^2$$
$$\cos\phi = \frac{R}{\sqrt{z_1^2 + R^2}}$$

Then:

$$dH = \frac{IR^2 d\alpha}{4\pi \sqrt{\left(z_1^2 + R^2\right)^3}}$$

and:

$$H = \frac{IR^2}{2\sqrt{\left(z_1^2 + R^2\right)^3}} \qquad [A/m]$$

For $z_1 = 0$, the magnetic field intensity in the plane of the loop at the loop *center* is:

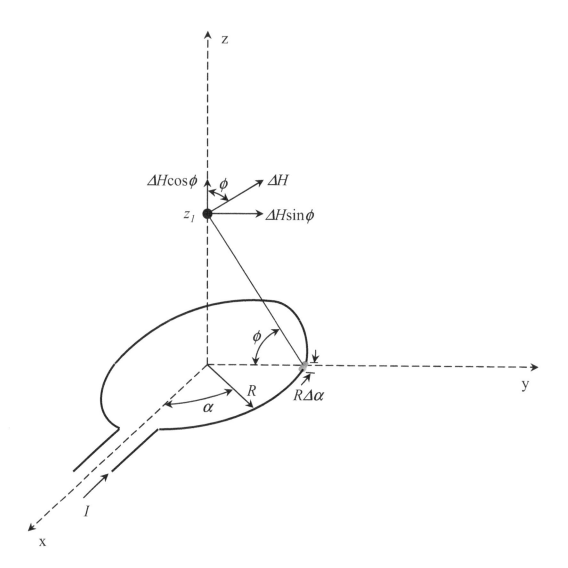

Figure A.9.3: Circular Loop Current

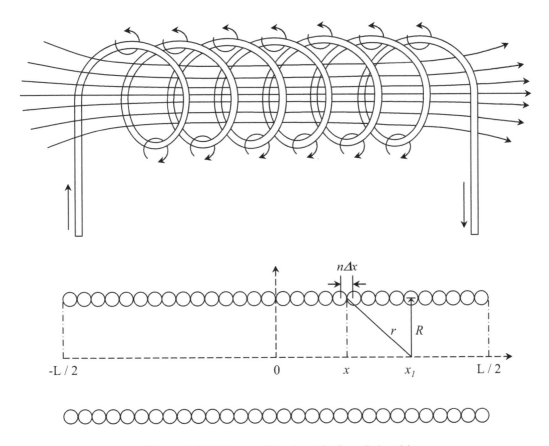

Figure A.9.4: Current-Carrying Air-Core Solenoid

$$H\big|_{x_1=0} = \frac{I}{2R} \qquad [A/m]$$

Example A.9.3

Consider the straight air-core solenoid of finite length illustrated in Figure A.9.4. We shall take the solenoid to consist of a series connection of current-carrying loops, n turns per meter with I amperes per turn. We wish to determine the strength of the H field along the central axis of the solenoid. We take advantage of the result obtained in Example A.9.2 and write:

$$dH = \frac{nIR^2\,dx}{2\sqrt{\left[(x_1 - x)^2 + R^2\right]^3}}$$

$$= \frac{nI\,dx}{2R\left[1 + \left(\dfrac{x - x_1}{R}\right)^2\right]^{\frac{3}{2}}}$$

Integrating from $x = -L/2$ to $x = +L/2$ we obtain:

$$H = \left(\frac{nI}{2}\right)\left(\frac{1}{\sqrt{\left(\dfrac{2R}{L - 2x_1}\right)^2 + 1}} + \frac{1}{\sqrt{\left(\dfrac{2R}{L + 2x_1}\right)^2 + 1}}\right)$$

For $x_1 = 0$ we have:

$$H\big|_{x_1=0} = \frac{nI}{\sqrt{\left(\dfrac{2R}{L}\right)^2 + 1}} \qquad [A/m]$$

For $x_1 = L/2$ we have:

$$H\big|_{x_1=\frac{L}{2}} = \frac{nI}{\sqrt{\left(\dfrac{2R}{L}\right)^2 + 4}} \qquad [A/m]$$

For $x_1 = 0$ and $L \to \infty$ we have:

$$H\big|_{\substack{x_1=0 \\ L\to\infty}} = nI \qquad [A/m]$$

A.9.3 Ampere's Circuital Law

Within two months after Oersted published a memoir on his observations, Andre Marie Ampere (1775–1836) presented a complete mathematical formulation concerning the nature of the mechanical force exerted by a current-carrying conductor upon a magnetic needle and the mechanical force exerted by two current-carrying conductors (see Section A.9.1). With his mathematical mind, Ampere resolved to work out a mathematical theory that would embrace not only all the phenomena of magnetism then known, but also the complete theory of the science of electrodynamics.

By 1825, Ampere, amongst others, had developed an analytical (vector calculus) formulation for the (invisible) magnetic field established by an electric current. Whether such a vector field actually exists is not critically important. What is important, though, is the fact that this vector field model has successfully permitted predictions of observable magnetic phenomena for over a century and therefore leads us to accept this as a valid model for analytical purposes.

We have, in differential form, Ampere's fundamental law:

$$curl\ \vec{H} = \nabla \times \vec{H} = \vec{J} \quad [A/m^2] \qquad (A.9.3)$$

which states that whenever there exists an electric conduction current density vector $\vec{J}$, a magnetic vector $\vec{H}$ field is created. Conversely, in order to establish a magnetic vector $\vec{H}$ field we must have a current density vector field (even the magnetic field about permanent magnets is attributable to a net alignment of electron spins in the magnetic material).

We recall from vector algebra that for:

$$\vec{C} = \vec{A} \times \vec{B} \qquad (A.9.4)$$

$\vec{C}$ lies in a plane perpendicular to the plane formed by the vectors $\vec{A}$ and $\vec{B}$. On this basis, we recognize that the vector field lies in a plane perpendicular to the current density vector $\vec{J}$. Hence:

Associated with every electric conduction current there exists a magnetic intensity ($\vec{H}$) vector field, which lies in a plane perpendicular to the current density ($\vec{J}$) vector.

Via Stoke's theorem we may convert Equation A.9.3 into the integral form:

$$\oint \vec{H} \cdot d\ell = i \quad [A] \qquad (A.9.5)$$

which is known as Ampere's Circuital Law and which is generally much easier to deal with in most practical engineering problems. In other words:

A line integral, taken around a *closed* path in the plane of the vector $\vec{H}$ field, is numerically equal to the magnitude of the electric current enclosed by the path of integration (i.e., equal to the current passing through the area whose perimeter is the *closed* path of integration).

Example A.9.4

Consider an element of an infinitely long current-carrying conductor in free space illustrated in Figure A.9.5. On the basis of geometric symmetry, loci of constant are circles in the plane perpendicular to the current. For the field outside the conductor we have:

$$\oint \vec{H} \cdot d\ell = H \oint d\ell = H(2\pi r) = i \quad [A]$$

and:

$$H = \frac{i}{2\pi r} \quad [A/m]$$

In Example A.9.1 we employed the Biot-Savart law to obtain the H field for a finite length of conductor, and if we let $L \to \infty$, we obtain the same result as above. We conclude that when we know, a priori, that loci of constant H are simple geometric figures (such that the closed line integral around the path of constant H can easily be evaluated), it is much simpler to employ Ampere's circuital law than the Biot-Savart law.

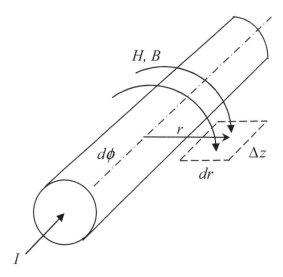

Figure A.9.5: Elemental Length of Infinitely Long Current-Carrying Conductor in Free Space

A.9.4   Magnetic Flux Density

Associated with the magnetic field intensity vector, there exists a magnetic flux density vector $\vec{B}$ according to:

$$\vec{B} = \mu\vec{H} \qquad [Wb/m^2] \qquad \text{(A.9.6)}$$

$$div\vec{B} = \nabla \cdot \vec{B} = 0 \qquad \text{(A.9.7)}$$

In other words,

> The magnetic flux density vector $\vec{B}$ field has no sources or sinks. Visualized in terms of lines of magnetic flux, flux lines are continuous and close upon themselves (unlike electrostatic flux lines, which emanate from positive charges and terminate upon negative charges).

The term μ (magnetic permeability) is a magnetic characteristic of the medium in which the magnetic field resides. It is common practice to write:

$$\mu = \mu_r\mu_0 \qquad [H/m] \qquad \text{(A.9.8)}$$

where:

$$\mu_0 = 4\pi \times 10^{-7} \quad [H/m]\text{[in free space]} \tag{A.9.9}$$

$$\mu_r \equiv \text{relative magnetic permeability} \tag{A.9.10}$$

It is of interest also to observe (since $div\ curl\ \vec{A} = 0$) that Equation A.9.3 leads to the conclusion that vector current density lines are also continuous and must close upon themselves (i.e., since $div\ curl\ \vec{H} = 0 = div\ \vec{J}$).

A.10   Faraday's Law

Michael Faraday (1791–1867), motivated by Oersted's discovery in 1819, became convinced, in 1824, that an electric current might be obtained by the motion of a magnet. After seven years of unsuccessful effort, he finally demonstrated the fact (1831) that a voltage (with consequent current) could be induced in a closed loop of wire by the relative motion between a magnetic field and the closed loop. Heinrich Friedrich Emil Lenz (1804–1865) discovered the fact (1834) that when the magnetic flux linking a closed circuit is changed, a current will flow in the closed circuit in a direction as to directly oppose the change in flux-linkages. In 1845, F.E. Neumann stated that the magnitude of the induced voltage is directly proportional to the time-rate-of-change of magnetic flux-linkages with the circuit. This relationship is expressed formally by:

$$\nabla \times \vec{E} = -\frac{d\vec{B}}{dt} \quad [V/m^2] \tag{A.10.1}$$

and in integral form:

$$e_i = \int \vec{E} \cdot d\ell = \frac{-d\left(\int\int \vec{B} \cdot d\vec{S}\right)}{dt} \quad [V] \tag{A.10.2}$$

Equation A.10.2 is actually quite profound and requires considerable expertise in vector calculus in order to be interpreted properly.

The magnetic flux through a cross-sectional area is rigorously defined by:

$$\phi = \int\int \vec{B} \cdot d\vec{S} \quad [Wb] \tag{A.10.3}$$

If the flux density happens to be constant (uniform) over the entire cross-sectional area in question (which will generally *not* be the case), then we have the simpler form:

$$\phi = BA \quad [Wb] \tag{A.10.4}$$

Therefore, on the surface one might interpret Equation A.10.2 as follows:

$$e_i = \frac{-d\left(\int\int \vec{B} \cdot d\vec{S}\right)}{dt} = -\frac{d\phi}{dt} \quad [V] \tag{A.10.5}$$

which would indeed be valid for a single-turn loop encircling a *total* magnetic flux ϕ.

Example A.10.1
Extending the results of Example A.9.4 we have, for the magnetic field in air external to the conductor element illustrated in Figure A.9.5:

$$B = \frac{\mu_0 i}{2\pi r} \quad [Wb/m^2]$$

and the magnetic flux about an elemental length Δz of conductor is:

$$\phi = \Delta z \int_{r_c}^{R} \frac{\mu_0 i}{2\pi r} \, dr = \frac{i \Delta z \mu_0}{2\pi} \ln \frac{R}{r_c} \quad [Wb]$$

where r_c is the radius of the conductor and R extends to infinity. Then:

$$e_i = -\left(\frac{\Delta z \mu_0}{2\pi} \ln \frac{R}{r_c} \right) \frac{di}{dt} = -L \frac{di}{dt} \quad [V]$$

which is the voltage drop along the length of conductor due to the time variation of the flux linking (surrounding) the conductor.

In Example A.10.1 it is observed that the flux-linkages are proportional to the current in the conductor and a term that reflects the permeability of the medium and the geometry of the conductor configuration. For the sake of convenience we introduce the practical concept of *inductance*, L, whereby L is a constant for a given conductor configuration as long as the permeability remains constant (no saturation).

Example A.10.2
Consider now the inside of a current-carrying conductor shown in Figure A.10.1. We have:

$$H = \frac{i}{2\pi r} \left(\frac{r}{r_c} \right)^2 \quad [A/m]$$

and:

$$\frac{d\phi}{\Delta z} = \frac{\mu_0 i}{2\pi r} \left(\frac{r}{r_c} \right)^2 \, dr \quad [Wb/m]$$

Now the tube of flux, $d\phi/\Delta z$, does *not* link (encircle, enclose) *all* of the current i. The tube of flux $d\phi/\Delta z$ induces a voltage only along those current filaments inside its perimeter. Therefore, we introduce:

$$\frac{d\lambda}{\Delta z} = \left(\frac{r}{r_c} \right)^2 \frac{d\phi}{\Delta z} \quad [V - s/m]$$

to represent the effective flux-linkages. Consequently:

$$\frac{\lambda}{\Delta z} = \int_{0}^{r_c} \frac{\mu_0 i}{2\pi r} \left(\frac{r}{r_c} \right)^4 \, dr = \frac{\mu_0 i}{8\pi} \quad [V - s/m]$$

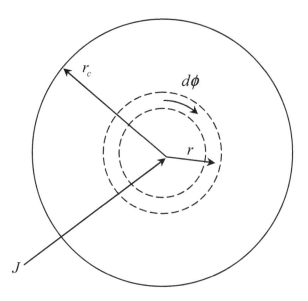

Figure A.10.1: Magnetic Field Inside Current-Carrying Conductor

and:

$$e_i = -\frac{\mu_0 i}{8\pi}\frac{di}{dt} = -L\frac{di}{dt} \qquad [V]$$

In Example A.10.2 we recognize that a time-varying closed tube of flux $d\phi$ induces a voltage only along those current filaments that lie inside its perimeter; i.e., $d\phi$ links (encircles) only a portion of the total current i. Therefore, we introduce the term $d\lambda$ for a tube of *flux-linkage*, as required by Faraday's law, in distinction to $d\phi$ for a tube of *flux*. In addition, it is helpful to employ units of *volt-seconds* when we are interested in *flux-linkages* as required by Faraday's law, and *Webers* when we are interested only in *total magnetic flux*, even though the units are physically equivalent.

Physicists appear to prefer to think in terms of "induced voltage" or "back EMF" for e_i:

$$e_i = -L\frac{di}{dt} \qquad [V] \qquad\qquad (A.10.6)$$

For electrical engineers, v is the voltage that would be measured at the terminals of an inductor for which:

$$v = L\frac{di}{dt} \qquad [V] \qquad\qquad (A.10.7)$$

with v considered as a voltage drop in the direction of current similar to an Ri drop across a resistor.

If we write Faraday's law in the form:

$$v = \frac{d\lambda}{dt} = L\frac{di}{dt} \qquad [V] \qquad\qquad (A.10.8)$$

with:

$$\lambda = \int v \, dt = Li \qquad [V \cdot s] \qquad \text{(A.10.9)}$$

we can avoid most pitfalls associated with the direct employment of Equation A.10.2.

Example A.10.3

Consider now two infinitely long parallel current-carrying conductors in free space. Let the spacing D between conductors be much greater than the radii r_c of the conductors. Extending the analysis in Example A.10.1, we have for the flux linking the first conductor:

$$\lambda_1 = \Delta z \int_{r_c}^{R} \frac{\mu_0 i_1}{2\pi r} \, dr + \Delta z \int_{D}^{R} \frac{\mu_0 i_2}{2\pi r} \, dr$$

$$= \Delta z \left[\left(\frac{\mu_0}{2\pi} \ln \frac{R}{r_c} \right) i_1 + \left(\frac{\mu_0}{2\pi} \ln \frac{R}{D} \right) i_2 \right]$$

and for the flux linking the second conductor:

$$\lambda_2 = \Delta z \left[\left(\frac{\mu_0}{2\pi} \ln \frac{R}{r_c} \right) i_2 + \left(\frac{\mu_0}{2\pi} \ln \frac{R}{D} \right) i_1 \right]$$

If the two currents are equal in magnitude but oppositely directed, then:

$$\lambda_1 = -\lambda_2 = \Delta z \left(\frac{\mu_0}{2\pi} \ln \frac{D}{r_c} \right) i \qquad [V \cdot s]$$

and:

$$e_{i1} = -e_{i2} = -\Delta z \left(\frac{\mu_0}{2\pi} \ln \frac{D}{r_c} \right) \frac{di}{dt} = -L \frac{di}{dt}$$

We observe that the inductance L in Example A.10.1 becomes infinitely large when R extends to infinity. This unrealistic result is due to the fact that we have considered only a single current-carrying conductor in free space, which violates the continuity of current law given in Section A.9.4. The inductance L in Example A.10.3, on the other hand, has a finite value because we accounted for a go and return current in the system.

Caution must be exercised, therefore, in interpreting the results of an analysis. Whereas the analysis in Example A.10.1 and the analysis via the Biot-Savart law in Example A.9.1 are mathematically correct, the results are only a portion of the complete result. The correct complete result is obtained only when the continuity of current law is also satisfied.

A.11   Magnetic Energy

Associated with a magnetic field there is magnetic energy. At any point in a magnetic field we have an energy density defined by:

$$w_{mag} = \frac{BH}{2} \qquad [J/m^3] = [N/m^2] \qquad \text{(A.11.1)}$$

For a perfectly uniform magnetic field we have:

$$W_{mag} = w_{mag} A l \quad [J] \qquad (A.11.2)$$

where A is the cross-sectional area through which the magnetic flux flows, and l is the length of the magnetic flux path. We also have:

$$\phi = BA = \mu_0 HA \quad [Wb] \qquad (A.11.3)$$

so that we may write:

$$W_{mag} = \frac{\phi^2}{2} \left(\frac{l}{\mu_0 A} \right) \quad [J] \qquad (A.11.4)$$

When the magnetic field is *not* uniform, the total magnetic energy within a given region must be evaluated upon the basis of *integration*.

Example A.11.1

Extending the results of Example A.10.1 we have, for the magnetic field in air *external* to the conductor element illustrated in Figure A.9.5:

$$B = \frac{\mu_0 i}{2\pi r} \quad [Wb/m^2]$$

so that:

$$w_{mag} = \frac{\mu_0}{2} \left(\frac{i}{2\pi r} \right)^2 \quad [J/m^3]$$

and for an elemental length of Δz of conductor:

$$W_{mag} = \Delta z \int_{r_c}^{R} w_{mag} \, 2\pi r \, dr$$

$$= \frac{i^2}{2} \left(\frac{\Delta z \, \mu_0}{2\pi} \ln \frac{R}{r_c} \right) \quad [J]$$

We may also write this in the forms:

$$W_{mag} = \frac{i^2 L}{2} = \frac{\phi^2}{2L} = \frac{\phi i}{2} \quad [J]$$

Example A.11.2

Extending the results of Example A.10.2 we have, for the magnetic field inside the current-carrying conductor:

$$B = \frac{\mu_0 i}{2\pi r}\left(\frac{r}{r_c}\right)^2 \quad [Wb/m^2]$$

so that:

$$w_{mag} = \frac{\mu_0}{2}\left(\frac{i}{2\pi r}\right)^2\left(\frac{r}{r_c}\right)^4 \quad [J/m^3]$$

and for an elemental length of Δz of conductor:

$$W_{mag} = \Delta z \int_{r_c}^{R} w_{mag}\, 2\pi r\, dr$$

$$= \frac{i^2}{2}\left(\frac{\Delta z\, \mu_0}{8\pi}\right) \quad [J]$$

We may also write this in the forms:

$$W_{mag} = \frac{i^2 L}{2} = \frac{\lambda^2}{2L} = \frac{\lambda i}{2} \quad [J]$$

In retrospect, the magnetic energy for Examples A.11.1 and A.11.2 could have been evaluated in another manner. We have found that we could write:

$$\lambda = Li \quad [V \cdot s] \tag{A.11.5}$$

where L is independent of i. According to Faraday's law we always have:

$$v = \frac{d\lambda}{dt} \quad [V] \tag{A.11.6}$$

Then:

$$p = vi = i\frac{d\lambda}{dt} \quad [W] \tag{A.11.7}$$

so that the associated energy is given by:

$$W_{mag} = \int p\, dt = \int i\, d\lambda = L\int i\, dt = \frac{Li^2}{2} \quad [J] \tag{A.11.8}$$

Example A.11.3

In evaluating the magnetic energy associated with the configuration in Example A.10.3 we prefer to employ the procedure outlined above. Thus:

$$p = v_1 i_1 + v_2 i_2$$

with:

$$v_1 = \frac{d\lambda_1}{dt} = L_{11}\frac{di_1}{dt} + L_{12}\frac{di_2}{dt}$$

$$v_2 = \frac{d\lambda_2}{dt} = L_{21}\frac{di_1}{dt} + L_{22}\frac{di_2}{dt}$$

Evaluating energy:

$$W_{mag} = \int (v_1 i_1 + v_2 i_2)\, dt$$

$$= \int (L_{11}i_1\, di_1 + L_{22}i_2\, di_2 + L_{12}i_2\, di_1 + L_{21}i_1\, di_2)$$

$$= \frac{L_{11}i_1^2}{2} + \frac{L_{22}i_2^2}{2} + L_{12}i_1 i_2 \qquad [J]$$

since $L_{12} = L_{21}$.

For the conditions in Example A.10.3 we have:

$$L_{11} = L_{22} = \frac{\Delta z\, \mu_0}{2\pi} \ln \frac{R}{r_c} \qquad [H]$$

$$L_{12} = L_{21} = \frac{\Delta z\, \mu_0}{2\pi} \ln \frac{R}{D} \qquad [H]$$

and $i_2 = -i_1 = i$ so that:

$$W_{mag} = i^2\, \frac{\Delta z\, \mu_0}{2\pi} \ln \frac{D}{r_c} \qquad [J]$$

A.12   Magnetic Forces

In a *conservative* system we have:

$$dW_{elect-in} = dW_{losses} + dW_{mag-stored} + dW_{mech-out} \qquad (A.12.1)$$

and in a *lossless conservative* system we have:

$$dW_{elect-in} = dW_{mag-stored} + dW_{mech-out} \qquad (A.12.2)$$

With:

$$dW_{elect-in} = vi\, dt = i\, d\lambda \qquad (A.12.3)$$

$$dW_{mag-stored} = \left(\frac{i\, d\lambda + \lambda\, di}{2} \right) \qquad (A.12.4)$$

so that:

$$dW_{mech-out} = \left(\frac{i\,d\lambda + \lambda\,di}{2} \right) \qquad\qquad (A.12.5)$$

In all events:

$$F = \frac{dW_{mech-out}}{dx} \qquad [N] \qquad\qquad (A.12.6)$$

Now the developed force F is independent of whether $di = 0$ or $d\lambda = 0$. We elect to have $di = 0$ so that:

$$F = \frac{dW_{mag-stored}}{dx} \qquad [N] \qquad\qquad (A.12.7)$$

According to the results from Example A.11.3 we have:

$$W_{mag-stored} = i^2 \frac{\Delta z\,\mu_0}{2\pi} \ln \frac{D}{r_c} \qquad [J] \qquad\qquad (A.12.8)$$

We determine the force acting upon the two parallel current-carrying conductors to be:

$$F = \frac{dW_{mag-stored}}{dD} = i^2 \frac{\Delta z\,\mu_0}{2\pi D} \qquad [N] \qquad\qquad (A.12.9)$$

The force will be in such a direction as to increase D. This is confirmed by Ampere's force law stated in Section A.9.1, "two parallel current elements, oppositely directed, will repel each other."

We may express the force in another form. The current in the first conductor produces a B field at the second conductor of magnitude:

$$B = \frac{\mu_0\,i}{2D} \qquad [Wb/m^2] \qquad\qquad (A.12.10)$$

which reacts with the current in the second conductor to produce a force:

$$F = B\,i\,\Delta z \qquad [N] \qquad\qquad (A.12.11)$$

A conductor, of length l, carrying a current i, located in a uniform B field, will experience a force $F = B\,i\,l$.

The right-hand rule will indicate the direction of F. Those familiar with electromagnetic field theory will recognize it as a generalized form of the Lorentz Force Equation.

A.13   References

[1] M. Abraham and R. Becker. *The Classical Theory of Electricity and Magnetism*. Hafner Pub. Co., Inc., New York, NY, 1949.

[2] S.S. Atwood. *Electric and Magnetic Fields*. John Wiley and Sons, Inc., New York, NY, 1941.

[3] L. V. Bewley. *Two-Dimensional Fields in Electrical Engineering*. Dover Publications, Inc., New York, NY.

[4] L. V. Bewley. *Flux Linkages and Electromagnetic Induction*. Dover Publications, Inc., New York, NY, 1964.

[5] B. I. Bleaney and B. Bleaney. *Electricity and Magnetism*. Oxford University Press, London, UK, 1965.

[6] C. P. Enz. *Pauli Lectures on Physics: 1 - Electrodynamics*. The MIT Press, Cambridge, MA, 1973.

[7] G. P. Harnwell. *Principles of Electricity and Electromagnetism*. McGraw-Hill Book Co.,Inc., New York, NY, 1949.

[8] William H. Hayt. *Engineering Electromaagnetics, 5th Edition*. McGraw-Hill Book Co., Inc., New York, NY, 1989.

[9] Oliver Heaviside. *Electromagnetic Theory, Volume I*. The Electrician Pub. Co. Ltd., London, UK, 1893.

[10] Oliver Heaviside. *Electromagnetic Theory, Volume II*. The Electrician Pub. Co. Ltd., London, UK, 1899.

[11] Oliver Heaviside. *Electromagnetic Theory, Volume III*. The Electrician Pub. Co. Ltd., London, UK, reprinted 1922.

[12] R. W. P. King. *Fundamental Electromagnetic Theory*. Dover Publications, Inc., New York, NY, 1963.

[13] K. Kupfmuller. *Einfuhrung in die theoretische Elektrotechnik*. Springer-Verlag, Berlin, Germany, 1957.

[14] M. Mason and W. Weaver. *The Electromagnetic Field*. Dover Pub- lications, Inc., New York, NY, 1929.

[15] J.C. Maxwell. *A Treatise on Electricity and Magnetism (Two Volumes)*. Dover Publications, Inc., New York, NY, 1954.

[16] J. Meixner. *Vorlesungen uber Maxwellsche Theorie*. Techn. U. Aachen, Aachen, Germany, 1954.

[17] P. Moon and D. E. Spencer. *Foundations of Electrodynamics*. D. Van Nostrand Company, Inc., Princeton, NJ, 1960.

[18] P. Moon and D. E. Spencer. *Field Theory For Engineers*. D. Van Nostrand Company, Inc., Princeton, NJ, 1961.

[19] F. Ollendorff. *Die Grundlagen der Hochfrequenztechnik*. Springer- Verlag, Berlin, Germany, 1926.

[20] F. Ollendorff. *Berechnung magnetischer Felder*. Springer-Verlag, Vienna, Austria, 1952.

[21] L. Page and N. I. Adams. *Principles of Electricity*. D. Van Nostrand Co., Inc., New York, NY, 1949.

[22] Brother Potamian and J. J. Walsh. *Makers Of Electricity*. Fordham University Press, New York, NY.

[23] H. H. Skilling. *Fundamentals of Electric Waves*. John Wiley and Sons, Inc., New York, NY, 1948.

[24] J. C. Slater and N. H. Frank. *Electromagnetism*. McGraw-Hill Book Co.,Inc., New York, NY, 1947.

[25] A. Sommerfeld. *Vorlesungen uber theoretische Physik - Elektro- dynamik, Vol. 3*. Ak. Verl.-Ges., Geest and Portig K.-G., Leipzig, Germany, 1954.

[26] J. A. Stratton. *Electromagnetic Theory*. McGraw-Hill Book Co., Inc., New York, NY, 1941.

[27] E. Weber. *Electromagnetic Theory*. Dover Publications, Inc., New York, NY, 1965.

[28] W. H. Westphal. *Physik - Ein Lehrbuch*. Springer-Verlag, Berlin, Germany, 1953.

Appendix B

Concept of Flux-Linkage and Inductance

According to Faraday's law, a voltage is induced in a circuit when there is a change of flux-linkages with the circuit. In many instances, the numerical value for flux-linkages is the same as for magnetic flux, at most multiplied by an obvious number of turns in a winding. The distinction, in concept, between flux and flux-linkages can, at other times, be considerably more subtle.

B.1   Faraday's Law

,

Faraday's Law states that:

$$v = \frac{d\lambda}{dt} \quad [V] \tag{B.1.1}$$

The magnetic field is due to a current, i, and for a purely magnetic circuit (i.e., no losses), the electrical power is given by:

$$p = vi = i\frac{d\lambda}{dt} \quad [W] \tag{B.1.2}$$

and the energy delivered to the magnetic circuit:

$$W_{mag} = \int i\frac{d\lambda}{dt} = \int i\,d\lambda \quad [J] \tag{B.1.3}$$

which represents the magnetic energy stored in the magnetic circuit.

Consider a current-carrying conductor segment shown in Figure B.1.1, wherein the current density J is constant over the cross section of the conductor, and for which we wish to determine the net flux-linkages within the conductor. At a radius $r < r_c$, we have by Ampere's Law:

$$\oint \vec{H} \cdot d\ell = 2\pi r H = i\left(\frac{r}{r_c}\right)^2 \quad [A] \tag{B.1.4}$$

and:

$$H = \frac{i}{2\pi r}\left(\frac{r}{r_c}\right)^2 \quad [A/m] \tag{B.1.5}$$

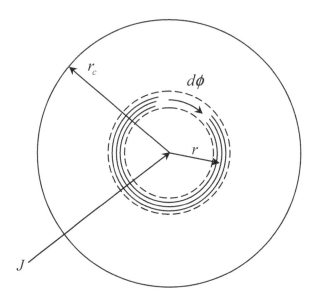

Figure B.1.1: Flux-Linkages within a Current-Carrying Conductor

Then:

$$B = \frac{\mu_0 i}{2\pi r} \left(\frac{r}{r_c}\right)^2 \qquad [Wb/m^2] \tag{B.1.6}$$

and, per unit length conductor:

$$\frac{d\phi}{\Delta z} = \frac{\mu_0 i}{2\pi r} \left(\frac{r}{r_c}\right)^2 \, dr \qquad [Wb/m] \tag{B.1.7}$$

If we integrate Equation B.1.7 from $r = 0$ to $r = r_c$, we obtain:

$$\frac{\phi}{\Delta z} = \frac{\mu_0 i}{2\pi} \int_0^{r_c} \left(\frac{r}{r_c^2}\right) \, dr = \frac{\mu_0 i}{4\pi} \qquad [Wb/m] \tag{B.1.8}$$

resulting in an *abstraction* with *no practical significance*.

The effective *flux-linkages* are given by:

$$\frac{d\lambda}{\Delta z} = \left(\frac{r}{r_c}\right)^2 \frac{d\phi}{\Delta z} \qquad [V \cdot s/m] \tag{B.1.9}$$

to account for the fact the $d\phi$ links (encircles, encloses) only the current within the $d\phi$ path. Then:

$$\frac{\lambda}{\Delta z} = \frac{\mu_0 i}{2\pi} \int_0^{r_c} \left(\frac{r^2}{r_c^4}\right) \, dr = \frac{\mu_0 i}{8\pi} \qquad [V \cdot s/m] \tag{B.1.10}$$

Because of the linear relationship between λ and i in Equation B.1.10, we have:

$$\frac{d\lambda}{\Delta z} = \frac{\mu_0}{8\pi} \, di \qquad [V \cdot s/m] \tag{B.1.11}$$

and on the basis of Equation B.1.3:

$$\frac{W_{mag}}{\Delta z} = \frac{\mu_0}{16\pi} i^2 = \frac{\lambda i}{2} \qquad [J/m] \tag{B.1.12}$$

B.2 Magnetic Energy Density Approach

Equation B.1.9 is based upon the *assertion* that $d\lambda$ represents the effective flux-linkages, consistent with Faraday's law, and not $d\phi$. The basis for this assertion is perhaps somewhat elusive, and we may avoid this confrontation entirely by proceeding from an energy density consideration.

Beginning with the general expression for magnetic energy density:

$$w_{mag} = \frac{BH}{2} = \frac{\mu_0}{2} H^2 \qquad [J/m^3] \tag{B.2.1}$$

and for our example (Equation B.1.5):

$$H = \frac{i}{2\pi r} \left(\frac{r}{r_c}\right)^2 \qquad [A/m]$$

and:

$$W_{mag} = \int_0^{r_c} w_{mag}\, 2\pi r\, dr \qquad [J] \tag{B.2.2}$$

we obtain:

$$\frac{W_{mag}}{\Delta z} = \frac{\mu_0}{16\pi} i^2 = \frac{\lambda i}{2} \qquad [J/m] \tag{B.2.3}$$

which is the same as Equation B.1.12. Our assertion concerning $d\lambda$ is therefore validated.

B.3 Inductance Concept

For magnetic fields in nonferromagnetic materials, there is generally a linear relationship between λ and i. It has therefore become customary to write:

$$\lambda = Li \qquad [V \cdot s] \tag{B.3.1}$$

for a singly excited magnetic circuit, where L [Henrys] is called the *inductance of the winding*.

For example, the inductance per unit length of a single current-carrying conductor in free space is:

$$L = \frac{\mu_0}{2\pi} \left(\frac{1}{4} + \ln \frac{R}{r_c}\right) = \frac{\mu_0}{2\pi} \ln \frac{r}{GMR} \qquad [H/m] \tag{B.3.2}$$

and the inductance for magnetic circuits for which the magnetic Ohm's law is applicable is:

$$L = N^2 \wp = \frac{N^2 \mu_0 A}{\ell} \qquad [H] \tag{B.3.3}$$

No matter how complex the expression for L may be, it is always directly proportional to the magnetic permeability of the material within which the magnetic field resides, the remainder being a function only of the geometry of the magnetic circuit (i.e., effective measurable lengths).

The inductance concept is of practical value in electric circuit analysis where a lumped L element, with an assigned numerical value, is employed in the circuit to reflect the appropriate magnetic circuit characteristics. Then, Faraday's law takes the form:

$$V = \frac{d\lambda}{dt} = \frac{d(Li)}{dt} = L\frac{di}{dt} + i\frac{dL}{dt} \qquad [V] \tag{B.3.4}$$

where $L\, di/dt$ represents the change in flux-linkages due to a time variation in current, and $i\, dL/dt$ represents the change in flux-linkages due to a time variation in L. Some authors refer to the former as voltage induced by *transformer action* and the latter as voltage induced by *flux-cutting action*. This distinction between the two is a matter of convenience rather than a matter of physical reality, as we show in Appendix D.

The associated magnetic energy for a *singly-excited* magnetic circuit may be written in the form:

$$W_{mag} = \frac{\lambda i}{2} = \frac{Li^2}{2} \quad [J] \quad \quad (\text{B.3.5})$$

The associated magnetic energy for a multiply-excited magnetic circuit may be written in the form:

$$W_{mag} = \frac{1}{2} \sum_{j=1}^{N} \lambda_j \, i_j$$

$$= \frac{1}{2} \sum_{j=1}^{N} i_j^2 \, L_{jj} + \frac{1}{2} \sum_{\substack{k=1 \\ k \neq j}}^{N} i_j \, i_k \, L_{jk} \quad [J] \quad \quad (\text{B.3.6})$$

Appendix C

Electromagnetic Fields above a Perfectly Conducting Plane

It is of interest to determine the electric charge distribution on a perfectly conducting plane due to a charged line above and parallel to it. Analogously, it is of interest to examine in detail the case of a single current-carrying conductor parallel to and above an infinite, perfectly conducting plane, and to determine the distribution of the return current in the plane.

C.1 Electrostatic Scalar Potential

We begin by considering a parallel conductor system in free space, illustrated in Figure C.1.1. The conductors are of identical dimensions and carry an equal charge of opposite polarity. Consider an additional parallel filament located at $P(x, y)$. The scalar potential at the filament is given by:

$$v(x, y) = \frac{q}{2\pi\epsilon_0} \ln \frac{d_{x2}}{d_{x1}} = \frac{q}{4\pi\epsilon_0} \ln \left(\frac{d_{x2}}{d_{x1}} \right)^2 \quad [V]$$

$$= \frac{q}{4\pi\epsilon_0} \ln \frac{x^2 + (h + y)^2}{x^2 + (h - y)^2} \quad [V] \tag{C.1.1}$$

We may rewrite this equation in the form:

$$e^{2\eta} = e^{\frac{4\pi\epsilon_0 \, v(x,y)}{q}} = \left(\frac{d_{x2}}{d_{x1}} \right)^2 \tag{C.1.2}$$

with:

$$\eta = \frac{2\pi\epsilon_0 \, v(x, y)}{q} \tag{C.1.3}$$

Then:

$$\left[x^2 + (h - y)^2 \right] e^{\eta} = \left[x^2 + (h + y)^2 \right] e^{-\eta} \tag{C.1.4}$$

and:

$$x^2 \left(e^{\eta} - e^{-\eta} \right) + y^2 \left(e^{\eta} - e^{-\eta} \right) - 2hy \left(e^{\eta} + e^{-\eta} \right) + h^2 \left(e^{\eta} - e^{-\eta} \right) = 0 \tag{C.1.5}$$

and finally:

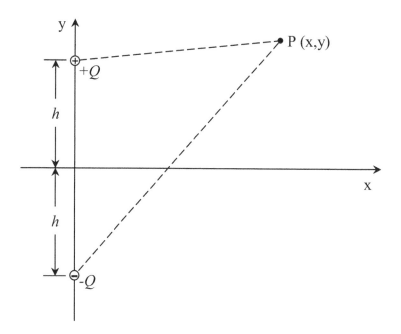

Figure C.1.1: Parallel-Charged Conductors in Free Space

$$x^2 + (y - h \coth \eta)^2 = (\coth^2 \eta - 1) h^2 = \left(\frac{h}{\sinh \eta}\right)^2 \qquad \text{(C.1.6)}$$

which represents a family of nonconcentric circles. The circles represent loci of *constant scalar potential*, and the separation distance normal to and between circles is a measure of the *electric field intensity*. Figure C.1.2 shows the circles of constant scalar potential overlaid on top of the field lines.

The scalar potential at the top conductor is given by:

$$v(0, h) = \frac{q}{2\pi\epsilon_0} \ln \frac{2h}{r} \qquad [V] \qquad \text{(C.1.7)}$$

so that:

$$\eta_1 = \frac{2\pi\epsilon_0 \, v(0, h)}{q} = \ln \frac{2h}{r} \qquad \text{(C.1.8)}$$

The scalar potential at the midplane is zero so that we might select:

$$\eta_x = \frac{\eta_1}{10}, \frac{2\eta_1}{10}, \frac{3\eta_1}{10}, \dots \frac{10\eta_1}{10} \qquad \text{(C.1.9)}$$

so that the set of circles represents 10% increments of the total scalar potential between the top conductor and the midplane.

C.2 Vector Electric Flux Density

Now consider the same parallel conductor system in free space, illustrated in Figure C.1.1. The conductors are of identical dimensions and carry an equal charge of opposite polarity. Consider an

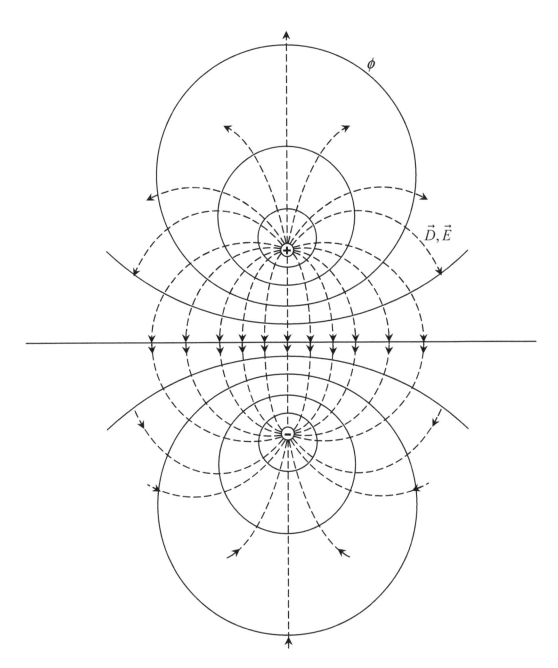

Figure C.1.2: Parallel-Charged Line Conductors in Free Space

additional parallel filament located at P(x,y). The net vector electric flux density at the filament is given by:

$$\vec{D}_{1x} = \frac{q}{2\pi\sqrt{x^2 + (h-y)^2}}\frac{x - j(h-y)}{\sqrt{x^2 + (h-y)^2}}$$

$$= \left(\frac{q}{2\pi}\right)\left[\frac{1}{x + j(h-y)}\right] \qquad [C/m^2] \tag{C.2.1}$$

$$\vec{D}_{2x} = \frac{-q}{2\pi\sqrt{x^2 + (h+y)^2}}\frac{x + j(h+y)}{\sqrt{x^2 + (h+y)^2}}$$

$$= \left(\frac{q}{2\pi}\right)\left[\frac{1}{-x + j(h+y)}\right] \qquad [C/m^2] \tag{C.2.2}$$

$$\vec{D}(x,y) = \vec{D}_{x1} + \vec{D}_{x2} = \left(\frac{q}{2\pi}\right)\left[\frac{2h}{2xy + j(x^2 - y^2 + h^2)}\right] \qquad [C/m^2] \tag{C.2.3}$$

From another point of view:

$$\vec{E}(x,y) = -grad\ v(x,y) = -\left[\frac{\partial v(x,y)}{\partial x} + j\frac{\partial v(x,y)}{\partial y}\right]$$

$$= \left(\frac{-q}{2\pi\epsilon_0}\right)\left[\frac{x + j(h+y)}{x^2 + (h+y)^2} - \frac{x - j(h-y)}{x^2 + (h-y)^2}\right]$$

$$= \left(\frac{-q}{2\pi\epsilon_0}\right)\left[\frac{1}{x - j(h+y)} - \frac{1}{x + j(h-y)}\right] \qquad [V/m] \tag{C.2.4}$$

so that:

$$\vec{E}(x,y) = \left(\frac{q}{2\pi\epsilon_0}\right)\left[\frac{2h}{2xy + j(x^2 - y^2 + h^2)}\right] \qquad [V/m] \tag{C.2.5}$$

which confirms Equation C.2.3 on the basis of:

$$\vec{D}(x,y) = \epsilon_0\vec{E}(x,y) \tag{C.2.6}$$

At the midplane between the two conductors we have:

$$\vec{D}(x,0) = -j\left(\frac{q}{2\pi}\right)\frac{2h}{(x^2 + h^2)} \qquad [C/m^2] \tag{C.2.7}$$

Now, according to Gauss's law, the integral taken around any circle shown in Figure C.1.2, for example, of the component of $D(x,y)$ normal to the circle is equal to the total charge per unit length enclosed within the path of integration. The midplane is, of course, one of these circles in the limit. Thus:

$$\int_{-\infty}^{+\infty}|D(x,0)|\ dx = \frac{q}{\pi}\int_{-\infty}^{+\infty}\frac{h}{x^2 + h^2}\ dx = \frac{q}{\pi}\left|\arctan\frac{x}{h}\right|_{-\infty}^{+\infty}$$

$$= \frac{q}{\pi}\left[\left(\frac{\pi}{2}\right) - \left(-\frac{\pi}{2}\right)\right] = q \qquad [C/m] \tag{C.2.8}$$

which verifies that Equation C.2.7 is correct.

Let us now insert a perfectly conducting metal sheet, infinite in area, and of infinitesimally small thickness 2δ, at the midplane between the two cylindrical conductors. Due to induction there will be charge separation within the metal sheet such that there will be a negative and positive charge distribution $q(x, \pm\delta)$ along the top and bottom surfaces, respectively, of the metal sheet as shown in Figure C.2.1. What must be the nature of $q(x, \pm\delta)$ such that $D(x, 0) = 0$? According to Gauss's law, the contribution of $q(x, \pm\delta)$ must be such that the net surface charge distribution is equal and opposite to the charge on the cylindrical conductor above it. Thus:

$$\int_{-\infty}^{+\infty} \left[|D(x, \pm\delta)| + q(x, \pm\delta) \right] dx = 0 \qquad [C/m] \tag{C.2.9}$$

so that we require:

$$q(x, \pm\delta) = -|D(x, \pm\delta)| = -\left(\frac{q}{\pi}\right) \frac{h}{x^2 + h^2} \qquad [C/m^2] \tag{C.2.10}$$

Observe that under the above conditions, the electric field above and below the thin metal sheet has not been disturbed since the net charge in the sheet is zero as viewed from either side of the metal sheet. If we consider the upper-charged conductor and the top-half thickness of the metal sheet, together with its charge distribution $q(x, \pm\delta)$, we have the case of a charged conductor above and parallel to a perfectly conducting plane with opposite charge distribution in the conducting plane.

Conversely, if we have the case of a charged conductor above and parallel to a perfectly conducting plane of infinite extent with induced charge in the conducting plane, we evaluate the electric field by placing an image conductor at an equal distance below the plane and carrying a charge of equal magnitude but of opposite sign.

C.3   Lines of Constant Electric Flux

Lines of electric flux have the same direction as the vector electric field intensity, $\vec{E}$. Since $\vec{E} = -grad\,v = -\nabla v$, we expect lines of constant electric flux to be normal (orthogonal) to the circles shown in Figure C.2.1. These are again a family of nonconcentric circles given by:

$$(x + h\cot\psi)^2 + y^2 = \left(\frac{h}{\sin\psi}\right)^2 \tag{C.3.1}$$

The two families of orthogonal circles defined by Equations C.1.6 and C.3.1 are characterized by the coordinates h and y and are called *bicylindrical coordinates* [1].

At $y = 0$ we have:

$$x = h\frac{(1 - \cos\psi)}{\sin\psi} = h\tan\frac{\psi}{2} \tag{C.3.2}$$

On the basis of Equation C.2.10 we have:

$$q(\Delta x, 0) = \int_0^{\Delta x} |D(x, 0)|\,dx = \frac{q}{\pi} \int_0^{\Delta x} \frac{h}{x^2 + h^2}\,dx$$
$$= \frac{q}{h} \left| \arctan\frac{x}{h} \right|_0^{\Delta x} \qquad [C] \tag{C.3.3}$$

so that:

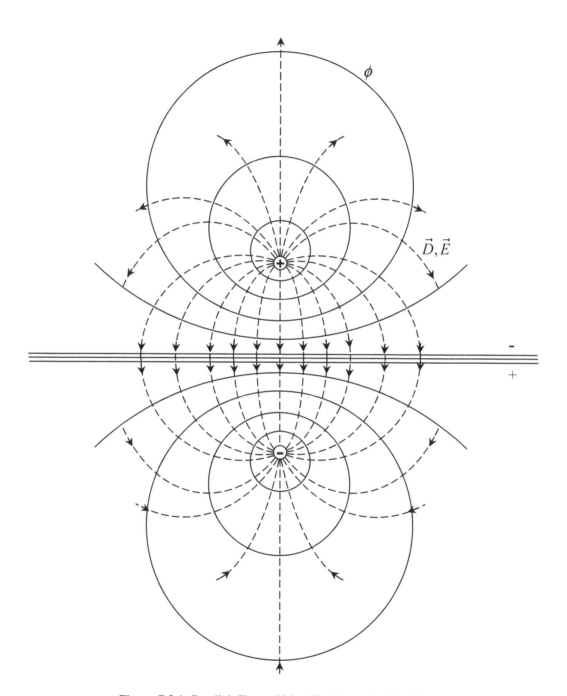

Figure C.2.1: Parallel-Charged Line Conductors in Free Space

$$\Delta x = h \tan \left[\frac{q(\Delta x, 0)}{q} \pi \right] = h \tan \frac{\psi}{2} \tag{C.3.4}$$

and:

$$\psi = \frac{q(\Delta x, 0)}{q} 2\pi \tag{C.3.5}$$

If $x \to \Delta\infty$:

$$q(\Delta\infty, 0) = \frac{q}{\pi} \left| \arctan \frac{x}{h} \right|_0^{\Delta\infty} = \frac{q}{\pi} \left(\frac{\pi}{2} \right) = \frac{q}{2} \qquad [C] \tag{C.3.6}$$

and when $\Delta x = \Delta h$:

$$q(\Delta h, 0) = \frac{q}{\pi} \left| \arctan \frac{x}{h} \right|_0^{\Delta h} = \frac{q}{\pi} \left(\frac{\pi}{4} \right) = \frac{q}{4} \qquad [C] \tag{C.3.7}$$

C.4 Magnetic Flux-Linkages

We begin by considering a parallel conductor system in free space, illustrated in Figure C.4.1. The conductors are of identical dimensions and carry equal currents in opposite directions. Consider an additional parallel filament located at P(x, y). The net flux linking the filament is given by:

$$\lambda_x = \frac{\mu_0 i}{2\pi} \ln \frac{d_{x2}}{d_{x1}} = \frac{\mu_0 i}{4\pi} \ln \left(\frac{d_{x2}}{d_{x1}} \right)^2 \qquad [V] \tag{C.4.1}$$

We may rewrite this equation in the form:

$$e^{2\eta} = e^{\frac{4\pi\lambda_x}{\mu_0 i}} = \left(\frac{d_{x2}}{d_{x1}} \right)^2 \tag{C.4.2}$$

with:

$$\eta = \frac{2\pi\lambda_x}{\mu_0 i} \tag{C.4.3}$$

Then:

$$\left[x^2 + (y - h)^2 \right] e^{\eta} = \left[x^2 + (y + h)^2 \right] e^{-\eta} \tag{C.4.4}$$

and:

$$x^2 \left(e^{-\eta} - e^{\eta} \right) + y^2 \left(e^{-\eta} - e^{\eta} \right) + 2hy \left(e^{-\eta} + e^{\eta} \right) + h^2 \left(e^{-\eta} - e^{\eta} \right) = 0 \tag{C.4.5}$$

and finally:

$$x^2 + (y - h \coth \eta)^2 = \left(\coth^2 \eta - 1 \right) h^2 = \left(\frac{h}{\sinh \eta} \right)^2 \tag{C.4.6}$$

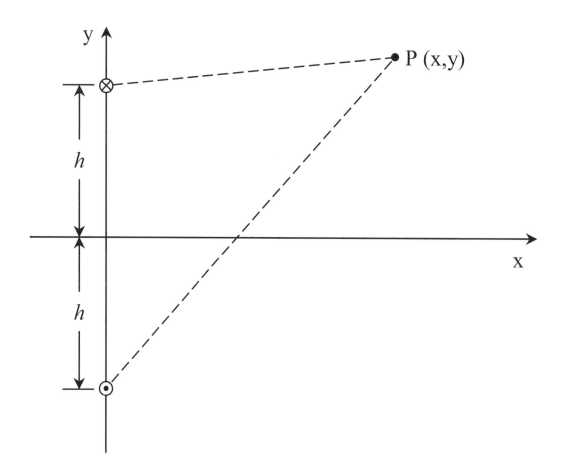

Figure C.4.1: Parallel Current-Carrying Conductors in Free Space

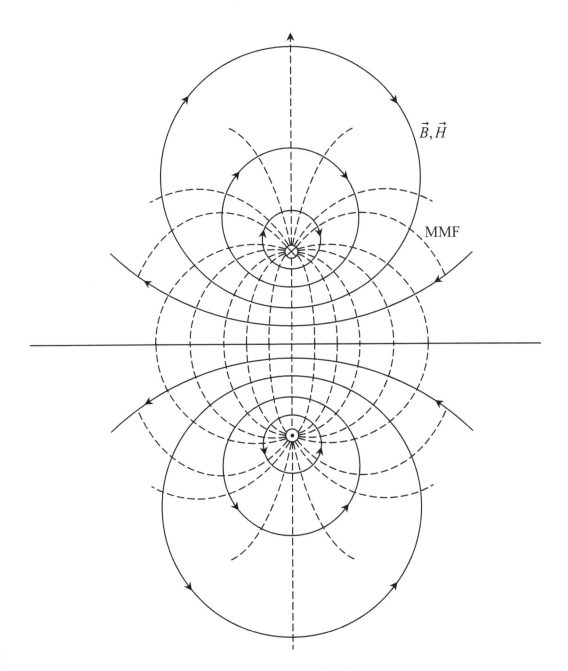

Figure C.4.2: Parallel Current-Carrying Line Conductors in Free Space

which represents a family of nonconcentric circles.

Since the circles represent loci of constant flux-linkages, each circle represents a magnetic line of flux. Thus, the circles also give the vector direction of the magnetic field intensity H, but the circles are not loci of constant magnitude of H. Figure C.4.2 illustrates the circles of flux with Magnetomotive Force (MMF) lines overlaid.

The total flux (per unit length of conductor) linking the top conductor is given by:

$$\lambda_1 = \frac{\mu_0 \, i}{2\pi} \, \ln \frac{2h}{GMR} \qquad [V \cdot s/m] \tag{C.4.7}$$

and if we ignore the flux-linkages inside the conductor, then:

$$\lambda_1 = \frac{\mu_0 \, i}{2\pi} \, \ln \frac{2h}{r} \qquad [V \cdot s/m] \tag{C.4.8}$$

so that:

$$\eta_1 = \frac{2\pi \lambda_2}{\mu_0 \, i} = \ln \frac{2h}{r} \tag{C.4.9}$$

We might select:

$$\eta_x = \frac{\eta_1}{10}, \, \frac{2\eta_1}{10}, \, \frac{3\eta_1}{10}, \cdots \frac{10\eta_1}{10} \tag{C.4.10}$$

so that between each adjacent set of circles represents 10% of the total flux-linkages.

C.5   Magnetic Field Intensity

Now consider the same parallel conductor system in free space, illustrated in Figure C.4.1. The conductors are of identical dimensions and carry equal currents in opposite directions. Consider an additional parallel filament located at P(x, y). The net vector magnetic field intensity at the filament is given by:

$$\vec{H}_{1x} = \frac{-ji}{2\pi \sqrt{x^2 + (h-y)^2}} \frac{x - j(h-y)}{\sqrt{x^2 + (h-y)^2}}$$

$$= \left(\frac{-ji}{2\pi} \right) \left[\frac{1}{x + j(h-y)} \right] \qquad [A/m] \tag{C.5.1}$$

$$\vec{H}_{2x} = \frac{+ji}{2\pi \sqrt{x^2 + (h+y)^2}} \frac{x + j(h+y)}{\sqrt{x^2 + (h+y)^2}}$$

$$= \left(\frac{-ji}{2\pi} \right) \left[\frac{1}{-x + j(h+y)} \right] \qquad [A/m] \tag{C.5.2}$$

$$\vec{H}(x,y) = \vec{H}_{x1} + \vec{H}_{x2} = \left(\frac{ji}{2\pi} \right) \left[\frac{2h}{2xy + j(x^2 - y^2 + h^2)} \right] \qquad [A/m] \tag{C.5.3}$$

At the midplane between the two conductors we have:

$$\vec{H}(x,0) = -\left(\frac{i}{2\pi} \right) \frac{2h}{(x^2 + h^2)} \qquad [A/m] \tag{C.5.4}$$

Now, according to Ampere's law, the line integral taken around any circle shown in Figure C.4.2 for example, is equal to the total current enclosed within the path of integration. The midplane is, of course, one of these circles in the limit. Thus:

$$\int_{+\infty}^{-\infty} |H(x,0)|\ dx = -\frac{i}{\pi} \int_{+\infty}^{-\infty} \frac{h}{x^2 + h^2}\ dx = -\frac{i}{\pi}\ \left| \arctan \frac{x}{h} \right|_{+\infty}^{-\infty}$$

$$= -\frac{i}{\pi}\left[\left(-\frac{\pi}{2}\right) - \left(+\frac{\pi}{2}\right)\right] = i \quad [A] \tag{C.5.5}$$

which only verifies that Equation C.5.4 is correct.

Let us now insert a perfectly conducting metal sheet, infinite in area, and of infinitesimally small thickness 2δ, at the midplane between the two cylindrical conductors. Let there be a go and return current distribution $j(x)$ along the top and bottom of the metal sheet as shown in Figure C.5.1. What must be the nature of $j(x)$ at $y = \pm\delta$ such that $H(x,0) = 0$? According to Ampere's law, the contribution of $j(x)$ must be such that:

$$-\int_{+\infty}^{-\infty} |H(x,0)| = \int_{-\infty}^{+\infty} j(x)\ dx = -i \quad [A] \tag{C.5.6}$$

so that we require:

$$j(x) = -|H(x,0)| = -\left(\frac{i}{\pi}\right)\frac{h}{x^2 + h^2} \quad [A/m] \tag{C.5.7}$$

Observe that under the above conditions, the magnetic field above and below the thin metal sheet has not been disturbed since the net current in the sheet is zero as viewed from either side of the metal sheet. If we consider the upper current-carrying conductor and the top-half thickness of the metal sheet, together with its current distribution $j(x)$, we have the case of a current-carrying conductor above and parallel to a perfectly conducting plane with return current in the conducting plane.

Conversely, if we have the case of a current-carrying conductor above and parallel to a perfectly conducting plane of infinite extent with return current in the conducting plane, we evaluate the magnetic field by placing an image conductor at an equal distance below the plane and carrying a current of equal magnitude but in opposite direction.

C.6 Bicylindrical Coordinates

The conformal mapping function:

$$z = -h \cot w \tag{C.6.1}$$

maps the z plane in Figure C.6.1 into the w plane in Figure C.6.2.
We have:

$$z = -h \cot w = -h\frac{\cos w}{\sin w} = -hj\frac{e^{j2w} + 1}{e^{j2w} - 1} \tag{C.6.2}$$

$$z = x + jy \tag{C.6.3}$$

$$w = u + jv \tag{C.6.4}$$

Then:

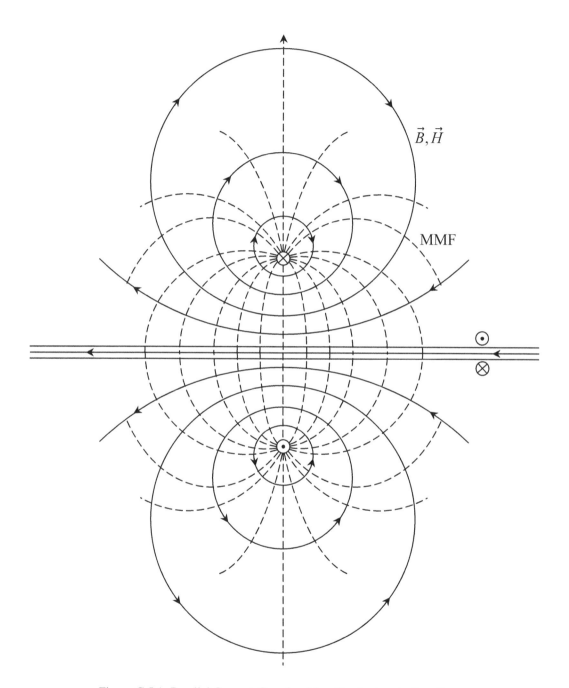

Figure C.5.1: Parallel Current-Carrying Line Conductors in Free Space

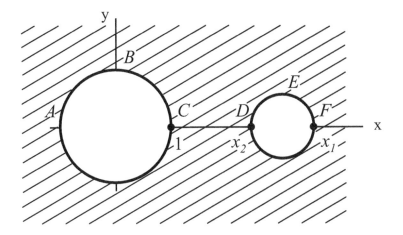

Figure C.6.1: Magnetic Field in the z Plane

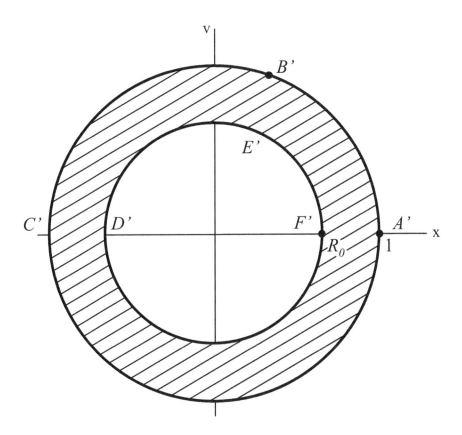

Figure C.6.2: Magnetic Field in the w Plane

$$\frac{z}{h} = \frac{-je^{2v}e^{j2u} - j}{e^{2v}e^{j2u} - 1} = \frac{-\sin 2u + j\left(e^{2v} + \cos 2u\right)}{\left(e^{2v} - \cos 2u\right) - j\sin 2u} \tag{C.6.5}$$

If we write:

$$\frac{-\sin 2u + j\left(e^{2v} + \cos 2u\right)}{\left(e^{2v} - \cos 2u\right) - j\sin 2u} = \frac{-a + jb}{c - ja} \tag{C.6.6}$$

Then:

$$\frac{-a + jb}{c - ja} = \frac{(-a + jb)(c + ja)}{a^2 + c^2} = \frac{-a(b + c) - j\left(a^2 - bc\right)}{a^2 + c^2} \tag{C.6.7}$$

where:

$$a(b + c) = 2e^{2v}\sin 2u \tag{C.6.8}$$
$$a^2 - bc = 1 - e^{4v} = -2e^{2v}\sinh 2v \tag{C.6.9}$$
$$a^2 + c^2 = 1 + e^{4v} - 2e^{2v}\cos 2u = 2e^{2v}\left(\cosh 2v - \cos 2u\right) \tag{C.6.10}$$

so that:

$$x = \frac{-h\sin 2u}{\cosh 2v - \cos 2u} \tag{C.6.11}$$

$$y = \frac{h\sinh 2v}{\cosh 2v - \cos 2u} \tag{C.6.12}$$

Then we may write:

$$\begin{bmatrix} x & 0 \\ y & -h \end{bmatrix}\begin{bmatrix} \cosh 2v \\ \sinh 2v \end{bmatrix} = \begin{bmatrix} x & -h \\ y & 0 \end{bmatrix}\begin{bmatrix} \cos 2u \\ \sin 2u \end{bmatrix} \tag{C.6.13}$$

so that:

$$\begin{bmatrix} \cosh 2v \\ \sinh 2v \end{bmatrix} = \left(\frac{1}{xh}\right)\begin{bmatrix} h & 0 \\ y & -x \end{bmatrix}\begin{bmatrix} x & -h \\ y & 0 \end{bmatrix}\begin{bmatrix} \cos 2u \\ \sin 2u \end{bmatrix}$$
$$= \left(\frac{1}{x}\right)\begin{bmatrix} x & -h \\ 0 & -y \end{bmatrix}\begin{bmatrix} \cos 2u \\ \sin 2u \end{bmatrix} \tag{C.6.14}$$

and:

$$\begin{bmatrix} \cos 2u \\ \sin 2u \end{bmatrix} = \left(\frac{1}{yh}\right)\begin{bmatrix} 0 & h \\ -y & x \end{bmatrix}\begin{bmatrix} x & 0 \\ y & -h \end{bmatrix}\begin{bmatrix} \cosh 2v \\ \sinh 2v \end{bmatrix}$$
$$= \left(\frac{1}{y}\right)\begin{bmatrix} y & -h \\ 0 & -x \end{bmatrix}\begin{bmatrix} \cosh 2v \\ \sinh 2v \end{bmatrix} \tag{C.6.15}$$

Then:

$$x^2\left(\cosh^2 2v - \sinh^2 2v\right) = x^2$$
$$= x^2\cos^2 2u - 2hx\sin 2u\cos 2u - y^2\sin^2 2u + h^2\sin^2 2u \tag{C.6.16}$$

so that:

$$(x + h\cot 2u)^2 + y^2 = \left(\frac{h}{\sin 2u}\right)^2 \tag{C.6.17}$$

and in terms of our variables η and ψ:

$$(x + h\cot \psi)^2 + y^2 = \left(\frac{h}{\sin \psi}\right)^2 \tag{C.6.18}$$

Similarly:

$$y^2 \left(\cos^2 2u - \sin^2 2u\right) = y^2$$
$$= y^2 \cosh^2 2v - 2hy \sinh 2v \cosh 2v + x^2 \sinh^2 2v + h^2 \sinh^2 2v \tag{C.6.19}$$

so that:

$$x^2 + (y - h\coth 2v)^2 = \left(\frac{h}{\sinh 2v}\right)^2 \tag{C.6.20}$$

and in terms of our variables η and ψ:

$$x^2 + (y - h\coth \eta)^2 = \left(\frac{h}{\sinh \eta}\right)^2 \tag{C.6.21}$$

C.7   References

[1] P. Moon and D. E. Spencer. *Field Theory For Engineers*. D. Van Nostrand Company, Inc., Princeton, NJ, 1961, pgs 361-368.

Appendix D

Carson's Earth-Return Correction Factors

In the introduction to his classic paper [1] published in 1926, Dr. John R. Carson wrote:

> The problem of wave propagation along a transmission system composed of an overhead wire parallel to the (plane) surface of the earth, in spite of great technical importance, does not appear to have been satisfactorily solved. While a complete solution of the actual problem is impossible, on account of the inequalities in the earth's surface and its lack of conductive homogeneity, the solution of the problem, where the actual earth is replaced by a plan homogeneous semi-infinite solid, is of considerable theoretical and practical interest. The solution of this problem is given in the present paper, together with formulas for calculating the inductive disturbances in neighboring transmission systems.

The pertinent results, adapted to units and nomenclature employed in this book, are presented in this appendix.

D.1   Definition of Terms

ρ_e	$\equiv$ resistivity of the earth $[\Omega \cdot m]$ (often written as $[\Omega/m^3]$)	(D.1.1)
f	$\equiv$ electrical frequency [Hz]	(D.1.2)
ω	$\equiv 2\pi f$	(D.1.3)
μ_0	$\equiv$ magnetic permeability of free space $= 4\pi \times 10^{-7}$ [H/m]	(D.1.4)
h_i	$\equiv$ height of conductor i above the earth's surface [m]	(D.1.5)
x_{ij}	$\equiv$ horizontal component of separation distance between conductors i and j [m]	(D.1.6)
d_{ij}	$\equiv \sqrt{x_{ij}^2 + (h_i - h_j)^2}$ [m]	(D.1.7)
D_{ij}	$\equiv \sqrt{x_{ij}^2 + (h_i + h_j)^2}$ [m]	(D.1.8)
GMR_i	$\equiv$ geometric mean radius of conductor i [m]	(D.1.9)
α	$= \dfrac{\mu_0\, \omega}{\rho_e}$ $[m^{-2}]$	(D.1.10)

$$p \qquad = x_{ij} \sqrt{\alpha} \qquad\qquad\qquad\qquad\qquad\qquad\qquad \text{(D.1.11)}$$
$$q \qquad = (h_i + h_j) \sqrt{\alpha} \qquad\qquad\qquad\qquad\qquad \text{(D.1.12)}$$
$$k \qquad = \sqrt{p^2 + q^2} \qquad\qquad\qquad\qquad\qquad\qquad \text{(D.1.13)}$$
$$k_{ii} \qquad = 2h_i \sqrt{\alpha} \qquad\qquad\qquad\qquad\qquad\qquad \text{(D.1.14)}$$
$$k_{ij} \qquad = D_{ij} \sqrt{\alpha} \qquad\qquad\qquad\qquad\qquad\qquad \text{(D.1.15)}$$
$$\theta_{ii} \qquad = 0 \quad [\text{rad}] \qquad\qquad\qquad\qquad\qquad\qquad \text{(D.1.16)}$$

$$\theta_{ij} \qquad = \arctan \frac{q}{p} \quad [\text{rad}] \qquad\qquad\qquad\qquad \text{(D.1.17)}$$

D.2   Carson's Results

Carson determined the influence of the earth upon line impedance to be:

$$\tilde{Z}_{ii-e} = \left(R_i + j \frac{\omega \mu_0}{2\pi} \ln \frac{2h_i}{GMR_i} \right)$$
$$+ \frac{\omega \mu_0}{\pi} \int_0^\infty \left[\sqrt{\beta^2 + j} - \beta \right] e^{-2\sqrt{\alpha}\, h_i\, \beta} \, d\beta \quad [\Omega/m] \qquad \text{(D.2.1)}$$

$$\tilde{Z}_{ij-e} = \left(j \frac{\omega \mu_0}{2\pi} \ln \frac{D_{ij}}{d_{ij}} \right)$$
$$+ \frac{\omega \mu_0}{\pi} \int_0^\infty \left[\sqrt{\beta^2 + j} - \beta \right] e^{-\sqrt{\alpha}\, (h_i + h_j)\beta} \cos \left(\sqrt{\alpha}\, x_{ij}\, \beta \right) d\beta \quad [\Omega/m] \qquad \text{(D.2.2)}$$

The major portion of his analysis involves the evaluation of the integral of the form:

$$P + jQ = \int_0^\infty \left[\sqrt{\beta^2 + j} - \beta \right] e^{-q\beta} \cos p\beta \, d\beta \qquad \text{(D.2.3)}$$

so that Equations D.2.1 and D.2.2 may be written as:

$$\tilde{Z}_{ii-e} = \left(R_i + j \frac{\omega \mu_0}{2\pi} \ln \frac{1}{GMR_i} \right)$$
$$+ \frac{\omega \mu_0}{\pi} \left[P_{ii} + j \left(Q_{ii} + \frac{1}{2} \ln \frac{k_{ii}}{\sqrt{\alpha}} \right) \right] \quad [\Omega/m] \qquad \text{(D.2.4)}$$

$$\tilde{Z}_{ij-e} = \left(j \frac{\omega \mu_0}{2\pi} \ln \frac{1}{d_{ij}} \right)$$
$$+ \frac{\omega \mu_0}{\pi} \left[P_{ij} + j \left(Q_{ii} + \frac{1}{2} \ln \frac{k_{ij}}{\sqrt{\alpha}} \right) \right] \quad [\Omega/m] \qquad \text{(D.2.5)}$$

Carson presents the following solutions:

For $k < 0.25$

$$P = \frac{\pi}{8} - \frac{k}{\sqrt{18}} \cos \theta + \frac{k^2}{16} \left[0.6728 + \ln \frac{2}{k} \right] \cos 2\theta + \frac{k^2 \theta}{16} \sin 2\theta \ldots \qquad \text{(D.2.6)}$$

$$Q = -0.0386 + \frac{1}{2} \ln \frac{2}{k} + \frac{k}{\sqrt{18}} \cos \theta \ldots \qquad \text{(D.2.7)}$$

For k > 5:

$$P = \frac{1}{k\sqrt{2}}\left(\cos\theta - \frac{\sqrt{2}}{k}\cos 2\theta + \frac{1}{k^2}\cos 3\theta + \frac{3}{k^4}\cos 5\theta \ldots\right) \qquad (D.2.8)$$

$$Q = \frac{1}{k\sqrt{2}}\left(\cos\theta - \frac{1}{k^2}\cos 3\theta + \frac{3}{k^4}\cos 5\theta \ldots\right) \qquad (D.2.9)$$

For k > 10:

$$P + jQ = \frac{1}{k^2\sqrt{2}}\left(k - \sqrt{2}\cos 2\theta\right) + j\frac{1}{k\sqrt{2}}\cos\theta \qquad (D.2.10)$$

D.3 Clarke's Series

Edith Clarke [2] presents a somewhat longer series, which gives P and Q to less than 1% error for values of k up to unity:

$$P = \frac{\pi}{8} + \left[-\frac{k}{\sqrt{18}}\cos\theta + \frac{k^2}{16}\left(0.6728 + \ln\frac{2}{k}\right)\cos 2\theta + \frac{k^2\theta}{16}\sin 2\theta\right]$$
$$+ \left[\frac{k^3}{45\sqrt{2}}\cos 3\theta - \frac{\pi k^4}{1536}\cos 4\theta\right] \qquad (D.3.1)$$

$$Q = \left[-0.0386 + \frac{1}{2}\ln\frac{2}{k}\right] + \left[\frac{k}{\sqrt{18}}\cos\theta - \frac{k^2\pi}{64}\cos 2\theta\right]$$
$$+ \left[\frac{k^3}{45\sqrt{2}}\cos 3\theta - \frac{k^4\theta}{384}\sin 4\theta - \frac{k^4}{384}\cos 4\theta\left(1.0895 + \ln\frac{2}{k}\right)\right] \qquad (D.3.2)$$

We shall write:

$$P = \frac{\pi}{8} + P' = 0.3927 + P' \qquad (D.3.3)$$

$$Q = \frac{1}{2}\ln\frac{k}{\sqrt{\alpha}} = -0.0386 + \frac{1}{4}\ln\frac{4}{\alpha} + Q' \qquad (D.3.4)$$

whereby P' and Q' are given by the sum of the remaining second and third bracketed terms in Equations D.3.1 and D.3.2.

Tables D.3.1 to D.3.2 present the results of evaluating P' and Q' using only the second bracketed term and using both second and third bracketed terms in Equations D.3.1 and D.3.2. Note that the third bracketed term contributes almost nothing for the range of k given in the tables.

D.4 Line Impedance

For $f = 60$ Hz and $\rho_e = 100\Omega/m^3$, we have $\alpha = 4.7374 \times 10^{-6}$ and $\sqrt{\alpha} = 2.1766 \times 10^{-3}$. Then, $\frac{1}{4}\ln\frac{4}{\alpha} = 3.4116$ and we have:

$$\theta = 0°$$

k	P'		Q'	
	One Term	Two Terms	One Term	Two Terms
.01	-0.0023	-0.0023	0.0024	0.0024
.02	-0.0046	-0.0046	0.0047	0.0047
.03	-0.0068	-0.0068	0.0070	0.0070
.04	-0.0090	-0.0090	0.0093	0.0094
.05	-0.0111	-0.0111	0.0117	0.0117
.06	-0.0132	-0.0132	0.0140	0.0140
.07	-0.0153	-0.0153	0.0163	0.0163
.08	-0.0173	-0.0173	0.0185	0.0185
.09	-0.0193	-0.0193	0.0208	0.0208
.10	-0.0213	-0.0213	0.0231	0.0231
.11	-0.0232	-0.0232	0.0253	0.0254
.12	-0.0251	-0.0251	0.0276	0.0276
.13	-0.0270	-0.0270	0.0298	0.0298
.14	-0.0289	-0.0289	0.0320	0.0321
.15	-0.0308	-0.0307	0.0343	0.0343
.16	-0.0326	-0.0325	0.0365	0.0365
.17	-0.0344	-0.0343	0.0387	0.0387
.18	-0.0362	-0.0361	0.0408	0.0409
.19	-0.0380	-0.0378	0.0430	0.0431
.20	-0.0397	-0.0396	0.0452	0.0453

For $f = 60$ Hz, $\rho_e = 100$ [Ω/m^3], and ignoring the P' and Q' terms, we have approximately:

$$P = 0.3927$$

$$Q + \frac{1}{2} \ln \frac{k}{\sqrt{\alpha}} = 3.3730$$

Table D.3.1: Evaluation of P' and Q', $\theta = 0°$

$$\theta = 30°$$

k	P'		Q'	
	One Term	Two Terms	One Term	Two Terms
.01	-0.0020	-0.0020	0.0020	0.0020
.02	-0.0040	-0.0040	0.0041	0.0041
.03	-0.0060	-0.0060	0.0061	0.0061
.04	-0.0079	-0.0079	0.0081	0.0081
.05	-0.0098	-0.0098	0.0101	0.0101
.06	-0.0117	-0.0117	0.0122	0.0122
.07	-0.0135	-0.0135	0.0142	0.0142
.08	-0.0154	-0.0154	0.0162	0.0162
.09	-0.0172	-0.0172	0.0182	0.0182
.10	-0.0190	-0.0190	0.0202	0.0202
.11	-0.0208	-0.0208	0.0222	0.0222
.12	-0.0225	-0.0225	0.0241	0.0241
.13	-0.0243	-0.0243	0.0261	0.0261
.14	-0.0260	-0.0260	0.0281	0.0281
.15	-0.0277	-0.0277	0.0301	0.0301
.16	-0.0294	-0.0294	0.0320	0.0320
.17	-0.0310	-0.0310	0.0340	0.0340
.18	-0.0327	-0.0327	0.0359	0.0360
.19	-0.0343	-0.0343	0.0379	0.0379
.20	-0.0360	-0.0360	0.0398	0.0398

For $f = 60$ Hz, $\rho_e = 100$ [Ω/m^3], and ignoring the P' and Q' terms, we have approximately:

$$P = 0.3927$$

$$Q + \frac{1}{2} \ln \frac{k}{\sqrt{\alpha}} = 3.3730$$

Table D.3.2: Evaluation of P' and Q', $\theta = 30^o$

$$\theta = 60°$$

k	P'		Q'	
	One Term	Two Terms	One Term	Two Terms
.01	-0.0012	-0.0012	0.0012	0.0012
.02	-0.0024	-0.0024	0.0024	0.0024
.03	-0.0036	-0.0036	0.0036	0.0036
.04	-0.0049	-0.0049	0.0048	0.0048
.05	-0.0061	-0.0061	0.0060	0.0060
.06	-0.0073	-0.0073	0.0072	0.0072
.07	-0.0086	-0.0086	0.0084	0.0084
.08	-0.0098	-0.0099	0.0096	0.0096
.09	-0.0111	-0.0111	0.0108	0.0108
.10	-0.0124	-0.0124	0.0120	0.0120
.11	-0.0136	-0.0136	0.0133	0.0132
.12	-0.0149	-0.0149	0.0145	0.0145
.13	-0.0162	-0.0162	0.0157	0.0157
.14	-0.0174	-0.0175	0.0170	0.0169
.15	-0.0187	-0.0187	0.0182	0.0182
.16	-0.0200	-0.0200	0.0195	0.0194
.17	-0.0212	-0.0213	0.0207	0.0207
.18	-0.0225	-0.0226	0.0220	0.0219
.19	-0.0238	-0.0239	0.0233	0.0232
.20	-0.0250	-0.0251	0.0246	0.0244

For $f = 60$ Hz, $\rho_e = 100$ [Ω/m^3], and ignoring the P' and Q' terms, we have approximately:

$$P = 0.3927$$

$$Q + \frac{1}{2}\ln\frac{k}{\sqrt{\alpha}} = 3.3730$$

Table D.3.3: Evaluation of P' and Q', $\theta = 60°$

$$P = \frac{\pi}{8} + P' = 0.3827 + P' \tag{D.4.1}$$

$$Q + \frac{1}{2} \ln \frac{k}{\sqrt{\alpha}} = 3.3730 + Q' \tag{D.4.2}$$

With these same assumptions, and for practical line configurations with $10 < h < 40$ we have $0.0348 < k < 0.1741$. According to Tables D.3.1 through D.3.3, we find that Q' is insignificant compared to 3.3730, so that we may ignore Q' altogether. Thus, for $f = 60$ Hz and $\rho_e = 100\Omega/m^3$:

$$Q_{ii} + \frac{1}{2} \ln \frac{k_{ii}}{\sqrt{\alpha}} = 3.3730 \tag{D.4.3}$$

$$Q_{ij} + \frac{1}{2} \ln \frac{k_{ij}}{\sqrt{\alpha}} = 3.3730 \tag{D.4.4}$$

and therefore, the expressions D.4.3 and D.4.4 are, in effect, independent of conductor height above ground and spacing between conductors.

According to Tables D.3.1 through D.3.3, we find that P' is not nearly as insignificant compared to 0.3927. However, from a practical engineering point of view, we recognize that these results are based upon the idealized constraints outlined in Dr. Carson's introductory paragraph. Furthermore, since Carson's earth-return correction factors are significant only when earth-return currents are present, there are inherently other inaccuracies involved in defining the resistances of ground wires, neutral wires, earthing electrodes, earth resistivity, etc. It has therefore become accepted practice to ignore P', so that, for $f = 60$ Hz and $\rho_e = 100\Omega/m^3$:

$$P_{ii} = 0.3927 \tag{D.4.5}$$
$$P_{ij} = 0.3927 \tag{D.4.6}$$

and therefore, the expressions D.4.5 and D.4.6 are, in effect, independent of conductor height above ground and spacing between conductors.

We may now write:

$$\tilde{Z}_{ii-e} = \left(R_i + \frac{R_e}{3} \right) + j \left(X_{ii} + \frac{X_e}{3} \right) \tag{D.4.7}$$

$$\tilde{Z}_{ij-e} = \frac{R_e}{3} + j \left(X_{ij} + \frac{X_e}{3} \right) \tag{D.4.8}$$

where:

$$X_{ii} = \frac{\omega\mu_0}{2\pi} \ln \frac{1}{GMR_i} \tag{D.4.9}$$

$$\tag{D.4.10}$$

$$X_{ij} = \frac{\omega\mu_0}{2\pi} \ln \frac{1}{d_{ij}} \tag{D.4.11}$$

$$\tag{D.4.12}$$

$$\frac{R_e}{3} = \frac{\omega\mu_0}{\pi} \frac{\pi}{8} \tag{D.4.13}$$

$$\tag{D.4.14}$$

$$\frac{X_e}{3} = \frac{\omega\mu_0}{\pi} \left(-0.0386 + \frac{1}{4} \ln \frac{4}{\alpha} \right) \tag{D.4.15}$$

and for $f = 60$ Hz and $\rho_e = 100\Omega/m^3$:

$$\frac{R_e}{3} = \frac{\omega\mu_0}{\pi}\frac{\pi}{8} = 0.05922 \quad [\Omega/km] \tag{D.4.16}$$

$$\frac{X_e}{3} = \frac{\omega\mu_0}{\pi}\left(-0.0386 + \frac{1}{4}\ln\frac{4}{\alpha}\right) = 0.5086 \quad [\Omega/km] \tag{D.4.17}$$

$$\frac{L_e}{3} = \frac{\mu_0}{\pi}\left(-0.0386 + \frac{1}{4}\ln\frac{4}{\alpha}\right) = 1.3492 \quad [mH/km] \tag{D.4.18}$$

D.5 References

[1] John R. Carson. *Wave Propagation in Overhead Wires with Ground Return.* Bell Systems Technical Journal, Volume 5, 1926, pgs. 539-554.

[2] Edith Clarke. *Circuit Analysis of AC Power Systems - Volume I.* John Wiley and Sons, Inc., New York, New York, 1943, pgs. 372-384.

Appendix E

Matrix Algebra

This appendix is dedicated to a concise review of some basic matrix algebra concepts and to the definition of some special matrix forms that will be employed throughout the book.

E.1 Matrix

A *matrix* is an *organized* rectangular *array* consisting of m rows and n columns. A matrix is not a mathematical operation, like the determinant, but is to be regarded generally as merely an array. The term *matrix* for such an array was suggested by J.J. Sylvester (1850). Hamilton (1853) employed a calculus of matrices in his work on linear vector functions, but Cayley (1858) developed the basic notions of matrix algebra without recognizing the relationship between his work and that of Hamilton. Little attention was paid to Cayley's work until Leguerre (1867) and Frobenius (1878) recognized matrix algebra as an important mathematical concept.

The elements of a matrix may be real or complex numbers, algebraic symbols, or functions. A *square* matrix may be written in the general format as:

$$[A] = \begin{bmatrix} a_{11} & a_{12} & a_{13} \\ a_{21} & a_{22} & a_{23} \\ a_{31} & a_{32} & a_{33} \end{bmatrix} = [a_{ij}] \tag{E.1.1}$$

where the double-subscript notation helps to locate the element as being in the ith row and j^{th} column.

General *rectangular* matrices may also take the form:

$$[a_{ij}] = \begin{bmatrix} a_{11} & a_{12} & a_{13} \\ a_{21} & a_{22} & a_{23} \\ a_{31} & a_{32} & a_{33} \\ a_{41} & a_{42} & a_{43} \end{bmatrix} , \quad [y_j] = \begin{bmatrix} y_1 \\ y_2 \\ y_3 \\ y_4 \end{bmatrix} , \quad [x_i] = \begin{bmatrix} x_1 & x_2 & x_3 & x_4 \end{bmatrix} \tag{E.1.2}$$

where the elements in row matrices and column matrices will be identified in terms of single-subscript notation. Single-row and single-column matrices are also referred to as *row vectors* and *column vectors*, respectively.

E.2 Order (Dimension) of a Matrix

The *order* or dimension of a matrix is a formal manner of designating the number of rows (m) and columns (n) of a general rectangular matrix. It is usually expressed in the form:

$$m \times n \tag{E.2.1}$$

For example, in E.1.2, the order of the matrices are, respectively:

$$4 \times 3 , \qquad 4 \times 1 , \qquad 1 \times 4 \tag{E.2.2}$$

For the square matrix in E.1.1, we may describe the order either as 3×3 or, since there can be no ambiguity, merely as a *third-order matrix*. Matrices are said to be compatible if they are of exactly the same order.

E.3   Matrix Addition

Matrix *addition* is defined as follows:

$$\begin{bmatrix} a_{11} & a_{12} \\ a_{21} & a_{22} \end{bmatrix} + \begin{bmatrix} b_{11} & b_{12} \\ b_{21} & b_{22} \end{bmatrix} = \begin{bmatrix} (a_{11} + b_{11}) & (a_{12} + b_{12}) \\ (a_{21} + b_{21}) & (a_{22} + b_{22}) \end{bmatrix} \tag{E.3.1}$$

Matrices can be added only if they are *compatible*. The order of the matrix addition may be reversed, i.e.:

$$[A] + [B] = [B] + [A] \tag{E.3.2}$$

E.4   Matrix Subtraction

Matrix *subtraction* is merely an extension of matrix addition. Thus:

$$\begin{bmatrix} a_{11} & a_{12} \\ a_{21} & a_{22} \end{bmatrix} - \begin{bmatrix} b_{11} & b_{12} \\ b_{21} & b_{22} \end{bmatrix} = \begin{bmatrix} (a_{11} - b_{11}) & (a_{12} - b_{12}) \\ (a_{21} - b_{21}) & (a_{22} - b_{22}) \end{bmatrix} \tag{E.4.1}$$

in symbolic notation, we write:

$$[A] - [B] = \quad [C] \tag{E.4.2}$$
$$-[B] + [A] = \quad [C] \tag{E.4.3}$$
$$[B] - [A] = -[C] \tag{E.4.4}$$

E.5   Equal Matrices

Two matrices are equal if every element of one matrix is equal to every corresponding element of the other.

E.6   Null (Zero) Matrix

A null (zero) matrix is defined as a matrix in which all elements are zero. We shall employ the symbolic notation [0] for the null matrix. Therefore, we have, in general:

$$[A] - [A] = [0] \tag{E.6.1}$$

On the basis of Equation E.4.2, we have:

$$[A] - [B] - [C] = [0] \tag{E.6.2}$$

and on the basis of Equation E.4.4, we have:

$$[B] - [A] + [C] = [0] \tag{E.6.3}$$

We observe that matrix addition and subtraction obey the ordinary rules of algebra.

E.7   Scalar Multiplication

In *scalar* multiplication of matrices, every element of the matrix is multiplied by the scalar quantity. This follows directly from the rule for scalar addition; thus:

$$[A] + [A] + [A] = (3)[A] \tag{E.7.1}$$

Conversely, a factor common to every element of a matrix may be withdrawn or removed as a common scalar factor. We point out that scalar multiplication of matrices is unlike scalar multiplication of determinants.

E.8   Matrix Multiplication

The multiplication of one matrix by another lies at the heart of matrix algebra and requires the introduction of some new concepts. Actually, the rules are quite simple, involving the cumulative multiplication of the elements of a row by the elements of a column.

Consider the following:

$$\begin{bmatrix} y_1 \\ y_2 \end{bmatrix} = \begin{bmatrix} a_{11} & a_{22} \\ a_{21} & a_{22} \end{bmatrix} \begin{bmatrix} x_{11} \\ x_{22} \end{bmatrix} = \begin{bmatrix} a_{11}x_{11} + a_{12}x_2 \\ a_{21}x_1 + a_{22}x_2 \end{bmatrix} \tag{E.8.1}$$

which, in expanded form, can be written as:

$$y_1 = a_{11}x_1 + a_{12}x_2 \tag{E.8.2}$$
$$y_2 = a_{21}x_1 + a_{22}x_2 \tag{E.8.3}$$

and which we recognize as a set of simultaneous linear algebraic equations.

In a more general sense, we define matrix multiplication as follows:

$$\begin{bmatrix} a_{11} & a_{12} \\ a_{21} & a_{22} \\ a_{31} & a_{32} \end{bmatrix} \begin{bmatrix} b_{11} & b_{12} \\ b_{21} & b_{22} \end{bmatrix} = \begin{bmatrix} a_{11}b_{11} + a_{12}b_{21} & a_{11}b_{12} + a_{12}b_{22} \\ a_{21}b_{11} + a_{22}b_{21} & a_{21}b_{12} + a_{22}b_{22} \\ a_{31}b_{11} + a_{32}b_{21} & a_{31}b_{12} + a_{32}b_{22} \end{bmatrix} \tag{E.8.4}$$

which is the accumulation of products of elements a_{ik} along left to right one row, by the elements b_{kj} down one column.

In a somewhat more concise format we may write:

$$[A][B] = [C] \tag{E.8.5}$$
$$[a_{ij}][b_{ij}] = [c_{ij}] \tag{E.8.6}$$

where each element of c_{ij} is defined by:

$$c_{ij} = \sum_{k=1}^{n} a_{ik} b_{kj} \tag{E.8.7}$$

n is the number of columns in the left (premultiplier) matrix and the number of rows in the right (postmultiplier) matrix, both of which must be the same for matrix multiplication to be defined.

E.8.1   Premultiplier and Postmultiplier

In Equations E.8.4 and E.8.7, note that there must be as many columns in the premulitplier as there are rows in the postmultiplier in order for multiplication to be possible. Thus, when the premultiplier is of order $m \times n$, then the order of the postmultiplier must be of the order $n \times p$. When this condition is satisfied, the matrices are said to be *conformable*. The product will be of the order $m \times p$. See Figure E.8.1 depicts this visually. When *any* (assuming that they are conformable) matrix expression is multiplied by a matrix on the left hand side of the expression, it is said to be premultiplied, which many also refer to as *left* multiplication. Conversely, when any matrix expression is multiplied by a matrix on the right hand side of the expression, it said to be postmultiplied, which is also referred to as *right* multiplication.

Left and right multiplication must be adhered to during algebraic manipulation as well. If one side of an algebraic expression is left multiplied, the expression is valid *only* if the other side is also left multiplied. In other words, one cannot left multiply one side of an expression and right multiply the other.

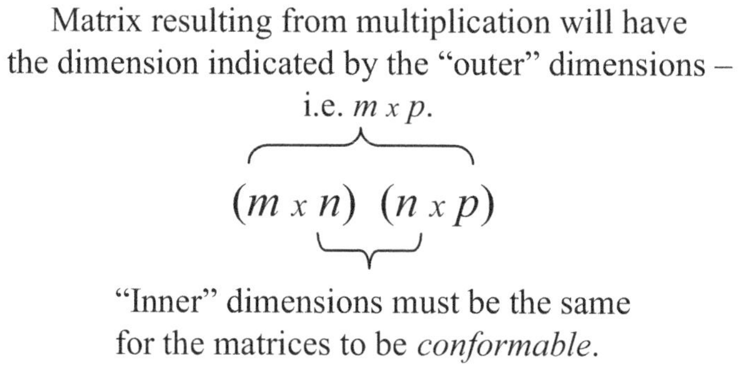

Matrix resulting from multiplication will have
the dimension indicated by the "outer" dimensions –
i.e. *m x p*.

$$(m \times n) \ (n \times p)$$

"Inner" dimensions must be the same
for the matrices to be *conformable*.

Figure E.8.1: Matrix Multiplication Dimensions

E.8.2   Order of Multiplication

In general, the same result will *not* be obtained if the order of the multiplication is reversed. Thus, in *general*:

$$[A]\,[B] \neq [B]\,[A] \tag{E.8.8}$$

In Equation E.8.4, reversing the order of multiplication is not possible since the two matrices would no longer be conformable. Even if the two matrices were square matrices of the same order, we might have, for example:

$$\begin{bmatrix} 1 & 2 \\ 3 & 4 \end{bmatrix} \begin{bmatrix} 2 & 1 \\ 2 & 3 \end{bmatrix} = \begin{bmatrix} 6 & 7 \\ 14 & 15 \end{bmatrix} \tag{E.8.9}$$

and reversing the order of multiplication:

$$\begin{bmatrix} 1 & 2 \\ 3 & 4 \end{bmatrix} \begin{bmatrix} 2 & 1 \\ 2 & 3 \end{bmatrix} = \begin{bmatrix} 5 & 8 \\ 11 & 16 \end{bmatrix} \tag{E.8.10}$$

In a formal manner, we say that matrix multiplication is not, in general, a *commutative* process.

E.8.3 Permutable (Commutable) Matrices

Permutable or *commutable* matrices are *square* matrices for which the result is not changed when the order of multiplication is reversed. The simplest example is when both matrices are identical:

$$[A][A] = [C] \tag{E.8.11}$$

However, they need not necessarily be identical. For example:

$$\begin{bmatrix} 1 & 2 \\ 3 & 4 \end{bmatrix} \begin{bmatrix} 1 & 4 \\ 6 & 7 \end{bmatrix} = \begin{bmatrix} 13 & 18 \\ 27 & 40 \end{bmatrix} \tag{E.8.12}$$

and:

$$\begin{bmatrix} 1 & 4 \\ 6 & 7 \end{bmatrix} \begin{bmatrix} 1 & 2 \\ 3 & 4 \end{bmatrix} = \begin{bmatrix} 13 & 18 \\ 27 & 40 \end{bmatrix} \tag{E.8.13}$$

Commutable matrices form a rather specialized set of matrices and although they exist, it is best practice not to reverse the order of an indicated multiplication.

E.9   Diagonal Matrix

We define a *diagonal* matrix as a *square* matrix whose off-diagonal elements, $a_{ij} = 0$, and which has at least one diagonal element, $a_{ii} \neq 0$.

E.9.1   Premultiplication by a Diagonal Matrix

Let:

$$[C] = \begin{bmatrix} d_{11} & 0 & 0 \\ 0 & d_{22} & 0 \\ 0 & 0 & d_{33} \end{bmatrix} \begin{bmatrix} b_{11} & b_{12} \\ b_{21} & b_{22} \\ b_{31} & b_{32} \end{bmatrix} = \begin{bmatrix} d_{12}b_{11} & d_{11}b_{12} \\ d_{22}b_{21} & d_{22}b_{22} \\ d_{33}b_{31} & d_{33}b_{32} \end{bmatrix} \tag{E.9.1}$$

In *premultiplication* by a diagonal matrix, every element in a row in the postmultiplier is multiplied by the diagonal element of the corresponding row of the premultiplier.

E.9.2   Postmultiplication by a Diagonal Matrix

Let:

$$[C] = \begin{bmatrix} b_{11} & b_{12} \\ b_{21} & b_{22} \\ b_{31} & b_{32} \end{bmatrix} \begin{bmatrix} d_{11} & 0 \\ 0 & d_{22} \end{bmatrix} = \begin{bmatrix} b_{11}d_{11} & b_{12}d_{22} \\ b_{21}d_{11} & b_{22}d_{22} \\ b_{31}d_{11} & b_{32}d_{22} \end{bmatrix} \tag{E.9.2}$$

In *postmultiplication* by a diagonal matrix, every element in a column in the premultiplier is multiplied by the diagonal element of the corresponding column in the postmultiplier.

E.9.3   Identity (Unit) Matrix

A special diagonal matrix in which all diagonal elements are equal to unity is called the *identity* (unit) matrix. We shall use the following nomenclature:

$$[I] = \begin{bmatrix} 1 & 0 & 0 & 0 & 0 & \cdots & \cdots & \cdots & 0 \\ 0 & 1 & 0 & 0 & 0 & \cdots & \cdots & \cdots & 0 \\ 0 & 0 & 1 & 0 & 0 & \cdots & \cdots & \cdots & 0 \\ 0 & 0 & 0 & 1 & 0 & \cdots & \cdots & \cdots & 0 \\ 0 & 0 & 0 & 0 & 1 & \cdots & \cdots & \cdots & 0 \\ \vdots & \vdots & \vdots & \vdots & \vdots & \ddots & \cdots & \cdots & 0 \\ \vdots & \vdots & \vdots & \vdots & \vdots & \cdots & \ddots & \cdots & 0 \\ \vdots & \vdots & \vdots & \vdots & \vdots & \cdots & \cdots & \ddots & 0 \\ 0 & 0 & 0 & 0 & 0 & 0 & 0 & 0 & 1 \end{bmatrix} \tag{E.9.3}$$

with the help of Equations E.9.1 and E.9.2, the reader may verify that:

$$[B] = [I]\,[B] = [B]\,[I] \tag{E.9.4}$$

While Equation E.9.4 is *always* valid, the reader is cautioned against jumping to conclusions about a slightly different form, such as:

$$[B] = [A]\,[B] \tag{E.9.5}$$

where it does *not* necessarily follow that $[A] = [I]$. For example, let:

$$[A] = \left(\frac{1}{2}\right) \begin{bmatrix} 1 & 1 & 1 \\ 1 & 1 & -1 \\ 1 & -1 & 1 \end{bmatrix} \tag{E.9.6}$$

and:

$$[B] = \begin{bmatrix} 2 & 1 & 1 \\ 1 & 2 & -1 \\ 1 & -1 & 2 \end{bmatrix} \tag{E.9.7}$$

for which:

$$[A]\,[B] = \left(\frac{1}{2}\right) \begin{bmatrix} 1 & 1 & 1 \\ 1 & 1 & -1 \\ 1 & -1 & 1 \end{bmatrix} \begin{bmatrix} 2 & 1 & 1 \\ 1 & 2 & -1 \\ 1 & -1 & 2 \end{bmatrix} = \begin{bmatrix} 2 & 1 & 1 \\ 1 & 2 & -1 \\ 1 & -1 & 2 \end{bmatrix} = [B] \tag{E.9.8}$$

but:

$$[A] \neq [I] \tag{E.9.9}$$

In a somewhat related manner, take $[B]$ as defined in Equation E.9.7 and let:

$$[C]\,[B] = [0] \tag{E.9.10}$$

Again, it does *not* necessarily follow that $[C] = [0]$. For example, let:

$$[C] = \begin{bmatrix} 1 & -1 & -1 \\ -1 & 1 & 1 \\ -1 & 1 & 1 \end{bmatrix} \tag{E.9.11}$$

The reader may easily verify that this will satisfy Equation E.9.10, but $[C] \neq [0]$.

E.9.4 Associative Property of Matrix Multiplication

The multiplication of matrices is *associative*. Hence:

$$[A][B]([C]) = [A]([B][C]) \qquad \text{(E.9.12)}$$

E.9.5 Distributive Property of Matrix Multiplication

The multiplication of matrices is *distributive*. Hence:

$$[A]([B] + [C]) = [A][B] + [A][C] \qquad \text{(E.9.13)}$$

E.10 Matrix Transposition

Given a matrix:

$$[A] = \begin{bmatrix} a_{11} & a_{12} \\ a_{21} & a_{22} \\ a_{31} & a_{32} \end{bmatrix} \qquad \text{(E.10.1)}$$

We define the *transpose* of this matrix as:

$$[A]^T = \begin{bmatrix} a_{11} & a_{21} & a_{31} \\ a_{12} & a_{22} & a_{32} \end{bmatrix} \qquad \text{(E.10.2)}$$

The *transpose* of a given matrix is obtained by taking the elements of the original matrix along a row and making them elements down a corresponding column. We may also express this as:

$$[a_{ij}]^T = [a_{ji}] \qquad \text{(E.10.3)}$$

E.10.1 Symmetrical Matrix

A matrix that is equal to its transpose is called a *symmetrical matrix*. Thus, a symmetrical matrix is necessarily a square matrix whose elements satisfy:

$$a_{ij} = a_{ji} \qquad \text{(E.10.4)}$$

The elements a_{ii} along the diagonal need not necessarily be equal.

E.10.2 Transpose of a Matrix Sum

The transpose of a sum of two matrices is equal to the sum of the transposes of the individual matrices:

$$([A] + [B])^T = [A]^T = [B]^T \qquad \text{(E.10.5)}$$

E.10.3   Transpose of a Matrix Product

The transpose of a product of matrices is equal to the product of the transposes of the individual matrices, with the order of multiplication reversed. Thus:

$$([A]\,[B]\,[C]\,[D]\,[E])^T = [E]^T\,[D]^T\,[C]^T\,[B]^T\,[A]^T \tag{E.10.6}$$

Consider the original product to consist of comformable rectangular matrices such that (expressed in terms of the *order* of the matrices):

$$(r \times c_1)\,(c_1 \times c_2)\,(c_2 \times c_3)\,(c_3 \times c_4)\,(c_4 \times c) = (r \times c) \tag{E.10.7}$$

For the transpose, we have:

$$(c \times c_4)\,(c_4 \times c_3)\,(c_3 \times c_2)\,(c_2 \times c_1)\,(c_1 \times r) = (c \times r) \tag{E.10.8}$$

Consider

$$[C] = [A]\,[B] \tag{E.10.9}$$

with

$$[C]^T = [B]^T\,[A]^T \tag{E.10.10}$$

Now let $[C]$ be a symmetrical matrix such that:

$$[C]^T = [C] \tag{E.10.11}$$

It does *not* follow that $[C] = [A]\,[B]$. For example:

$$[C] = [A]\,[B] = \begin{bmatrix} 1 & 1 \\ 2 & 3 \end{bmatrix} \begin{bmatrix} 1 & 2 \\ 3 & 9 \end{bmatrix} = \begin{bmatrix} 4 & 11 \\ 11 & 31 \end{bmatrix} \tag{E.10.12}$$

and:

$$[C]^T = [B]^T\,[A]^T = \begin{bmatrix} 1 & 3 \\ 2 & 9 \end{bmatrix} \begin{bmatrix} 1 & 2 \\ 1 & 3 \end{bmatrix} = \begin{bmatrix} 4 & 11 \\ 11 & 31 \end{bmatrix} \tag{E.10.13}$$

but:

$$[B]\,[A] = \begin{bmatrix} 1 & 2 \\ 3 & 9 \end{bmatrix} \begin{bmatrix} 1 & 1 \\ 2 & 3 \end{bmatrix} = \begin{bmatrix} 5 & 7 \\ 21 & 30 \end{bmatrix} \tag{E.10.14}$$

However, if $[A]$, $[B]$, and $[C]$ are *all* symmetrical matrices, then $[A]\,[B] = [B]^T\,[A]^T = [B]\,[A]$. For example:

$$[C] = [A]\,[B] = \begin{bmatrix} 1 & 1 \\ 1 & 2 \end{bmatrix} \begin{bmatrix} 1 & 2 \\ 2 & 3 \end{bmatrix} = \begin{bmatrix} 3 & 5 \\ 5 & 8 \end{bmatrix} \tag{E.10.15}$$

and:

$$[C]^T = [B]^T\,[A]^T = \begin{bmatrix} 1 & 2 \\ 2 & 3 \end{bmatrix} \begin{bmatrix} 1 & 1 \\ 1 & 2 \end{bmatrix} = \begin{bmatrix} 3 & 5 \\ 5 & 8 \end{bmatrix} \tag{E.10.16}$$

and it is left to the reader to verify that $[B]^T\,[A]^T = [B]\,[A]$

Finally, the product of two symmetrical matrices does *not necessarily* result in a symmetric matrix. For example:

$$\begin{bmatrix} 1 & 1 \\ 1 & 2 \end{bmatrix} \begin{bmatrix} 1 & 2 \\ 2 & 2 \end{bmatrix} = \begin{bmatrix} 3 & 4 \\ 5 & 6 \end{bmatrix} \tag{E.10.17}$$

E.11   Complex Conjugate of a Matrix

When the elements of a matrix are complex (i.e,. having real and imaginary components), the conjugate of the matrix is the matrix obtained by replacing each element of the original matrix by its complex conjugate. Thus, let:

$$[\tilde{C}] = [A] + j\,[B] \tag{E.11.1}$$

Then:

$$[\tilde{C}]^{*} = [A] - j\,[B] \tag{E.11.2}$$

It should be apparent that if $[\tilde{C}]^{*} = [\tilde{C}]$, then $[\tilde{C}]$ is a real matrix.

E.11.1   Complex Conjugate of a Matrix Sum

The complex conjugate of a matrix sum is equal to the sum of the complex conjugates of the individual matrices.

E.11.2   Complex Conjugate of a Matrix Product

The complex conjugate of a matrix product is equal to the product of the complex conjugates of the individual matrices with the order of multiplication remaining unchanged. For example, let:

$$\begin{aligned}[\tilde{E}] &= ([A] + j\,[B])\,([C] + j\,[D]) \\ &= ([A]\,[C] - [B]\,[D]) + j\,([A]\,[D] + [B]\,[C]) \end{aligned} \tag{E.11.3}$$

and:

$$\begin{aligned}[\tilde{E}]^{*} &= ([A]\,[C] - [B]\,[D]) - j\,([A]\,[D] + [B]\,[C]) \\ &= ([A] - j\,[B])\,([C] - j\,[D]) \end{aligned} \tag{E.11.4}$$

E.11.3   Product of a Matrix by Its Complex Conjugate

In ordinary algebra, a complex number multiplied by its complex conjugate results in a *real number*. A complex matrix, multiplied by its complex conjugate, does *not*, in general, result in a *real matrix*. Thus:

$$\begin{aligned}[\tilde{C}][\tilde{C}]^{*} &= ([A] + j\,[B])\,([A] - j\,[B]) \\ &= ([A]\,[A] + [B]\,[B]) + j\,([B]\,[A] - [A]\,[B]) \end{aligned} \tag{E.11.5}$$

E.12   Partitioned Matrices

Given:

$$[C] = [A]\,[B] = \begin{bmatrix} a_{11} & a_{12} & a_{13} \\ a_{21} & a_{22} & a_{23} \\ a_{31} & a_{32} & a_{33} \\ a_{41} & a_{42} & a_{43} \end{bmatrix} \begin{bmatrix} b_{11} & b_{12} & b_{13} & b_{14} & b_{15} \\ b_{21} & b_{22} & b_{23} & b_{24} & b_{25} \\ b_{31} & b_{32} & b_{33} & b_{34} & b_{35} \end{bmatrix} \tag{E.12.1}$$

Sometimes it is more convenient to subdivide the matrix into submatrices. For example, we may write:

$$[A] = \left[\begin{array}{c|c} A_{11} & A_{12} \\ \hline A_{21} & A_{22} \end{array} \right] = \left[\begin{array}{cc|c} a_{11} & a_{12} & a_{13} \\ a_{21} & a_{22} & a_{23} \\ \hline a_{31} & a_{32} & a_{33} \\ a_{41} & a_{42} & a_{43} \end{array} \right] \qquad \text{(E.12.2)}$$

with:

$$[A_{11}] = \left[\begin{array}{cc} a_{11} & a_{12} \\ a_{21} & a_{22} \end{array} \right] \quad [A_{12}] = \left[\begin{array}{c} a_{13} \\ a_{23} \end{array} \right]$$

$$[A_{21}] = \left[\begin{array}{cc} a_{31} & a_{32} \\ a_{41} & a_{42} \end{array} \right] \quad [A_{22}] = \left[\begin{array}{c} a_{33} \\ a_{43} \end{array} \right] \qquad \text{(E.12.3)}$$

We say that $[A]$ has been partitioned into submatrices. If $[B]$ in Equation E.12.1 is also partitioned, this must be done in a manner that the submatrices to be multiplied are conformable. This requires a partition line between the second and third rows in $[B]$. Whether or not vertical partition lines are employed in $[B]$ is optional.

Let us take, for example:

$$[B] = \left[\begin{array}{c|c} B_{11} & B_{12} \\ \hline B_{21} & B_{22} \end{array} \right] = \left[\begin{array}{ccc|cc} b_{11} & b_{12} & b_{13} & b_{14} & b_{15} \\ b_{21} & b_{22} & b_{23} & b_{24} & b_{25} \\ \hline b_{31} & b_{32} & b_{33} & b_{34} & b_{35} \end{array} \right] \qquad \text{(E.12.4)}$$

Then:

$$[C] = \left[\begin{array}{c|c} A_{11} & A_{12} \\ \hline A_{21} & A_{22} \end{array} \right] \left[\begin{array}{c|c} B_{11} & B_{12} \\ \hline B_{21} & B_{22} \end{array} \right] = \left[\begin{array}{c|c} C_{11} & C_{12} \\ \hline C_{21} & C_{22} \end{array} \right]$$

$$= \left[\begin{array}{c|c} A_{11}B_{11} + A_{12}B_{21} & A_{11}B_{12} + A_{12}B_{22} \\ \hline A_{21}B_{11} + A_{22}B_{21} & A_{21}B_{12} + A_{22}B_{22} \end{array} \right] \qquad \text{(E.12.5)}$$

with:

$$[C_{11}] = \left[\begin{array}{ccc} c_{11} & c_{12} & c_{13} \\ c_{21} & c_{22} & c_{23} \end{array} \right] \quad [C_{12}] = \left[\begin{array}{cc} c_{14} & c_{15} \\ c_{24} & c_{25} \end{array} \right]$$

$$[C_{21}] = \left[\begin{array}{ccc} c_{31} & c_{32} & c_{33} \\ c_{41} & c_{42} & c_{43} \end{array} \right] \quad [C_{22}] = \left[\begin{array}{cc} c_{34} & c_{35} \\ c_{44} & c_{45} \end{array} \right] \qquad \text{(E.12.6)}$$

E.13   Determinants

The determinant of a square matrix is a defined *mathematical operation* involving the elements of a matrix. For a 2 x 2 matrix we have:

$$|A| = \left| \begin{array}{cc} a_{11} & a_{12} \\ a_{21} & a_{22} \end{array} \right| = a_{11}a_{22} - a_{21}a_{12} \qquad \text{(E.13.1)}$$

For a 3 x 3 matrix we have:

$$|A| = \begin{vmatrix} a_{11} & a_{12} & a_{13} \\ a_{21} & a_{22} & a_{23} \\ a_{31} & a_{32} & a_{33} \end{vmatrix}$$

$$= a_{11}a_{22}a_{33} + a_{21}a_{13}a_{32} + a_{31}a_{12}a_{23} - a_{11}a_{23}a_{32} - a_{21}a_{12}a_{33} - a_{31}a_{22}a_{13}$$

$$= a_{11}\left(a_{22}a_{33} - a_{23}a_{32}\right) - a_{21}\left(a_{12}a_{33} - a_{13}a_{32}\right) + a_{31}\left(a_{12}a_{23} - a_{13}a_{22}\right)$$

$$= a_{11}\begin{vmatrix} a_{22} & a_{23} \\ a_{32} & a_{33} \end{vmatrix} - a_{21}\begin{vmatrix} a_{12} & a_{13} \\ a_{32} & a_{33} \end{vmatrix} + a_{31}\begin{vmatrix} a_{12} & a_{13} \\ a_{22} & a_{23} \end{vmatrix} \qquad (E.13.2)$$

Now the second-order determinants are exactly the *subdeterminants* that remain after the row and column containing the coefficient elements a_{11}, a_{21}, and a_{31}, respectively, are eliminated. We can therefore abbreviate the expression in Equation E.13.2 by introducing the concepts of *minors* and *cofactors*.

A *minor* of a determinant is defined as the subdeterminant remaining after a row and column are eliminated from the original determinant. We employ the notation:

$$|A_{ij}| \qquad (E.13.3)$$

for the minor formed by eliminating row i and column j from $|A|$.

A *cofactor* (sometimes called a *signed minor*) is defined as the minor of a determinant with an appropriate positive or negative sign according to:

$$|CF_{ij}| = (-1)^{i+j} |A_{ij}| \qquad (E.13.4)$$

On the basis of these definitions, we may write the denominator of Equation E.13.2 as follows:

$$|A| = a_{11}|A_{11}| - a_{21}|A_{21}| + a_{31}|A_{31}|$$

$$= a_{11}|CF_{11}| + a_{21}|CF_{21}| + a_{31}|CF_{31}|$$

$$= \sum_{i=1}^{3} a_{i1}|CF_{i1}| = \sum_{i=1}^{3} (-1)^{i+1} a_{i1}|A_{i1}| \qquad (E.13.5)$$

In this instance, the evaluation of $|A|$ has been carried out in terms of expansion in cofactors down the first *column*. If carried out in terms of cofactors down the second and third *columns* respectively, we have:

$$|A| = \sum_{i=1}^{3} a_{i2}|CF_{i2}| = \sum_{i=1}^{3} a_{i3}|CF_{i3}| \qquad (E.13.6)$$

It is left for the reader to verify that the expansion in terms of cofactors along any one of the three rows will lead to:

$$|A| = \sum_{j=1}^{3} a_{1j} \, |CF_{1j}| = \sum_{j=1}^{3} a_{2j} \, |CF_{2j}| = \sum_{j=1}^{3} a_{3j} \, |CF_{3j}| \qquad \text{(E.13.7)}$$

In general, for a third-order determinant:

$$|A| = \sum_{i=1}^{3} (-1)^{i+j} \, a_{ij} \, |A_{ij}| = \sum_{j=1}^{3} (-1)^{i+j} \, a_{ij} \, |A_{ij}| \qquad \text{(E.13.8)}$$

We shall *define* an nth order determinant on the basis of expansion of minors as:

$$|A| = \sum_{i=1}^{n} (-1)^{i+j} \, a_{ij} \, |A_{ij}| = \sum_{j=1}^{n} (-1)^{i+j} \, a_{ij} \, |A_{ij}| \qquad \text{(E.13.9)}$$

where the expansion may proceed along any one row or any one column. It is left for the reader to verify that $|A|$ will consist of $n!$ products, each product consisting of n elements a_{ij}, none of which have a common row or column [Hint: In each product, the first element can be chosen in (n) ways, the second in $(n-1)$ ways, etc.].

The value of the *third*-order determinant, consisting of only six triple products, is given in Equation E.13.2. None of the elements within each triple product has a common row or column. This permits a simple scheme for evaluation purposes:

First, form an augmented matrix (3 x 5) by copying the first two columns to the fourth and fifth columns:

$$\begin{vmatrix} a_{11} & a_{12} & a_{13} \\ a_{21} & a_{22} & a_{23} \\ a_{31} & a_{32} & a_{33} \end{vmatrix} \rightarrow \begin{vmatrix} a_{11} & a_{12} & a_{13} & a_{11} & a_{12} \\ a_{21} & a_{22} & a_{23} & a_{21} & a_{22} \\ a_{31} & a_{32} & a_{33} & a_{31} & a_{32} \end{vmatrix}$$

Now, *add* the products of *complete* diagonals going from left to right and *subtract* the products of *complete* diagonals going from right to left.

$$(+) \quad \begin{vmatrix} a_{11} & a_{12} & a_{13} & a_{11} & a_{12} \\ a_{21} & a_{22} & a_{23} & a_{21} & a_{22} \\ a_{31} & a_{32} & a_{33} & a_{31} & a_{32} \end{vmatrix}$$

$$(-) \quad \begin{vmatrix} a_{11} & a_{12} & a_{13} & a_{11} & a_{12} \\ a_{21} & a_{22} & a_{23} & a_{21} & a_{22} \\ a_{31} & a_{32} & a_{33} & a_{31} & a_{32} \end{vmatrix} \qquad \text{(E.13.10)}$$

which evaluates as:

$$(+) \ (a_{11}a_{22}a_{33} + a_{12}a_{23}a_{31} + a_{13}a_{21}a_{32})$$
$$(-) \ (a_{13}a_{22}a_{31} + a_{23}a_{32}a_{11} + a_{21}a_{12}a_{33}) \tag{E.13.11}$$

It should be noted that this algorithm will not work for higher order determinants – i.e., the expression in Equation E.13.9 must be used. For example, for a fourth-order determinant, this algorithm would deliver only eight quadruple products when, in fact, twenty-four such products are required.

We shall also define the adjoint of $[A]$ as a matrix where elements are the corresponding cofactors of $|A|$. Suppose that we have a matrix $[A] = [a_{ij}]$ with associated determinant $|A|$ and cofactors $(-1)^{i+j} |A_{ij}|$. Then the matrix:

$$[A'] = [a'_{ij}] = \left[(-1)^{i+j} |A_{ij}| \right] = [|CF_{ij}|] \tag{E.13.12}$$

is called the *adjoint* of $[A]$. Thus, the adjoint of $[A]$ is a matrix whose elements are the corresponding cofactors of $[A]$.

For a 3 x 3 matrix we have for the adjoint of $[A]$:

$$[A'] = \begin{bmatrix} CF_{11} & CF_{12} & CF_{13} \\ CF_{21} & CF_{22} & CF_{23} \\ CF_{31} & CF_{32} & CF_{33} \end{bmatrix} \tag{E.13.13}$$

E.14 Inverse (Reciprocal) Matrix

We define the *inverse* of $[A]$ as:

$$[A]^{-1} = \frac{1}{|A|} [A']^T \tag{E.14.1}$$

where $[A']^T$ is the transpose of the adjoint matrix. In expanded form:

$$[A]^{-1} = \frac{1}{|A|} \begin{bmatrix} |A_{11}| & -|A_{21}| & |A_{31}| & \cdots & \cdots \\ -|A_{12}| & |A_{22}| & -|A_{32}| & \cdots & \cdots \\ |A_{13}| & -|A_{23}| & |A_{33}| & \cdots & \cdots \\ \vdots & \vdots & \vdots & \cdots & \cdots \\ \vdots & \vdots & \vdots & \cdots & \cdots \end{bmatrix} \tag{E.14.2}$$

Thus, the *inverse* of a *square* matrix is the *transposed* adjoint of the matrix, multiplied by a scalar equal to the reciprocal of the determinant of the matrix. Then:

$$[A] [A]^{-1} = [A]^{-1} [A] = [I] \tag{E.14.3}$$

The reader may verify that when a column of $[A']^T$ is multiplied by the corresponding row in $[A]$, the result will be $|A|$ per Equation E.13.8. We state without proof that when a column of $[A']^T$ is multiplied by any other row of the $[A]$, the result will be zero.

Observe that a matrix inverse exists only if $|A|$ is nonzero. The inverse of a 2 x 2 matrix is easily remembered:

$$\begin{bmatrix} a & b \\ c & d \end{bmatrix}^{-1} = \frac{1}{ad - bc} \begin{bmatrix} d & -b \\ -c & a \end{bmatrix} \tag{E.14.4}$$

and for a partitioned square matrix:

$$\begin{bmatrix} A & B \\ C & D \end{bmatrix}^{-1} = \begin{bmatrix} P & Q \\ R & S \end{bmatrix} \tag{E.14.5}$$

with:

$$[P] = \left([A] - [B][D]^{-1}[C]\right)^{-1} = [A]^{-1} + [A]^{-1}[B][S][C][A]^{-1} \tag{E.14.6}$$

$$[S] = \left([D] - [C][A]^{-1}[B]\right)^{-1} = [D]^{-1} + [D]^{-1}[C][P][B][D]^{-1} \tag{E.14.7}$$

$$[Q] = [P][B][D]^{-1} = [A]^{-1}[B][S] \tag{E.14.8}$$

$$[R] = [D]^{-1}[C][P] = [S][C][A]^{-1} \tag{E.14.9}$$

The inverse of a partitioned matrix offers some advantage when the classical cofactor method is to be employed. Observe that the matrices to be inverted are of smaller order than the original matrix.

In many instances, inversion is done in a digital computer environment. In this situation, or even in manual situations, it may be more convenient to use Gaussian Elimination to find the inverse. We now illustrate the algorithm to do this to find the inverse of a 3 x 3 matrix that we will call $[A]$:

First, form an augmented matrix by inserting an identity matrix of the same order as the original matrix, $[A]$:

$$\begin{bmatrix} a_{11} & a_{12} & a_{13} \\ a_{21} & a_{22} & a_{23} \\ a_{31} & a_{32} & a_{33} \end{bmatrix} \rightarrow \left[\begin{array}{ccc|ccc} a_{11} & a_{12} & a_{13} & 1 & 0 & 0 \\ a_{21} & a_{22} & a_{23} & 0 & 1 & 0 \\ a_{31} & a_{32} & a_{33} & 0 & 0 & 1 \end{array} \right] \tag{E.14.10}$$

Next, perform Gaussian elimination on the rows of the new matrix such that the original matrix is transformed into the identity matrix. A new 3 x 3 matrix will evolve where the augmented identity matrix was originally positioned.

$$\left[\begin{array}{ccc|ccc} 1 & 0 & 0 & a^{-1}_{11} & a^{-1}_{12} & a^{-1}_{13} \\ 0 & 1 & 0 & a^{-1}_{21} & a^{-1}_{22} & a^{-1}_{23} \\ 0 & 0 & 1 & a^{-1}_{31} & a^{-1}_{32} & a^{-1}_{33} \end{array} \right] \tag{E.14.11}$$

This new matrix is the inverse of $[A]$:

$$[A]^{-1} = \begin{bmatrix} a^{-1}_{11} & a^{-1}_{12} & a^{-1}_{13} \\ a^{-1}_{21} & a^{-1}_{22} & a^{-1}_{23} \\ a^{-1}_{31} & a^{-1}_{32} & a^{-1}_{33} \end{bmatrix} \tag{E.14.12}$$

E.15 Matrix Reduction

Given a system of equations as follows:

$$\begin{bmatrix} y_1 \\ y_2 \\ 0 \end{bmatrix} = \begin{bmatrix} a_{11} & a_{12} & a_{13} \\ a_{21} & a_{22} & a_{23} \\ a_{31} & a_{32} & a_{33} \end{bmatrix} \begin{bmatrix} x_1 \\ x_2 \\ x_3 \end{bmatrix} \qquad (E.15.1)$$

which may be written in an expanded form as:

$$y_1 = a_{11}x_1 + a_{12}x_2 + a_{13}x_3$$
$$y_2 = a_{21}x_1 + a_{22}x_2 + a_{23}x_3$$
$$0 = a_{31}x_1 + a_{32}x_2 + a_{33}x_3 \qquad (E.15.2)$$

Solving the last equation for x_3 we have:

$$x_3 = -\frac{a_{31}}{a_{33}}x_1 - \frac{a_{32}}{a_{33}}x_2 \qquad (E.15.3)$$

and substituting this into the first two equations gives:

$$y_1 = \left(a_{11} - \frac{a_{13}a_{31}}{a_{33}} \right) x_1 + \left(a_{12} - \frac{a_{13}a_{32}}{a_{33}} \right) x_2$$

$$= b_{11}x_1 + b_{12}x_2 \qquad (E.15.4)$$

$$y_1 = \left(a_{21} - \frac{a_{23}a_{31}}{a_{33}} \right) x_1 + \left(a_{22} - \frac{a_{23}a_{32}}{a_{33}} \right) x_2$$

$$= b_{21}x_1 + b_{22}x_2 \qquad (E.15.5)$$

and in matrix form:

$$\begin{bmatrix} y_1 \\ y_2 \end{bmatrix} = \begin{bmatrix} b_{11} & b_{12} \\ b_{21} & b_{22} \end{bmatrix} \begin{bmatrix} x_1 \\ x_2 \end{bmatrix} \qquad (E.15.6)$$

with:

$$b_{ii} = a_{ii} - \frac{a_{ik}a_{ki}}{a_{kk}} \qquad (E.15.7)$$

$$b_{ij} = a_{ij} - \frac{a_{ik}a_{kj}}{a_{kk}} \qquad (E.15.8)$$

$$b_{ji} = a_{ji} - \frac{a_{jk}a_{ki}}{a_{kk}} \qquad (E.15.9)$$

Observe that if $a_{ij} = a_{ji}$, then $b_{ij} = b_{ji}$, and the calculation need be performed only once.

If there are several rows of zeros in the $[Y]$ matrix, we may employ the partitioned matrix form and write:

$$\left[\begin{array}{c} Y_a \\ \hline 0 \end{array} \right] = \left[\begin{array}{c|c} A & B \\ \hline C & D \end{array} \right] \left[\begin{array}{c} X_a \\ X_b \end{array} \right] \qquad (E.15.10)$$

and after eliminating the rows of zeros, and their associated columns, we have the result:

$$[Y_a] = \left([A] - [B][D]^{-1}[C]\right)[X_a] \tag{E.15.11}$$

It is apparent that we may obtain the result in Equation E.15.10 either by block matrix operation, or we may obtain the same result by eliminating one row and column at a time according to Equation E.15.7. The latter approach is, of course, the easier to carry out.

E.16   Numerical Matrix Inversion

Given a matrix of the form

$$[Y] = [A][X] \tag{E.16.1}$$

for which we seek the solution:

$$[X] = [A]^{-1}[Y] \tag{E.16.2}$$

Consider $[Y]$, $[A]$, and $[X]$ to be submatrices of a larger system of equations:

$$\left[\frac{X}{0}\right] = \left[\begin{array}{c|c} 0 & I \\ \hline I & -A \end{array}\right]\left[\frac{Y}{X}\right] \tag{E.16.3}$$

Then, on the basis of Equation E.15.11 we have directly:

$$[X] = \left([0] - [I][-A]^{-1}[I]\right)[Y]$$
$$= [A]^{-1}[Y] \tag{E.16.4}$$

Thus, by eliminating the rows and columns in the matrix:

$$\left[\begin{array}{c|c} 0 & I \\ \hline I & -A \end{array}\right] \tag{E.16.5}$$

we obtain $[A]^{-1}$. It follows directly that if we write:

$$\left[\begin{array}{c|c} 0 & I \\ \hline I & A \end{array}\right] \tag{E.16.6}$$

then by eliminating rows and columns in the matrix we obtain $-[A]^{-1}$.

This method is directly related to the inversion method presented in Section E.14.

E.17   Special Matrix Forms

Summation Matrix

$$[\Sigma] = \begin{bmatrix} 1 & 1 & 1 \\ 1 & 1 & 1 \\ 1 & 1 & 1 \end{bmatrix} \qquad\qquad [\Sigma]^n = \left(3^{n-1}\right)[\Sigma] \tag{E.17.1}$$

Rotation Matrix

$$\text{Forward:} \quad [R] = \begin{bmatrix} 0 & 1 & 0 \\ 0 & 0 & 1 \\ 1 & 0 & 0 \end{bmatrix}$$

$$\text{Backward:} \quad [R]^{-1} = \begin{bmatrix} 0 & 0 & 1 \\ 1 & 0 & 0 \\ 0 & 1 & 0 \end{bmatrix}$$

$$[R]^2 = [R]^{-1} \qquad [R]^3 = [I] \qquad \text{(E.17.2)}$$

Transposition Matrix

$$[T] = [T]^{-1} = \begin{bmatrix} 1 & 0 & 0 \\ 0 & 0 & 1 \\ 0 & 1 & 0 \end{bmatrix} \qquad\qquad [T]^2 = [I] \qquad \text{(E.17.3)}$$

Symmetrical Component Transformation

$$[\tilde{S}] = \left(\frac{1}{3}\right) \begin{bmatrix} 1 & 1 & 1 \\ 1 & a & a^2 \\ 1 & a^2 & a \end{bmatrix} \qquad [\tilde{S}]^{-1} = \begin{bmatrix} 1 & 1 & 1 \\ 1 & a^2 & a \\ 1 & a & a^2 \end{bmatrix}$$

$$[\tilde{S}]^{-1} = (3)[\tilde{S}]^* \qquad [\tilde{S}]^2 = \left(\frac{1}{3}\right)[T] \qquad [\tilde{S}]^{-2} = (3)[T] \qquad (\text{E.17.4})$$

Phase Quantities

$$[\tilde{V}_\phi] = \begin{bmatrix} \tilde{V}_a \\ \tilde{V}_b \\ \tilde{V}_c \end{bmatrix} \qquad\qquad [\tilde{I}_\phi] = \begin{bmatrix} \tilde{I}_a \\ \tilde{I}_b \\ \tilde{I}_c \end{bmatrix}$$

$$[\tilde{Z}_\phi] = \begin{bmatrix} \tilde{Z}_{aa} & \tilde{Z}_{ab} & \tilde{Z}_{ac} \\ \tilde{Z}_{ba} & \tilde{Z}_{bb} & \tilde{Z}_{bc} \\ \tilde{Z}_{ca} & \tilde{Z}_{cb} & \tilde{Z}_{cc} \end{bmatrix} \qquad\qquad [\tilde{Z}_\phi]^T = [\tilde{Z}_\phi] \qquad (\text{E.17.5})$$

Symmetrical Component Quantities

$$[\tilde{V}_c] = \begin{bmatrix} \tilde{V}_{a0} \\ \tilde{V}_{a1} \\ \tilde{V}_{a2} \end{bmatrix} = \begin{bmatrix} \tilde{V}_0 \\ \tilde{V}_1 \\ \tilde{V}_2 \end{bmatrix} = [\tilde{S}][\tilde{V}_\phi] \qquad [\tilde{I}_c] = \begin{bmatrix} \tilde{I}_{a0} \\ \tilde{I}_{a1} \\ \tilde{I}_{a2} \end{bmatrix} = \begin{bmatrix} \tilde{I}_0 \\ \tilde{I}_1 \\ \tilde{I}_2 \end{bmatrix} = [\tilde{S}][\tilde{I}_\phi]$$

$$[\tilde{Z}_c] = \begin{bmatrix} \tilde{Z}_{00} & \tilde{Z}_{01} & \tilde{Z}_{02} \\ \tilde{Z}_{10} & \tilde{Z}_{11} & \tilde{Z}_{12} \\ \tilde{Z}_{20} & \tilde{Z}_{21} & \tilde{Z}_{22} \end{bmatrix} = [\tilde{S}][\tilde{Z}_\phi][\tilde{S}]^{-1} \qquad (E.17.6)$$

$$[\tilde{Z}_c]^T = \left([\tilde{S}]^T\right)^{-1} [\tilde{Z}_\phi]^T [\tilde{S}]^T$$
$$= [\tilde{S}]^{-1} [\tilde{Z}_\phi][\tilde{S}]$$
$$= [\tilde{S}]^* [\tilde{Z}_\phi] \left([\tilde{S}]^{-1}\right)^* \neq [\tilde{Z}_c] \qquad (E.17.7)$$

$$[\tilde{V}_\phi] = [\tilde{Z}_\phi][\tilde{I}_\phi]$$
$$[\tilde{V}_c] = [\tilde{S}][\tilde{V}_\phi] = [\tilde{S}][\tilde{Z}_\phi][\tilde{S}]^{-1}[\tilde{S}][\tilde{I}_\phi]$$
$$= [\tilde{Z}_c][\tilde{I}_c] \qquad (E.17.8)$$

$$\tilde{S}_\phi = P_\phi + jQ_\phi = [\tilde{V}_\phi]^T [\tilde{I}_\phi]^*$$
$$[\tilde{V}_c]^T [\tilde{I}_c]^* = [\tilde{V}_\phi]^T [\tilde{S}][\tilde{S}]^* [\tilde{I}_\phi]^*$$
$$= \left(\frac{1}{3}\right) [P_\phi + jQ_\phi] \qquad (E.17.9)$$

Appendix F

Magnetic Energy in Transformers

We shall show in this appendix that to a high degree of accuracy, the total magnetic energy stored in a practical power transformer with n windings on a common closed magnetic core is stored in the leakage flux paths. From short-circuit tests between the $n(n-2)/2$ pairs of windings, we can determine the various L_{j-k} (or $\wp_{j-k}$ when referred to a per turn basis), and given the current in each winding, we can obtain the total magnetic stored energy, no matter how these windings may be externally connected.

If these n windings are reconnected to an N-winding transformer where $N < n$, the total magnetic energy is stored in $N(N-1)/2$ equivalent leakage flux paths. In this manner, we may determine the $N(N-1)/2$ equivalent short-circuit reactances for the reconnected transformer.

F.1   Short-Circuit Permeance of a Two-Winding Transformer

In a two-winding transformer, it is conceptually convenient to define the total flux linking a winding in terms of mutual and leakage flux components. Thus, for sinusoidal excitation, and in terms of rms phasor notation:

$$\left[\begin{array}{c} \dfrac{\tilde{\Lambda}_1}{N_1} \\ \dfrac{\tilde{\Lambda}_2}{N_2} \end{array} \right] = \left[\begin{array}{cc} (\wp_{\ell 1} + \wp_m) & \wp_m \\ \wp_m & (\wp_{\ell 2} + \wp_m) \end{array} \right] = \left[\begin{array}{c} N_1 \tilde{I}_1 \\ N_2 \tilde{I}_2 \end{array} \right] \tag{F.1.1}$$

$$\frac{\tilde{\Lambda}_1}{N_1} = \wp_{\ell 1} N_1 \tilde{I}_1 + \wp_m \left(N_1 \tilde{I}_1 + N_2 \tilde{I}_2 \right) \tag{F.1.2}$$

$$\frac{\tilde{\Lambda}_2}{N_2} = \wp_{\ell 2} N_2 \tilde{I}_2 + \wp_m \left(N_1 \tilde{I}_1 + N_2 \tilde{I}_2 \right) \tag{F.1.3}$$

$$\left(\frac{\tilde{\Lambda}_1}{N_1} - \frac{\tilde{\Lambda}_2}{N_2} \right) = \wp_{\ell 1} N_1 \tilde{I}_1 - \wp_{\ell 2} N_2 \tilde{I}_2 \tag{F.1.4}$$

If we write:

$$\tilde{\epsilon} = N_1 \tilde{I}_1 + N_2 \tilde{I}_2 \tag{F.1.5}$$

then:

$$\left(\frac{\tilde{\Lambda}_1}{N_1} - \frac{\tilde{\Lambda}_2}{N_2} \right) = \left(\wp_{\ell 1} + \wp_{\ell 2} \right) N_1 \tilde{I}_1 - \wp_{\ell 2} \, \tilde{\epsilon} \tag{F.1.6}$$

If the second winding is short-circuited, we have:

$$\frac{\tilde{\Lambda}_1}{N_1^2 \, \tilde{I}_1} = \wp_{\ell 1} + \wp_{\ell 2} - \wp_{\ell 2} \, \frac{\tilde{\epsilon}}{N_1 \tilde{I}_1} \tag{F.1.7}$$

Now the permeance of the iron core in practical power transformers is so much larger than the permeance of the leakage flux paths that the magnetizing ampere turns in Equation F.1.5 are generally less than 1% of the rated ampere-turns $N_1 I_1$ so that ignoring the last term in Equation F.1.7 causes an acceptable error of less than 0.5%. We may therefore write:

$$\frac{\tilde{\Lambda}_1}{N_1^2 \, \tilde{I}_1} = \wp_{1-2} = \wp_{\ell 1} + \wp_{\ell 2} \tag{F.1.8}$$

We observe that $\wp_{\ell 1}$ and $\wp_{\ell 2}$ cannot be separately identified from such a short-circuit test, which reminds us that dividing the total flux into mutual and leakage flux components was done quite arbitrarily and merely for the sake of convenience. That this division in not reflected in practical tests on transformers should not be surprising. Indeed, an actual short-circuit test on a power transformer would reflect winding resistance in addition to the short-circuit reactance, but this is of no interest to us in this appendix.

F.2   Multiwinding Transformers

In a multiwinding transformer, it no longer makes sense to define $\wp_{\ell 1}$, $\wp_{\ell 2}$, $\wp_{\ell 3}$, etc. For example, in concentric winding construction, the leakage flux between the inside winding and the outside winding must necessarily link some of the intermediate windings and therefore no longer conforms to our definition of leakage flux.

Consider a three-winding transformer for which we have:

$$\begin{bmatrix} \dfrac{\tilde{\Lambda}_1}{N_1} \\[2mm] \dfrac{\tilde{\Lambda}_2}{N_2} \\[2mm] \dfrac{\tilde{\Lambda}_3}{N_3} \end{bmatrix} = \begin{bmatrix} \wp_{11} & \wp_{12} & \wp_{13} \\ \wp_{21} & \wp_{22} & \wp_{23} \\ \wp_{31} & \wp_{32} & \wp_{33} \end{bmatrix} \begin{bmatrix} N_1 \tilde{I}_1 \\ N_2 \tilde{I}_2 \\ N_3 \tilde{I}_3 \end{bmatrix} \tag{F.2.1}$$

Based upon our analysis of the two-winding transformer, intuitively we feel that the $\wp_{jj}$ are somewhat larger than the $\wp_{jk}$. However, we also feel that $\wp_{12}$ is not necessarily identical to $\wp_{13}$. We must therefore embark on a variation of the analysis given in the previous section.

We now consider a single pair of windings, say #1 and #3, and write:

$$\tilde{\epsilon}_{13} = N_1 \tilde{I}_1 + N_3 \tilde{I}_3 \tag{F.2.2}$$

Then:

$$\frac{\tilde{\Lambda}_1}{N_1} = \wp_{11}\, N_1 \tilde{I}_1 + \wp_{13}\, N_3 \tilde{I}_3 = (\wp_{11} - \wp_{13})\, N_1 \tilde{I}_1 + \wp_{13}\, \tilde{\epsilon}_{13} \tag{F.2.3}$$

$$\frac{\tilde{\Lambda}_3}{N_3} = \wp_{31}\, N_1 \tilde{I}_1 + \wp_{33}\, N_3 \tilde{I}_3 = (\wp_{33} - \wp_{13})\, N_3 \tilde{I}_3 + \wp_{13}\, \tilde{\epsilon}_{13}$$

$$= -(\wp_{33} - \wp_{13})\, N_1 \tilde{I}_1 + (\wp_{33} - \wp_{13})\, \tilde{\epsilon}_{13} + \wp_{13}\, \tilde{\epsilon}_{13} \tag{F.2.4}$$

and:

$$\left(\frac{\tilde{\Lambda}_1}{N_1} - \frac{\tilde{\Lambda}_3}{N_3}\right) = (\wp_{11} + \wp_{33} - 2\wp_{13})\, N_1\, \tilde{I}_1 - (\wp_{33} - \wp_{13})\, \tilde{\epsilon}_{13} \tag{F.2.5}$$

Then:

$$\wp_{1-3} = (\wp_{11} + \wp_{33} - 2\wp_{13}) - (\wp_{33} - \wp_{13})\frac{\tilde{\epsilon}_{13}}{N_1 \tilde{I}_1} \tag{F.2.6}$$

and with an acceptably small error:

$$\wp_{1-3} = (\wp_{11} + \wp_{33} - 2\wp_{13}) \tag{F.2.7}$$

Thus, the short-circuit permeance between a pair of windings of a multiwinding transformer may be related to the permeance elements of the permeance matrix in Equation F.2.1, and in general:

$$\wp_{i-j} = (\wp_{ii} + \wp_{jj} - 2\wp_{ij}) \tag{F.2.8}$$

F.3 Magnetic Energy – Phasor Notation

When we deal with sinusoidal voltages and currents, it is generally most convenient to represent these quantities as rms phasors. Thus, we have, for example:

$$\tilde{V} = (R + jX)\, \tilde{I} \tag{F.3.1}$$

$$\tilde{S} = \tilde{V}\tilde{I}^* = P + jQ \tag{F.3.2}$$

$$P = R\tilde{I}\tilde{I}^* = \left|I^2\right| R \tag{F.3.3}$$

$$Q = X\tilde{I}\tilde{I}^* = \left|I^2\right| X = 2\omega\, \frac{\left|I^2\right| L}{2} = 2\omega\, \frac{\tilde{\Lambda}\tilde{I}^*}{2} \tag{F.3.4}$$

Thus, we have, in terms of phasor notation:

$$W_{mag} = \frac{\tilde{\Lambda}\tilde{I}^*}{2} = \frac{\tilde{\Lambda}^*\tilde{I}}{2} \qquad [J] \tag{F.3.5}$$

F.4   Stored Magnetic Energy in Transformers

The total magnetic energy stored in electromagnetic devices is given by the form:

$$W_{mag} = \frac{1}{2} \sum_{j=1}^{n} \tilde{\Lambda}_j \, \tilde{I}_j^* \qquad [J] \qquad\qquad (\text{F.4.1})$$

In the case of a transformer, we have a system of linear equations that may be written as:

$$\left[\frac{\tilde{\Lambda}}{N} \right] = [\wp] \left[N\tilde{I}^* \right] \qquad [V \cdot s/turn] \qquad\qquad (\text{F.4.2})$$

Then:

$$W_{mag} = \frac{1}{2} \left[N\tilde{I}^* \right]^T \left[\frac{\tilde{\Lambda}}{N} \right] = \frac{1}{2} \left[N\tilde{I}^* \right]^T [\wp] \left[N\tilde{I}^* \right]$$

$$= \frac{1}{2} \sum_{j=1}^{n} N_j^2 \, |I_j|^2 \, \wp_{jj} + \frac{1}{2} \sum_{j=1}^{n} \sum_{\substack{k=1 \\ k \neq j}}^{n} N_j \tilde{I}_j^* N_k \tilde{I}_k \, \wp_{jk} \qquad [J] \qquad\qquad (\text{F.4.3})$$

We prefer to employ the permeances established from short-circuit tests so that we have, on the basis of Equation F.2.8:

$$\wp_{jk} = \frac{1}{2} \left(\wp_{jj} + \wp_{kk} - \wp_{j-k} \right) \qquad\qquad (\text{F.4.4})$$

so that:

$$W_{mag} = \frac{1}{2} \sum_{j=1}^{n} N_j^2 \, |I_j|^2 \, \wp_{jj} + \frac{1}{4} \sum_{j=1}^{n} \sum_{\substack{k=1 \\ k \neq j}}^{n} N_j \tilde{I}_j^* N_k \tilde{I}_k \, \wp_{jj}$$

$$+ \frac{1}{4} \sum_{j=1}^{n} \sum_{\substack{k=1 \\ k \neq j}}^{n} N_j \tilde{I}_j^* N_k \tilde{I}_k \, \wp_{kk} - \frac{1}{4} \sum_{j=1}^{n} \sum_{\substack{k=1 \\ k \neq j}}^{n} N_j \tilde{I}_j^* N_k \tilde{I}_k \, \wp_{j-k} \qquad [J] \qquad (\text{F.4.5})$$

If we write:

$$\tilde{\epsilon} = \sum_{k=1}^{n} N_k \tilde{I}_k = N_j \tilde{I}_j + \sum_{\substack{k=1 \\ k \neq j}}^{n} N_k \tilde{I}_k \qquad\qquad (\text{F.4.6})$$

then the second term on the right-hand side of Equation F.4.5 may be written as:

$$\frac{1}{4} \sum_{j=1}^{n} N_j \tilde{I}_j^* \left(\tilde{\epsilon} - N_j \tilde{I}_j \right) \wp_{jj} = \frac{1}{4} \sum_{j=1}^{n} N_j \tilde{I}_j^* \, \tilde{\epsilon} \, \wp_{jj} - \frac{1}{4} \sum_{j=1}^{n} N_j^2 \, |I_j|^2 \, \wp_{jj} \qquad (\text{F.4.7})$$

It is left for the reader to show that third term on the right-hand side of Equation F.4.5 is simply the conjugate of the second term. Thus, Equation F.4.5 may be written as:

$$W_{mag} = -\frac{1}{4} \sum_{j=1}^{n} \sum_{\substack{k=1 \\ k \neq j}}^{n} N_j \tilde{I}_j^* N_k \tilde{I}_k \, \wp_{j-k} + \frac{1}{4} \sum_{j=1}^{n} N_j \left(\tilde{\epsilon} \tilde{I}_j^* - \tilde{\epsilon}^* \tilde{I}_j \right) \wp_{jj} \qquad [J] \qquad (\text{F.4.8})$$

The last term in Equation F.4.8 becomes vanishingly small. We can demonstrate this approximately if we assume all $\wp_{jj}$ to be about equal in magnitude so that we may write:

$$\frac{1}{4}\sum_{j=1}^{n} N_j \left(\tilde{\epsilon}\tilde{I}_j^* + \tilde{\epsilon}^*\tilde{I}_j\right)\wp_{jj} = \frac{\tilde{\epsilon}\,\wp_{jj}}{4}\sum_{j=1}^{n} N_j\tilde{I}_j^* = \frac{\tilde{\epsilon}^*\,\wp_{jj}}{4}\sum_{j=1}^{n} N_j\tilde{I}_j$$

$$= \frac{|\epsilon|^2\,\wp_{jj}}{2} \to 0 \qquad \text{(F.4.9)}$$

Thus, with negligible error, we determine that all of the magnetic energy stored within a multi-winding transformer resides in the leakage flux paths. The magnetic energy stored within a multi-winding transformer may be expressed in terms of the permeances associated with the short-circuit reactances between the various pairs of transformer windings as:

$$W_{mag} = -\frac{1}{4}\sum_{j=1}^{n}\sum_{\substack{k=1\\k\neq j}}^{n} N_j\tilde{I}_j^* N_k\tilde{I}_k\,\wp_{j-k} \qquad [J] \qquad \text{(F.4.10)}$$

Equation F.4.10 may also be written in the form:

$$W_{mag} = -\frac{1}{4}\sum_{j=1}^{n}\sum_{k=j+1}^{n} N_j N_k \left(\tilde{I}_j^*\tilde{I}_k + \tilde{I}_j\tilde{I}_k^*\right)\wp_{j-k} \qquad [J] \qquad \text{(F.4.11)}$$

in which case we see more clearly that W_{mag} is always a real number, as it should be, since we have neglected all real losses.

F.5   Per Unit Quantities

We now wish to establish the relationship between physical quantities employed in Equations F.4.10 and F.4.11, and the more commonly known per unit quantities $\bar{X}_{j-k}$.

Since all windings have been assumed to have a common core, we have a single MVA_{base}, and voltage bases according to:

$$\frac{V_{j\,base}}{N_j} = \frac{V_{k\,base}}{N_k} = \cdots \qquad \text{(F.5.1)}$$

and current bases according to:

$$N_j I_{j\,base} = N_k I_{k\,base} = \cdots \qquad \text{(F.5.2)}$$

and impedance bases according to:

$$\frac{Z_{j\,base}}{N_j^2} = \frac{Z_{k\,base}}{N_k^2} = \cdots \qquad \text{(F.5.3)}$$

In all cases, we have:

$$VA_{base} = \left(\frac{V_{base}}{N}\right)(NI_{base}) = (NI_{base})^2\left(\frac{Z_{base}}{N^2}\right) \qquad \text{(F.5.4)}$$

If we multiply both sides of Equation F.4.10 by 2ω, we obtain:

$$Q_{mag} = -\frac{1}{2} \sum_{j=1}^{n} \sum_{\substack{k=1 \\ k \neq j}}^{n} N_j \tilde{I}_j^* N_k \tilde{I}_k X_{j-k} \qquad [VAR] \qquad \text{(F.5.5)}$$

where X_{j-k} is on a per turn basis. After dividing both sides of Equation F.5.5 by the base quantities, we have:

$$\bar{Q}_{mag} = -\frac{1}{2} \sum_{j=1}^{n} \sum_{\substack{k=1 \\ k \neq j}}^{n} \tilde{\bar{I}}_j^* \tilde{\bar{I}}_k \bar{X}_{j-k} \qquad \text{(F.5.6)}$$

which may also be written in the form:

$$\bar{Q}_{mag} = -\frac{1}{2} \sum_{j=1}^{n} \sum_{k=j+1}^{n} \left(\tilde{\bar{I}}_j^* \tilde{\bar{I}}_k + \tilde{\bar{I}}_j \tilde{\bar{I}}_k^* \right) \bar{X}_{j-k} \qquad \text{(F.5.7)}$$

Appendix G

Exciting Current in Three-Legged Core-Type Transformer

From a no-load test on a single-phase transformer, we determine the exciting current, the core losses, the magnetizing current, and the turns ratio of the transformer. Occasionally, these tests are conducted over a range of values for applied voltage, and the data plotted graphically. Via an appropriate curve-fitting process, this data may be expressed as a mathematical function suitable for analytic purposes.

Because of the inherent dissymmetry in the core structure of a three-legged core-type transformer, the exciting currents observed in the three phases are not identical. Whether, in an analysis, the identities of the three currents should be maintained or whether some mean value will be satisfactory is a matter of engineering judgment. The following detailed examination of the three exciting currents should be helpful in making such a decision.

G.1   Analysis

Consider the three-legged core-type transformer shown in Figure G.1.1. Then:

$$\widetilde{MMF}_{AB} = N\tilde{I}_a - (\Re_{leg} + 2\Re_{yoke})\,\tilde{\Phi}_a$$
$$\widetilde{MMF}_{AB} = N\tilde{I}_b - (\Re_{leg})\,\tilde{\Phi}_b$$
$$\widetilde{MMF}_{AB} = N\tilde{I}_c - (\Re_{leg} + 2\Re_{yoke})\,\tilde{\Phi}_c \qquad (G.1.1)$$

If we add all three equations, we obtain:

$$3\widetilde{MMF}_{AB} = \left(N\tilde{I}_a + N\tilde{I}_b + N\tilde{I}_c\right) - \Re_{leg}\left(\tilde{\Phi}_a + \tilde{\Phi}_b + \tilde{\Phi}_c\right) - 2\Re_{yoke}\left(\tilde{\Phi}_a + \tilde{\Phi}_c\right) \qquad (G.1.2)$$

For an ungrounded wye connection we must have:

$$\left(\tilde{I}_a + \tilde{I}_b + \tilde{I}_c\right) = 0 \qquad (G.1.3)$$

If we apply a balanced set of positive sequence line-to-line voltages, then by Faraday's Law we must have:

$$\left(\tilde{\Phi}_{a1} + \tilde{\Phi}_{b1} + \tilde{\Phi}_{c1}\right) = 0 \qquad (G.1.4)$$

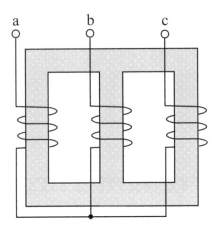

Figure G.1.1: Three-Legged Core-Type Transformer

for the positive sequence winding fluxes. There could conceivably be an additional zero sequence winding flux, but for our purposes we may assume this to be negligibly small since this flux would have to leave the core and return over an air path. Substituting Equations G.1.3 and G.1.4 into Equation G.1.2, we obtain:

$$\widetilde{MMF}_{AB} = \frac{2}{3}\Re_{yoke}\tilde{\Phi}_{b1} \tag{G.1.5}$$

Then, substituting Equation G.1.5 into Equation G.1.1, we have:

$$N\tilde{I}_{a1} = \left(\Re_{leg} + 2\Re_{yoke}\right)\tilde{\Phi}_{a1} + \frac{2}{3}\Re_{yoke}\tilde{\Phi}_{b1}$$

$$N\tilde{I}_{b1} = \left(\Re_{leg} + 2\Re_{yoke}\right)\tilde{\Phi}_{b1} + \frac{2}{3}\Re_{yoke}\tilde{\Phi}_{b1} - 2\Re_{yoke}\tilde{\Phi}_{b1}$$

$$N\tilde{I}_{c1} = \left(\Re_{leg} + 2\Re_{yoke}\right)\tilde{\Phi}_{c1} + \frac{2}{3}\Re_{yoke}\tilde{\Phi}_{b1} \tag{G.1.6}$$

G.2 Fundamental Frequency Magnetizing Current

Let:

$$\frac{\Re_{leg}}{\Re_{yoke}} = k \tag{G.2.1}$$

so that we may write Equation G.1.6 in the form:

$$\tilde{I}_{a1} = \frac{\Re_{yoke}}{3N}\left[\left(3k+6\right)\tilde{\Phi}_{a1} + 2\tilde{\Phi}_{b1}\right]$$

$$\tilde{I}_{b1} = \frac{\Re_{yoke}}{3N}\left[\left(3k+6\right)\tilde{\Phi}_{b1} + 2\tilde{\Phi}_{b1} - 6\tilde{\Phi}_{b1}\right]$$

$$\tilde{I}_{c1} = \frac{\Re_{yoke}}{3N}\left[\left(3k+6\right)\tilde{\Phi}_{c1} + 2\tilde{\Phi}_{b1}\right] \tag{G.2.2}$$

Now let:

$$\tilde{\Phi}_{a1} = 1 \qquad \tilde{\Phi}_{b1} = \tilde{a}^2 \qquad \tilde{\Phi}_{c1} = \tilde{a} \qquad \text{(G.2.3)}$$

so that:

$$\tilde{I}_{a1} = \left(\frac{\Re_{yoke}}{3N}\right)\left[(3k+6) + 2\tilde{a}^2\right]$$

$$\tilde{I}_{b1} = \left(\frac{\Re_{yoke}}{3N}\right)\left[(3k+6)\,\tilde{a}^2 + 2\tilde{a}^2 - 6\tilde{a}^2\right]$$

$$\tilde{I}_{c1} = \left(\frac{\Re_{yoke}}{3N}\right)\left[(3k+6)\,\tilde{a} + 2\tilde{a}^2\right] \qquad \text{(G.2.4)}$$

or:

$$\tilde{I}_{a1} = \left(\frac{\Re_{yoke}}{3N}\right)\left[(3k+4) - 2\tilde{a}\right]$$

$$\tilde{I}_{b1} = \left(\frac{\Re_{yoke}}{3N}\right)\left[(3k+2)\right]\tilde{a}^2$$

$$\tilde{I}_{c1} = \left(\frac{\Re_{yoke}}{3N}\right)\left[(3k+4) - 2\tilde{a}^2\right]\tilde{a} \qquad \text{(G.2.5)}$$

and:

$$\tilde{I}_{a1} = \left(\frac{\Re_{yoke}}{3N}\right)\left[3k+5 - j\sqrt{3}\right]$$

$$\tilde{I}_{b1} = \left(\frac{\Re_{yoke}}{3N}\right)\left[3k+2\right]\tilde{a}^2$$

$$\tilde{I}_{c1} = \left(\frac{\Re_{yoke}}{3N}\right)\left[3k+5 + j\sqrt{3}\right]\tilde{a} \qquad \text{(G.2.6)}$$

For small transformers, we might typically have:

$$\frac{\Re_{leg}}{\Re_{yoke}} = 0.4 \qquad \text{(G.2.7)}$$

Then:

$$\tilde{I}_{a1} = \left(\frac{\Re_{yoke}}{3N}\right)\left[6.2 - j\sqrt{3}\right] = 6.44\angle{-15.61°}\left(\frac{\Re_{yoke}}{3N}\right)$$

$$\tilde{I}_{b1} = \left(\frac{\Re_{yoke}}{3N}\right)[3.2]\,\tilde{a}^2 = 3.20\angle{-120.00°}\left(\frac{\Re_{yoke}}{3N}\right)$$

$$\tilde{I}_{c1} = \left(\frac{\Re_{yoke}}{3N}\right)\left[6.2 + j\sqrt{3}\right]\tilde{a} = 6.44\angle{135.61°}\left(\frac{\Re_{yoke}}{3N}\right) \qquad \text{(G.2.8)}$$

Thus, the fundamental frequency component of the magnetizing current in the outer legs can be about twice that in the center leg.

G.3   Third Harmonic Magnetizing Current

In order to maintain a sinusoidal winding flux in the core, a third harmonic component of magnetizing current must also be able to flow. If the third harmonic component of current cannot flow, a third harmonic component of flux will be developed. Conversely, a third harmonic component of flux, if it exists, will result in a third harmonic voltage, which will attempt to cause third harmonic currents to flow.

In order to determine the magnitude of the third harmonic component of magnetizing current, we shall assume that a third harmonic component of flux exists.

Based upon Equation G.2.3, we may now write:

$$\tilde{\Phi}_{a3} = \tilde{\Phi}_{b3} = \tilde{\Phi}_{c3} = h_3 \qquad \text{(G.3.1)}$$

whereby all triplen harmonic fluxes are in phase with one another (recall that the third harmonic is a zero sequence harmonic). Then, based on Equation G.2.2, we have for the third harmonics:

$$\tilde{I}_{a3} = \left(\frac{\Re_{yoke}}{3N}\right)[3k + 6 + 2]\,h_3$$

$$\tilde{I}_{b3} = \left(\frac{\Re_{yoke}}{3N}\right)[3k + 6 - 4]\,h_3$$

$$\tilde{I}_{c3} = \left(\frac{\Re_{yoke}}{3N}\right)[3k + 6 + 2]\,h_3 \qquad \text{(G.3.2)}$$

Now the terms $(3k + 6)$, being equal and in phase, cannot cause a third harmonic current to flow. However, the remaining terms, representing fluxes and/or associated voltages, will cause a third harmonic magnetizing current component to flow in the outer legs, returning into the center leg. Thus, for $k = 0.4$, we have effectively:

$$\tilde{I}_{a3} = \frac{\Re_{yoke}}{3N}\left[2h_3\right]$$

$$\tilde{I}_{b3} = \frac{\Re_{yoke}}{3N}\left[-4h_3\right]$$

$$\tilde{I}_{c3} = \frac{\Re_{yoke}}{3N}\left[2h_3\right] \tag{G.3.3}$$

G.4   Fifth Harmonic Magnetizing Current

In order to maintain a sinusoidal winding flux in the core, a fifth harmonic component of magnetizing current must also be able to flow. If the fifth harmonic component of current cannot flow, a fifth harmonic component of flux will be developed. Conversely, a fifth harmonic component of flux, if it exists, will result in a fifth harmonic voltage, which will attempt to cause fifth harmonic currents to flow.

In order to determine the magnitude of the fifth harmonic component of magnetizing current, we shall assume that a fifth harmonic component of flux exists. Thus:

$$\tilde{\Phi}_{a5} = h_5 \qquad\qquad \tilde{\Phi}_{b5} = \tilde{a}h_5 \qquad\qquad \tilde{\Phi}_{c5} = \tilde{a}^2 h_5 \tag{G.4.1}$$

The fifth harmonic appears effectively as a negative sequence set of fluxes, so that:

$$\tilde{I}_{a5} = \left(\frac{\Re_{yoke}}{3N}\right)\left[(3k+6) + 2\tilde{a}\right]h_5$$

$$\tilde{I}_{b5} = \left(\frac{\Re_{yoke}}{3N}\right)\left[(3k+6)\,\tilde{a} + 2\tilde{a} - 6\tilde{a}\right]h_5$$

$$\tilde{I}_{c5} = \left(\frac{\Re_{yoke}}{3N}\right)\left[(3k+6)\,\tilde{a}^2 + 2\tilde{a}\right]h_5 \tag{G.4.2}$$

or:

$$\tilde{I}_{a5} = \left(\frac{\Re_{yoke}}{3N}\right)\left[(3k+4) - 2\tilde{a}^2\right]h_5$$

$$\tilde{I}_{b5} = \left(\frac{\Re_{yoke}}{3N}\right)\left[(3k+2)\right]\tilde{a}h_5$$

$$\tilde{I}_{c5} = \left(\frac{\Re_{yoke}}{3N}\right)\left[(3k+4) - 2\tilde{a}\right]\tilde{a}^2 h_5 \tag{G.4.3}$$

and:

$$\tilde{I}_{a5} = \left(\frac{\mathfrak{R}_{yoke}}{3N}\right)\left[3k + 5 + j\sqrt{3}\right]h_5$$

$$\tilde{I}_{b5} = \left(\frac{\mathfrak{R}_{yoke}}{3N}\right)\left[3k + 2\right]\tilde{a}h_5$$

$$\tilde{I}_{c5} = \left(\frac{\mathfrak{R}_{yoke}}{3N}\right)\left[3k + 5 - j\sqrt{3}\right]\tilde{a}^2 h_5 \qquad\qquad \text{(G.4.4)}$$

Then, on the basis of Equation G.2.8, we may write directly:

$$\tilde{I}_{a5} = \left(\frac{\mathfrak{R}_{yoke}}{3N}\right)\left[6.2 + j\sqrt{3}\right]h_5 = 6.44 h_5 \angle 15.61^o \left(\frac{\mathfrak{R}_{yoke}}{3N}\right)$$

$$\tilde{I}_{b5} = \left(\frac{\mathfrak{R}_{yoke}}{3N}\right)\left[3.2\right]\tilde{a}h_5 = 3.20 h_5 \angle 120.00^o \left(\frac{\mathfrak{R}_{yoke}}{3N}\right)$$

$$\tilde{I}_{c5} = \left(\frac{\mathfrak{R}_{yoke}}{3N}\right)\left[6.2 - j\sqrt{3}\right]\tilde{a}^2 h_5 = 6.44 h_5 \angle -135.61^o \left(\frac{\mathfrak{R}_{yoke}}{3N}\right) \qquad \text{(G.4.5)}$$

G.5   Comparison with Test Results

Figures G.5.1 and G.5.2 show oscillograms of the exciting currents for a three-legged core-type transformer. For analytical purposes, $\mathfrak{R}_{leg}/\mathfrak{R}_{yoke}$ was taken to be 0.4, as is perhaps typical for this transformer. Via a little trial and error, the following values were selected:

$$h_3 = 0.20 \qquad\qquad h_5 = 0.09 \qquad\qquad\qquad \text{(G.5.1)}$$

for giving such a reasonable match for Figure G.5.2. Figure G.5.2 also shows the results of the analysis neglecting higher-order harmonics.

The results of the analysis compare quite well with the test results, considering that the input data was only estimated, and the core loss was not included. Certainly, they confirm the validity of the theory presented in this appendix. The following equations were employed in the analysis:

$$i_a = + 6.44 \sin(\omega t - 15.61^o) - 2(.2)\sin(3\omega t)$$
$$\qquad + 6.44(.09)\sin(5\omega t + 15.61^o)$$

$$i_b = + 3.20 \sin(\omega t - 120^o) + 4(.2)\sin(3\omega t)$$
$$\qquad + 3.20(.09)\sin(5\omega t + 120^o)$$

$$i_c = + 6.44 \sin(\omega t + 135.61^o) - 2(.2)\sin(3\omega t)$$
$$\qquad + 6.44(.09)\sin(5\omega t - 135.61^o) \qquad\qquad \text{(G.5.2)}$$

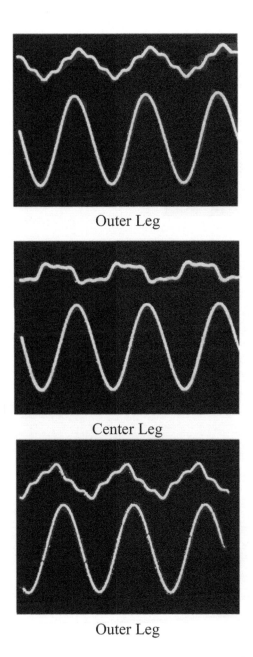

Outer Leg

Center Leg

Outer Leg

Figure G.5.1: Exciting Currents for a Three-Legged Core-Type Transformer, V = 700 Volts rms

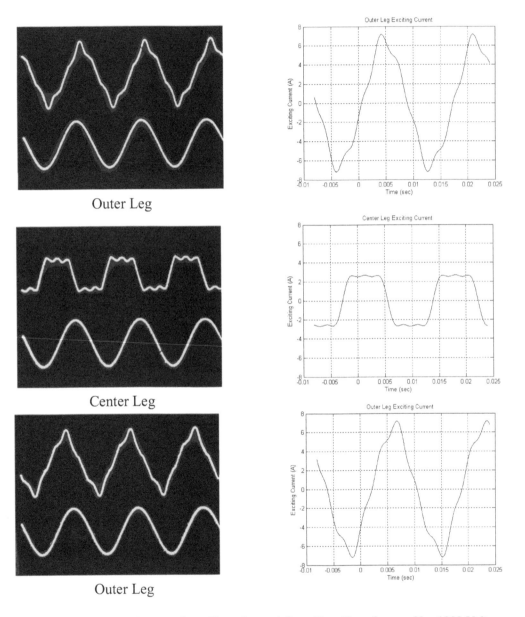

Outer Leg

Center Leg

Outer Leg

Figure G.5.2: Exciting Currents for a Three-Legged Core-Type Transformer, V = 1000 Volts rms

Appendix H

Hyperbolic Functions

The power series expansion for the exponential functions are:

$$e^{\theta} = 1 + \theta + \frac{\theta^2}{2!} + \frac{\theta^3}{3!} + \frac{\theta^4}{4!} + \cdots$$

$$e^{-\theta} = 1 - \theta + \frac{\theta^2}{2!} - \frac{\theta^3}{3!} + \frac{\theta^4}{4!} - \cdots$$

H.1 Hyperbolic Functions

For the hyperbolic functions, we have:

$$\cosh \theta = \frac{e^{\theta} + e^{-\theta}}{2} = 1 + \frac{\theta^2}{2!} + \frac{\theta^4}{4!} + \cdots \tag{H.1.1}$$

$$\sinh \theta = \frac{e^{\theta} - e^{-\theta}}{2} = \theta + \frac{\theta^3}{3!} + \frac{\theta^5}{5!} + \cdots \tag{H.1.2}$$

Let $\theta = \sqrt{\tilde{z}\tilde{y}}\,\ell$ and $\theta^2 = \tilde{z}\ell\,\tilde{y}\ell = \tilde{Z}\tilde{Y}$. Then, we have:

369

$$\cosh \sqrt{\tilde{z}\tilde{y}}\,\ell = 1 + \frac{\tilde{Z}\tilde{Y}}{2} + \frac{1}{6}\left(\frac{\tilde{Z}\tilde{Y}}{2}\right)^2 + \cdots \tag{H.1.3}$$

$$\sinh \sqrt{\tilde{z}\tilde{y}}\,\ell = \sqrt{\tilde{Z}\tilde{Y}}\left[1 + \frac{1}{3}\frac{\tilde{Z}\tilde{Y}}{2} + \frac{1}{30}\left(\frac{\tilde{Z}\tilde{Y}}{2}\right)^2 + \cdots\right] \tag{H.1.4}$$

$$\sqrt{\frac{\tilde{z}}{\tilde{y}}}\,\sinh \sqrt{\tilde{z}\tilde{y}}\,\ell = \tilde{Z}\left[1 + \frac{1}{3}\frac{\tilde{Z}\tilde{Y}}{2} + \frac{1}{30}\left(\frac{\tilde{Z}\tilde{Y}}{2}\right)^2 + \cdots\right] \tag{H.1.5}$$

$$\sqrt{\frac{\tilde{y}}{\tilde{z}}}\,\sinh \sqrt{\tilde{z}\tilde{y}}\,\ell = \tilde{Y}\left[1 + \frac{1}{3}\frac{\tilde{Z}\tilde{Y}}{2} + \frac{1}{30}\left(\frac{\tilde{Z}\tilde{Y}}{2}\right)^2 + \cdots\right] \tag{H.1.6}$$

H.2   Pure Imaginary Arguments

If $\tilde{z}$ and $\tilde{y}$ are purely imaginary, both with $+j$, then we have on the basis of:

$$\cosh j\theta = \cos \theta \tag{H.2.1}$$
$$\sinh j\theta = j \sin \theta \tag{H.2.2}$$

the following expressions:

$$\cos \sqrt{\tilde{z}\tilde{y}}\,\ell = 1 - \frac{ZY}{2} + \frac{1}{6}\left(\frac{ZY}{2}\right)^2 - \cdots \tag{H.2.3}$$

$$j\sqrt{\frac{\tilde{z}}{\tilde{y}}}\,\sin \sqrt{\tilde{z}\tilde{y}}\,\ell = jZ\left[1 - \frac{1}{3}\frac{ZY}{2} + \frac{1}{30}\left(\frac{ZY}{2}\right)^2 - \cdots\right] \tag{H.2.4}$$

$$j\sqrt{\frac{\tilde{y}}{\tilde{z}}}\,\sin \sqrt{\tilde{z}\tilde{y}}\,\ell = jY\left[1 - \frac{1}{3}\frac{ZY}{2} + \frac{1}{30}\left(\frac{ZY}{2}\right)^2 - \cdots\right] \tag{H.2.5}$$

where Z and Y now represent absolute magnitudes.

Appendix I

Equivalent Networks

In analyzing the performance of electric power systems we may be interested in identifying the performance of each and every element in the network (e.g., load flows) in which case we need to deal with large systems of equations. Very often, however, we are interested only in the response at some finite locations in the power system. For example, we might be interested in the magnitude of fault currents and voltages at some point in the system without needing to know how the currents and voltages distribute throughout the system. Or we may be interested only in the terminal performance characteristics of a transformer or rotating machine without requiring to know, in detail, what goes on inside the machine.

In the latter instances we may simplify the analysis considerably by employing relatively simple equivalent networks to represent large portions of the system. We shall briefly review some fundamental concepts associated with equivalent networks from the point of view of the electric power engineer, including some concepts not found in most modern texts on electric circuit analysis.

I.1 Two-Port Networks

We shall consider a two-port (two-terminal pair, four-pole) network shown in Figure I.1.1, in which the box represents any complex linear passive electrical network connected between the two pairs of terminals. We shall demonstrate that the electrical performance of the two-port network is given by:

$$\begin{bmatrix} \tilde{V}_s \\ \tilde{I}_s \end{bmatrix} = \begin{bmatrix} \tilde{A} & \tilde{B} \\ \tilde{C} & \tilde{D} \end{bmatrix} \begin{bmatrix} \tilde{V}_r \\ \tilde{I}_r \end{bmatrix} \tag{I.1.1}$$

where the voltages and currents may be expressed in terms of phasors, real-time, or operational form, and the A, B, C, D parameters correspondingly in terms of complex impedances, derivatives and/or integrals, or operational form.

There are three elementary two-port networks, illustrated in Figures I.1.2 to I.1.4, on the basis of which many other practical two-port networks can easily be constructed.

(1) For a single-series impedance element we have:

$$\begin{bmatrix} \tilde{V}_s \\ \tilde{I}_s \end{bmatrix} = \begin{bmatrix} 1 & \tilde{Z} \\ 0 & 1 \end{bmatrix} \begin{bmatrix} \tilde{V}_r \\ \tilde{I}_r \end{bmatrix} \tag{I.1.2}$$

(2) For a single-shunt admittance element we have:

$$\begin{bmatrix} \tilde{V}_s \\ \tilde{I}_s \end{bmatrix} = \begin{bmatrix} 1 & 0 \\ \tilde{Y} & 1 \end{bmatrix} \begin{bmatrix} \tilde{V}_r \\ \tilde{I}_r \end{bmatrix} \tag{I.1.3}$$

371

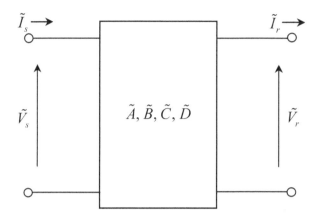

Figure I.1.1: Two-Port Network

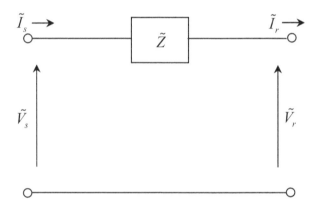

Figure I.1.2: Single-Series Impedance Element

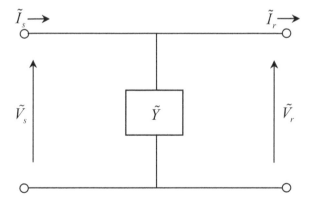

Figure I.1.3: Single-Shunt Admittance Element

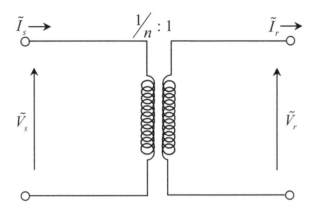

Figure I.1.4: Ideal Transformer

(3) For an ideal transformer we have:

$$
\begin{bmatrix} \tilde{V}_s \\ \tilde{I}_s \end{bmatrix} = \begin{bmatrix} \dfrac{1}{n} & 0 \\ 0 & n \end{bmatrix} \begin{bmatrix} \tilde{V}_r \\ \tilde{I}_r \end{bmatrix}
\tag{I.1.4}
$$

Observe that the determinant $\tilde{A}\tilde{D} - \tilde{B}\tilde{C} = 1.0$ in all three cases. We therefore directly have the inverse:

$$
\begin{bmatrix} \tilde{V}_s \\ \tilde{I}_s \end{bmatrix} = \begin{bmatrix} \tilde{A} & \tilde{B} \\ \tilde{C} & \tilde{D} \end{bmatrix} \begin{bmatrix} \tilde{V}_r \\ \tilde{I}_r \end{bmatrix}
\qquad
\begin{bmatrix} \tilde{V}_r \\ \tilde{I}_r \end{bmatrix} = \begin{bmatrix} \tilde{D} & -\tilde{B} \\ -\tilde{C} & \tilde{A} \end{bmatrix} \begin{bmatrix} \tilde{V}_s \\ \tilde{I}_s \end{bmatrix}
\tag{I.1.5}
$$

I.2 Gamma and Reverse Gamma Networks

By combining the elementary networks, we may obtain more complex networks. For example, the series impedance element in Figure I.1.2 may be preceded by the shunt admittance element in Figure I.1.3 to form the gamma network shown in Figure I.2.1. The appropriate $ABCD$ matrix is easily obtained by premultiplying the series impedance transfer matrix by the shunt admittance transfer matrix:

$$
\begin{bmatrix} \tilde{V}_s \\ \tilde{I}_s \end{bmatrix} = \begin{bmatrix} 1 & \tilde{Z} \\ \tilde{Y} & \left(1 + \tilde{Z}\tilde{Y}\right) \end{bmatrix} \begin{bmatrix} \tilde{V}_r \\ \tilde{I}_r \end{bmatrix}
\tag{I.2.1}
$$

Alternatively, the shunt admittance element in Figure I.1.3 may be preceded by the series impedance element in Figure I.1.2 to form the reverse gamma network shown in Figure I.2.2. The appropriate $ABCD$ matrix is easily obtained by premultiplying the shunt admittance transfer matrix by the series impedance transfer matrix:

$$
\begin{bmatrix} \tilde{V}_s \\ \tilde{I}_s \end{bmatrix} = \begin{bmatrix} \left(1 + \tilde{Z}\tilde{Y}\right) & \tilde{Z} \\ \tilde{Y} & 1 \end{bmatrix} \begin{bmatrix} \tilde{V}_r \\ \tilde{I}_r \end{bmatrix}
\tag{I.2.2}
$$

Observe that the determinant $\tilde{A}\tilde{D} - \tilde{B}\tilde{C} = 1.0$ remains true.

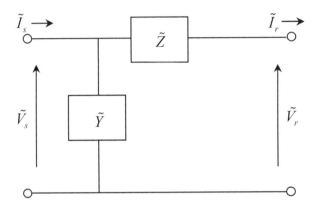

Figure I.2.1: Gamma Network

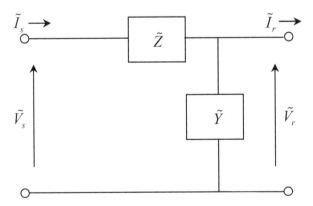

Figure I.2.2: Reverse Gamma Network

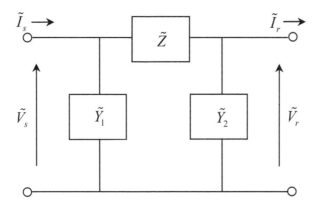

Figure I.3.1: Delta Network in Terms of Series Impedance and Shunt Admittance

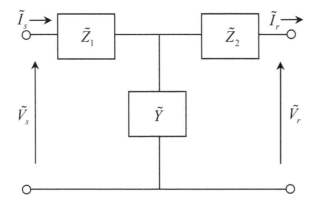

Figure I.3.2: Wye Network in Terms of Series Impedance and Shunt Admittance

I.3 Delta and Wye Networks

Three elementary networks may be concatenated to form delta (commonly referred to as a π network) and wye (commonly referred to as a T network) networks whereby the resultant $ABCD$ matrices are easily obtained as triple products of the associated elementary network matrices as shown in Figures I.3.1 and I.3.2. For the delta network we have:

$$
\begin{bmatrix} \tilde{V}_s \\ \tilde{I}_s \end{bmatrix} = \begin{bmatrix} \left(1 + \tilde{Z}\tilde{Y}_2\right) & \tilde{Z} \\ \left(\tilde{Y}_1 + \tilde{Y}_2 + \tilde{Y}_1\tilde{Z}\tilde{Y}_2\right) & \left(1 + \tilde{Z}\tilde{Y}_1\right) \end{bmatrix} \begin{bmatrix} \tilde{V}_r \\ \tilde{I}_r \end{bmatrix}
\tag{I.3.1}
$$

and for the wye network we have:

$$
\begin{bmatrix} \tilde{V}_s \\ \tilde{I}_s \end{bmatrix} = \begin{bmatrix} \left(1 + \tilde{Z}_1\tilde{Y}\right) & \left(\tilde{Z}_1 + \tilde{Z}_2 + \tilde{Z}_1\tilde{Y}\tilde{Z}_2\right) \\ \tilde{Y} & \left(1 + \tilde{Z}_2\tilde{Y}\right) \end{bmatrix} \begin{bmatrix} \tilde{V}_r \\ \tilde{I}_r \end{bmatrix}
\tag{I.3.2}
$$

If the admittances are expressed as impedances, then the delta and wye networks, and their associated matrices, take the forms shown in Figures I.3.3 and I.3.4. For the delta network we have:

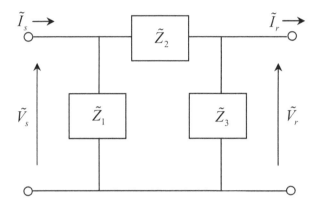

Figure I.3.3: Delta Network in Terms of Impedance Only

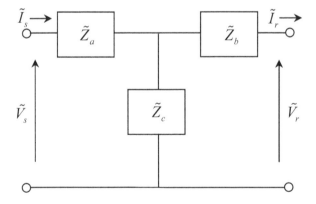

Figure I.3.4: Wye Network in Terms of Impedance Only

$$\begin{bmatrix} \tilde{V}_s \\ \tilde{I}_s \end{bmatrix} = \begin{bmatrix} \left(1 + \dfrac{\tilde{Z}_2}{\tilde{Z}_3}\right) & \tilde{Z}_2 \\ \left(\dfrac{\tilde{Z}_1 + \tilde{Z}_2 + \tilde{Z}_3}{\tilde{Z}_3 \tilde{Z}_1}\right) & \left(1 + \dfrac{\tilde{Z}_2}{\tilde{Z}_1}\right) \end{bmatrix} \begin{bmatrix} \tilde{V}_r \\ \tilde{I}_r \end{bmatrix} \tag{I.3.3}$$

and for the wye network we have:

$$\begin{bmatrix} \tilde{V}_s \\ \tilde{I}_s \end{bmatrix} = \begin{bmatrix} \left(1 + \dfrac{\tilde{Z}_a}{\tilde{Z}_c}\right) & \left(\dfrac{\tilde{Z}_a \tilde{Z}_b + \tilde{Z}_b \tilde{Z}_c + \tilde{Z}_c \tilde{Z}_a}{\tilde{Z}_c}\right) \\ \dfrac{1}{\tilde{Z}_c} & \left(1 + \dfrac{\tilde{Z}_b}{\tilde{Z}_c}\right) \end{bmatrix} \begin{bmatrix} \tilde{V}_r \\ \tilde{I}_r \end{bmatrix} \tag{I.3.4}$$

We may convert from one to the other by simply equating matrix elements:

(1) Converting from delta to wye

$$\tilde{Z}_a = \frac{\tilde{Z}_1 \tilde{Z}_2}{\tilde{Z}_1 + \tilde{Z}_2 + \tilde{Z}_3}$$

$$\tilde{Z}_b = \frac{\tilde{Z}_2 \tilde{Z}_3}{\tilde{Z}_1 + \tilde{Z}_2 + \tilde{Z}_3}$$

$$\tilde{Z}_c = \frac{\tilde{Z}_3 \tilde{Z}_1}{\tilde{Z}_1 + \tilde{Z}_2 + \tilde{Z}_3} \tag{I.3.5}$$

(2) Converting from wye to delta:

$$\tilde{Z}_1 = \frac{\tilde{Z}_a \tilde{Z}_b + \tilde{Z}_b \tilde{Z}_c + \tilde{Z}_c \tilde{Z}_a}{\tilde{Z}_b}$$

$$\tilde{Z}_2 = \frac{\tilde{Z}_a \tilde{Z}_b + \tilde{Z}_b \tilde{Z}_c + \tilde{Z}_c \tilde{Z}_a}{\tilde{Z}_c}$$

$$\tilde{Z}_3 = \frac{\tilde{Z}_a \tilde{Z}_b + \tilde{Z}_b \tilde{Z}_c + \tilde{Z}_c \tilde{Z}_a}{\tilde{Z}_a} \tag{I.3.6}$$

I.4   Modal Component Forms

When N identical $ABCD$ networks are concatenated (connected in series) the resultant $ABCD$ matrix is the Nth power of the $ABCD$ matrix of the individual network. The evaluation can be expedited considerably if we first diagonalize the $ABCD$ matrix by means of a modal (eigenvector) analysis. We begin with:

$$\begin{bmatrix} \tilde{V}_s \\ \tilde{I}_s \end{bmatrix} = \begin{bmatrix} \tilde{A} & \tilde{B} \\ \tilde{C} & \tilde{D} \end{bmatrix} \begin{bmatrix} \tilde{V}_r \\ \tilde{I}_r \end{bmatrix} \qquad \tilde{A}\tilde{D} - \tilde{B}\tilde{C} = 1.0 \tag{I.4.1}$$

Then we write:

$$\left[\begin{array}{cc} \left(\tilde{A} - \tilde{\lambda} \right) & \tilde{B} \\ \tilde{C} & \left(\tilde{D} - \tilde{\lambda} \right) \end{array} \right]^{-1} = \frac{1}{\tilde{\Delta}} \left[\begin{array}{cc} \left(\tilde{D} - \tilde{\lambda} \right) & -\tilde{B} \\ -\tilde{C} & \left(\tilde{A} - \tilde{\lambda} \right) \end{array} \right] \tag{I.4.2}$$

for which the determinant must be zero. Therefore:

$$\tilde{\Delta} = \left(\tilde{A} - \tilde{\lambda} \right) \left(\tilde{D} - \tilde{\lambda} \right) - \tilde{B}\tilde{C} = \tilde{\lambda}^2 - \tilde{\lambda} \left(\tilde{A} + \tilde{D} \right) + 1.0 = 0 \tag{I.4.3}$$

$$\tilde{\lambda} = \left(\frac{\tilde{A} + \tilde{D}}{2} \right) \pm \sqrt{\left(\frac{\tilde{A} + \tilde{D}}{2} \right)^2 - 1.0} \tag{I.4.4}$$

For the sake of typographical simplicity, we recall the identity:

$$\cosh^2 \tilde{\theta} - \sinh^2 \tilde{\theta} = 1.0 \tag{I.4.5}$$

and introduce:

$$\left(\frac{\tilde{A} + \tilde{D}}{2} \right) = \cosh \tilde{\theta} = \left(\frac{e^{\tilde{\theta}} + e^{-\tilde{\theta}}}{2} \right) \tag{I.4.6}$$

$$\sqrt{\left(\frac{\tilde{A} + \tilde{D}}{2} \right)^2 - 1.0} = \sinh \tilde{\theta} = \left(\frac{e^{\tilde{\theta}} - e^{-\tilde{\theta}}}{2} \right) \tag{I.4.7}$$

so that:

$$\tilde{\lambda} = \cosh \tilde{\theta} \pm \sinh \tilde{\theta} \tag{I.4.8}$$

Thus, the eigenvalues are:

$$\tilde{\lambda}_1 = \cosh \tilde{\theta} + \sinh \tilde{\theta} = e^{\tilde{\theta}} \tag{I.4.9}$$

$$\tilde{\lambda}_1 = \cosh \tilde{\theta} - \sinh \tilde{\theta} = e^{-\tilde{\theta}} \tag{I.4.10}$$

and:

$$\left(\tilde{A} + \tilde{D} \right) = \tilde{\lambda}_1 + \tilde{\lambda}_2 \tag{I.4.11}$$

$$\left(\tilde{A} - \tilde{\lambda}_2 \right) = - \left(\tilde{D} - \tilde{\lambda}_1 \right) \tag{I.4.12}$$

$$\tilde{\lambda}_1 \tilde{\lambda}_2 = 1.0 \tag{I.4.13}$$

If we designate the modal transformation by $\left[\tilde{M} \right]$, then:

$$\left[\tilde{M} \right] \left[\tilde{M} \right]^{-1} = [I] \tag{I.4.14}$$

and specifically:

$$\left(\frac{1}{\tilde{d}}\right)\left[\begin{array}{cc} \left(\tilde{A}-\tilde{\lambda}_2\right) & \tilde{B} \\ \tilde{C} & \left(\tilde{D}-\tilde{\lambda}_1\right) \end{array}\right]\left[\begin{array}{cc} \left(\tilde{D}-\tilde{\lambda}_1\right) & -\tilde{B} \\ -\tilde{C} & \left(\tilde{A}-\tilde{\lambda}_2\right) \end{array}\right]=\left[\begin{array}{cc} 1.0 & 0 \\ 0 & 1.0 \end{array}\right] \tag{I.4.15}$$

so that:

$$\tilde{d}=\left(\tilde{A}-\tilde{\lambda}_2\right)\left(\tilde{D}-\tilde{\lambda}_1\right)-\tilde{B}\tilde{C}=\tilde{\lambda}_1\tilde{\lambda}_2-\left(\tilde{A}\tilde{\lambda}_1+\tilde{D}\tilde{\lambda}_2\right)+1.0$$

$$=\left(\tilde{D}-\tilde{\lambda}_1\right)\left(\tilde{\lambda}_1-\tilde{\lambda}_2\right)=\left(\tilde{\lambda}_2-\tilde{A}\right)\left(\tilde{\lambda}_1-\tilde{\lambda}_2\right) \tag{I.4.16}$$

for compactness, let us define:

$$\tilde{S}=\left(\tilde{D}-\tilde{\lambda}_1\right)$$
$$\tilde{T}=\left(\tilde{A}-\tilde{\lambda}_2\right)$$

Then:

$$\left(\frac{1}{\tilde{d}}\right)\left[\begin{array}{cc} \tilde{T} & \tilde{B} \\ \tilde{C} & \tilde{S} \end{array}\right]\left[\begin{array}{cc} \tilde{A} & \tilde{B} \\ \tilde{C} & \tilde{D} \end{array}\right]\left[\begin{array}{cc} \tilde{S} & -\tilde{B} \\ -\tilde{C} & \tilde{T} \end{array}\right]$$

$$=\left(\frac{1}{\tilde{d}}\right)\left[\begin{array}{cc} \tilde{T} & \tilde{B} \\ \tilde{C} & \tilde{S} \end{array}\right]\left[\begin{array}{cc} \left(1-\tilde{A}\tilde{\lambda}_1\right) & -\tilde{B}\tilde{\lambda}_2 \\ -\tilde{C}\tilde{\lambda}_1 & \left(1-\tilde{D}\tilde{\lambda}_2\right) \end{array}\right]$$

$$=\left(\frac{1}{\tilde{d}}\right)\left[\begin{array}{cc} \tilde{T} & \tilde{B} \\ \tilde{C} & \tilde{S} \end{array}\right]\left[\begin{array}{cc} \left(\tilde{\lambda}_2-\tilde{A}\right) & -\tilde{B} \\ -\tilde{C} & \left(\tilde{\lambda}_1-\tilde{D}\right) \end{array}\right]\left[\begin{array}{cc} \tilde{\lambda}_1 & 0 \\ 0 & \tilde{\lambda}_2 \end{array}\right]$$

$$=\left(\frac{1}{\tilde{d}}\right)\left[\begin{array}{cc} \tilde{T} & \tilde{B} \\ \tilde{C} & \tilde{S} \end{array}\right]\left[\begin{array}{cc} \tilde{S} & -\tilde{B} \\ -\tilde{C} & \tilde{T} \end{array}\right]\left[\begin{array}{cc} \tilde{\lambda}_1 & 0 \\ 0 & \tilde{\lambda}_2 \end{array}\right]$$

$$=\left[\begin{array}{cc} \tilde{\lambda}_1 & 0 \\ 0 & \tilde{\lambda}_2 \end{array}\right] \tag{I.4.17}$$

and:

$$\left[\begin{array}{cc} \tilde{A} & \tilde{B} \\ \tilde{C} & \tilde{D} \end{array}\right]=\left(\frac{1}{\tilde{d}}\right)\left[\begin{array}{cc} \tilde{S} & -\tilde{B} \\ -\tilde{C} & \tilde{T} \end{array}\right]\left[\begin{array}{cc} \tilde{\lambda}_1 & 0 \\ 0 & \tilde{\lambda}_2 \end{array}\right]\left[\begin{array}{cc} \tilde{T} & \tilde{B} \\ \tilde{C} & \tilde{S} \end{array}\right] \tag{I.4.18}$$

I.5 Powers of ABCD Matrices

In terms of symbolic matrix notation, let:

$$\left[\tilde{H}\right]=\left[\begin{array}{cc} \tilde{A} & \tilde{B} \\ \tilde{C} & \tilde{D} \end{array}\right] \tag{I.5.1}$$

If N such matrices are concatenated, then the resultant $ABCD$ matrix takes the form:

$$\left[\tilde{H}\right]^{N} = \left[\tilde{H}\right]\left[\tilde{H}\right]\left[\tilde{H}\right]\cdots\cdots\cdots\left[\tilde{H}\right] \tag{I.5.2}$$

which would be a rather tedious process of multiplication.

For $\left[\tilde{H}\right]$, let $\left[\tilde{M}\right]$ and $\left[\tilde{\lambda}\right]$ be the associated modal transformation and eigenvalue matrices respectively. Then:

$$\left[\tilde{H}\right]^{N} = \left[\tilde{M}\right]^{-1}\left[\tilde{\lambda}\right]\left[\tilde{M}H\right]\left[\tilde{M}\right]^{-1}\left[\tilde{\lambda}\right]\left[\tilde{M}H\right]\cdots\cdots\left[\tilde{M}\right]^{-1}\left[\tilde{\lambda}\right]\left[\tilde{M}H\right] = \left[\tilde{M}\right]^{-1}\left[\tilde{\lambda}\right]^{N}\left[\tilde{M}H\right] \tag{I.5.3}$$

as the $\left[\tilde{M}\right]^{-1}\left[\tilde{M}\right]$ products in the middle of the expression all simplify to the identity matrix. Such an operation would involve raising each of the two eigenvalues to the Nth power and one triple matrix multiplication. Recalling the definition (again for compactness) of $\tilde{S}$ and $\tilde{T}$:

$$\tilde{S} = \left(\tilde{D} - \tilde{\lambda}_1\right)$$

$$\tilde{T} = \left(\tilde{A} - \tilde{\lambda}_2\right)$$

Then:

$$\left[\begin{array}{cc} \tilde{A} & \tilde{B} \\ \tilde{C} & \tilde{D} \end{array}\right]^{N} = \left(\frac{1}{\tilde{d}}\right)\left[\begin{array}{cc} \tilde{S} & -\tilde{B} \\ -\tilde{C} & \tilde{T} \end{array}\right]\left[\begin{array}{cc} \tilde{\lambda}_1^{N} & 0 \\ 0 & \tilde{\lambda}_2^{N} \end{array}\right]\left[\begin{array}{cc} \tilde{T} & \tilde{B} \\ \tilde{C} & \tilde{S} \end{array}\right]$$

$$= \left(\frac{1}{\tilde{d}}\right)\left[\begin{array}{cc} \tilde{\lambda}_1^{N}\tilde{S} & -\tilde{\lambda}_2^{N}\tilde{B} \\ -\tilde{\lambda}_1^{N}\tilde{C} & \tilde{T} \end{array}\right]\left[\begin{array}{cc} \tilde{T} & \tilde{B} \\ \tilde{C} & \tilde{S} \end{array}\right]$$

$$= \left(\frac{1}{\tilde{d}}\right)\left[\begin{array}{cc} \tilde{\lambda}_1^{N}\tilde{S}\tilde{T} - \tilde{\lambda}_2^{N}\tilde{B}\tilde{C} & \tilde{B}\tilde{S}\left(\tilde{\lambda}_1^{N} - \tilde{\lambda}_2^{N}\right) \\ \tilde{C}\tilde{T}\left(\tilde{\lambda}_2^{N} - \tilde{\lambda}_1^{N}\right) & \tilde{\lambda}_2^{N}\tilde{S}\tilde{T} - \tilde{\lambda}_1^{N}\tilde{B}\tilde{C} \end{array}\right]$$

$$= \left(\frac{1}{\tilde{d}}\right)\left(\tilde{\lambda}_1^{N} - \tilde{\lambda}_2^{N}\right)\tilde{S}\left[\begin{array}{cc} \tilde{T} & \tilde{B} \\ \tilde{C} & \tilde{S} \end{array}\right] + \left[\begin{array}{cc} \tilde{\lambda}_2^{N} & 0 \\ 0 & \tilde{\lambda}_1^{N} \end{array}\right] \tag{I.5.4}$$

Recall that $\tilde{d} = \tilde{S}\left(\tilde{\lambda}_1 - \tilde{\lambda}_2\right)$. Hence:

$$\frac{1}{\tilde{d}} = \frac{1}{\tilde{S}\left(\tilde{\lambda}_1 - \tilde{\lambda}_2\right)}$$

therefore:

$$\left[\begin{array}{cc} \tilde{A} & \tilde{B} \\ \tilde{C} & \tilde{D} \end{array}\right]^{N} = \left(\frac{\tilde{\lambda}_1^{N} - \tilde{\lambda}_2^{N}}{\tilde{\lambda}_1 - \tilde{\lambda}_2}\right)\left[\begin{array}{cc} \tilde{T} & \tilde{B} \\ \tilde{C} & \tilde{S} \end{array}\right] + \left[\begin{array}{cc} \tilde{\lambda}_2^{N} & 0 \\ 0 & \tilde{\lambda}_1^{N} \end{array}\right]$$

$$= \left(\frac{\tilde{\lambda}_1^{N} - \tilde{\lambda}_2^{N}}{\tilde{\lambda}_1 - \tilde{\lambda}_2}\right)\left[\begin{array}{cc} \tilde{A} & \tilde{B} \\ \tilde{C} & \tilde{D} \end{array}\right] - \left(\frac{\tilde{\lambda}_1^{N}\tilde{\lambda}_2 - \tilde{\lambda}_2^{N}\tilde{\lambda}_1}{\tilde{\lambda}_1 - \tilde{\lambda}_2}\right)\left[\begin{array}{cc} 1 & 0 \\ 0 & 1 \end{array}\right] \tag{I.5.5}$$

Per Equations I.4.9 and I.4.10, we have:

$$\tilde{\lambda}_1 = e^{\tilde{\theta}} \qquad \tilde{\lambda}_2 = e^{-\tilde{\theta}} \tag{I.5.6}$$

so that:

$$\tilde{\lambda}_1 - \tilde{\lambda}_2 = 2 \sinh \tilde{\theta} \qquad \tilde{\lambda}_1^N - \tilde{\lambda}_2^N = 2 \sinh N\tilde{\theta} \tag{I.5.7}$$

Per Equation I.4.13, we have:

$$\tilde{\lambda}_1 \tilde{\lambda}_2 = 1.0 \tag{I.5.8}$$

so that:

$$\tilde{\lambda}_1^N \tilde{\lambda}_2 - \tilde{\lambda}_2^N \tilde{\lambda}_1 = \tilde{\lambda}_1^{N-1} - \tilde{\lambda}_2^{N-1} = 2 \sinh (N-1)\tilde{\theta} \tag{I.5.9}$$

Then:

$$\begin{bmatrix} \tilde{A} & \tilde{B} \\ \tilde{C} & \tilde{D} \end{bmatrix}^N = \frac{\sinh N\tilde{\theta}}{\sinh \tilde{\theta}} \begin{bmatrix} \tilde{A} & \tilde{B} \\ \tilde{C} & \tilde{D} \end{bmatrix} - \frac{\sinh (N-1)\tilde{\theta}}{\sinh \tilde{\theta}} \begin{bmatrix} 1 & 0 \\ 0 & 1 \end{bmatrix} \tag{I.5.10}$$

with:

$$\cosh \tilde{\theta} = \frac{\tilde{A} + \tilde{D}}{2} \qquad \sinh \tilde{\theta} = \sqrt{\cosh^2 \tilde{\theta} - 1} \tag{I.5.11}$$

I.6   Helmholtz Equivalent Generator Theorem

In 1853, Heinrich Helmholtz published a paper proposing an equivalent generator concept in which the voltage and current at some terminal pair within a large complex electrical network could be determined without the need for solving a large system of equations based on Kirchhoff's laws. We shall verify this from a somewhat different point of view than that taken by Helmholtz.

Given a large electrical network, shown in Figure I.6.1, consisting of voltage and current sources and linear conductors. We wish to apply an impedance between some terminal pair $x - y$ within this network and are only interested in the voltage across and the current through this impedance.

Let $\tilde{V}_{o.c.}$ be the voltage between the terminal pair before the impedance is inserted. Then, Figure I.6.2(a) is equivalent to the condition of an impedance $\tilde{Z}_r$ connected between the pair of terminals, and Figure I.6.2(b) is equivalent to the condition before the impedance is inserted so that $\tilde{I}'_r$ is zero.

Observe that in Figure I.6.2(b), all voltage and current sources, except one, are active, and in Figure I.6.2(c), only one voltage source is active. Then by the principle of superposition we have:

$$\tilde{I}_r = \tilde{I}'_r + \tilde{I}''_r = \tilde{I}''_r = \frac{\tilde{V}_{o.c.}}{\tilde{Z}_r + \tilde{Z}_{eq}} \tag{I.6.1}$$

where $\tilde{Z}_{eq}$ is the impedance of the large system as seen from the terminal pair. Figure I.6.3(a) illustrates one possible equivalent generator network.

When $\tilde{Z}_r = 0$, we have a short circuit between the terminal pair in which case:

$$\tilde{I}_{s.c} = \frac{\tilde{V}_{o.c.}}{\tilde{Z}_{eq}} \tag{I.6.2}$$

Therefore, we may also write:

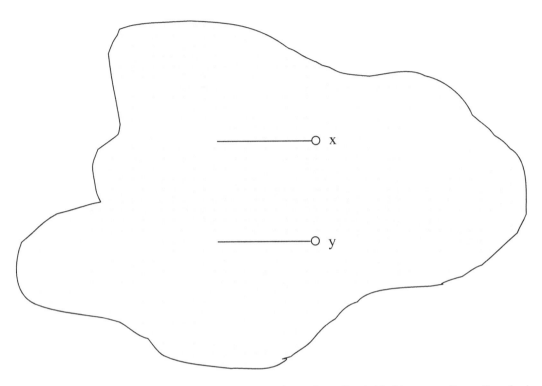

Figure I.6.1: Existing Network to Which an Impedance Is to Be Added between Some Terminal Pair $x - y$

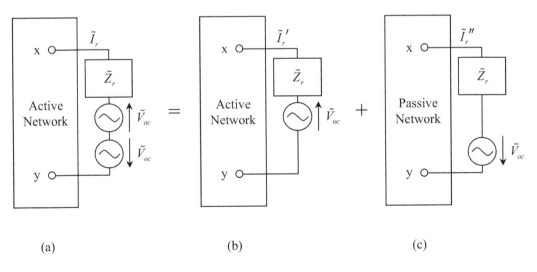

Figure I.6.2: Superposition Networks

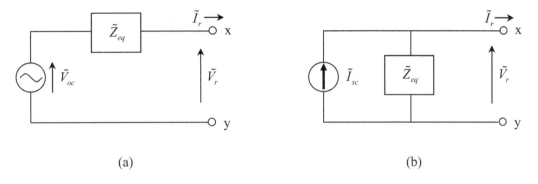

(a)   (b)

Figure I.6.3: Equivalent Generator Networks

$$\tilde{I}_r = \frac{\tilde{I}_{s.c.}\tilde{Z}_{eq}}{\tilde{Z}_r + \tilde{Z}_{eq}} \tag{I.6.3}$$

Figure I.6.3(b) illustrates a second possible equivalent generator network.

At this point in time, the reader may wonder why equivalent voltage source networks are referred to as *Helmholtz* equivalents rather than *Thevenin* equivalents. L. Charles Thevenin published a paper entitled "Sur un nouveau theoreme d'electricite dynamique" in which he described the equivalent generator network in Figure I.6.3(a) and which is known in the American literature as Thevenin's equivalent network. This was, however, published in 1883, some 30 years after Helmholtz's paper. Additionally, Helmholtz's paper is generally considered more rigorous and provided a mathematical proof, whereas Thevenin's paper presented the network equivalent without proof. In 1926 E. L. Norton of the Bell Telephone Laboratories published a paper in which he described the equivalent generator network in Figure I.6.3(b) and which is known in the American literature as a Norton equivalent network.

A discussion to the paper at Reference [1] provides a terse summation of the historical context of this subject and provides further references for those interested. The authors of this text will be happy to provide it upon request.

I.7   Constant Source Networks

The equivalent generator networks in Figure I.6.3 will be significantly different depending upon whether the original system consists of constant voltage sources or constant current sources. Consider, for example, the complete equivalent circuit for a two-winding transformer shown in Figures I.7.1 and I.7.2, where typical per unit values for the impedances are indicated.

In Figure I.7.1, the transformer is energized from a constant voltage source as is typical for power transformers and potential instrument transformers. Then for rated applied voltage (1.0 in per unit):

$$\tilde{V}_{o.c.} = 1.0 \tag{I.7.1}$$

$$\tilde{I}_{s.c.} = 10.0 \tag{I.7.2}$$

$$\tilde{Z}_{eq} = 0.10 \tag{I.7.3}$$

In Figure I.7.2, the same transformer is energized from a constant current source as is typical for current instrument transformers. Then, for rated applied current (1.0 in per unit):

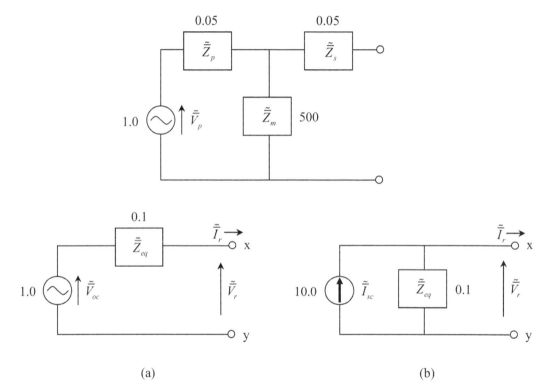

Figure I.7.1: Equivalent Circuit for a Two-Winding Transformer and Equivalent Generator Networks for *Constant Voltage* Source

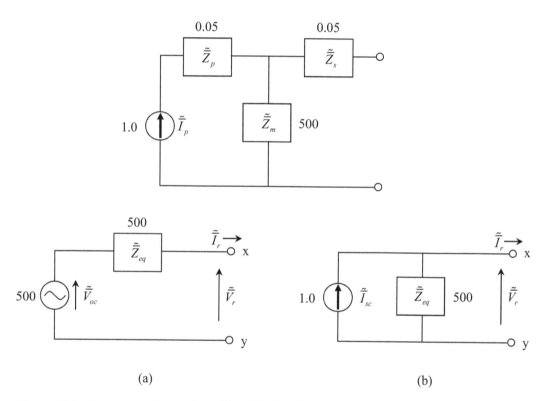

Figure I.7.2: Equivalent Circuit for a Two-Winding Transformer and Equivalent Generator Networks for *Constant Current* Source

$$\tilde{\tilde{V}}_{o.c.} = 500.0 \tag{I.7.4}$$

$$\tilde{\tilde{I}}_{s.c.} = 1.0 \tag{I.7.5}$$

$$\tilde{\tilde{Z}}_{eq} = 500.0 \tag{I.7.6}$$

This case clearly illustrates the hazard of not short-circuiting the secondary terminals of an instrument current transformer before removing an ammeter or current relay from the secondary terminals.

I.8 References

[1] Colin Adamson and S. M. El-Soki. *The General Form of Some Common Network Theorems.* AIEE Trans., Part I, Comm. and El., Vol. 78, 1959.

Appendix J

Y-Δ Relationships

Consider the circuit arrangement shown in Figure J.3.1. Given the line-to-earth voltages, we may uniquely define the line-to-line voltages. Similarly, given the delta-currents, we may uniquely define the line-currents. The inverse relationships are, however, indeterminate from a rigorous point of view. Nonetheless, from a practical point of view, useful inverse relationships may be obtained as established in the following analysis.

J.1 Kirchhoff's Voltage Law

According to Kirchhoff's voltage law we may write:

$$
\begin{aligned}
\tilde{V}_{ae} - \tilde{V}_{ab} - \tilde{V}_{be} &= 0 \\
\tilde{V}_{be} - \tilde{V}_{bc} - \tilde{V}_{ce} &= 0 \\
\tilde{V}_{ce} - \tilde{V}_{ca} - \tilde{V}_{ae} &= 0
\end{aligned}
\tag{J.1.1}
$$

$$
\begin{bmatrix} \tilde{V}_{ab} \\ \tilde{V}_{bc} \\ \tilde{V}_{ca} \end{bmatrix} =
\begin{bmatrix} 1 & -1 & 0 \\ 0 & 1 & -1 \\ -1 & 0 & 1 \end{bmatrix}
\begin{bmatrix} \tilde{V}_{ae} \\ \tilde{V}_{be} \\ \tilde{V}_{ce} \end{bmatrix}
\tag{J.1.2}
$$

and in terms of symbolic matrix notation:

$$
\left[\tilde{V}_{delta} \right] = [T_v] \left[\tilde{V}_{wye} \right]
\tag{J.1.3}
$$

J.2 Kirchhoff's Current Law

According to Kirchhoff's current law we may write:

$$
\begin{aligned}
\tilde{I}_a - \tilde{I}_{ab} + \tilde{I}_{ca} &= 0 \\
\tilde{I}_b - \tilde{I}_{bc} + \tilde{I}_{ab} &= 0 \\
\tilde{I}_c - \tilde{I}_{ca} + \tilde{I}_{bc} &= 0
\end{aligned}
\tag{J.2.1}
$$

$$
\begin{bmatrix} \tilde{I}_a \\ \tilde{I}_b \\ \tilde{I}_c \end{bmatrix} =
\begin{bmatrix} 1 & 0 & -1 \\ -1 & 1 & 0 \\ -0 & -1 & 1 \end{bmatrix}
\begin{bmatrix} \tilde{I}_{ab} \\ \tilde{I}_{bc} \\ \tilde{I}_{ca} \end{bmatrix}
\tag{J.2.2}
$$

and in terms of symbolic notation:

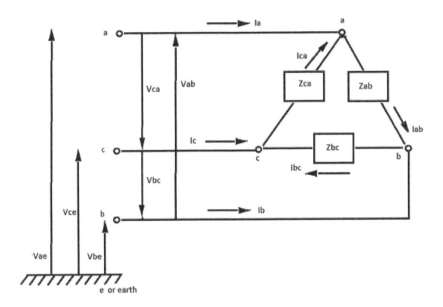

Figure J.3.1: Delta Circuit

$$\left[\tilde{I}_{wye}\right] = [T_i]\left[\tilde{I}_{delta}\right] \tag{J.2.3}$$

J.3   Inverse of $[T_v]$ and $[T_i]$

Since the determinants of $[T_v]$ and $[T_i]$ are equal to zero, no formal inverse exists. However, we can write:

$$\begin{bmatrix} 1 & -1 & 0 \\ 0 & 1 & -1 \\ -1 & 0 & 1 \end{bmatrix} = \lim_{\varepsilon \to 0} \begin{bmatrix} 1 & -1 & \varepsilon \\ \varepsilon & 1 & -1 \\ -1 & \varepsilon & 1 \end{bmatrix} \tag{J.3.1}$$

$$\begin{bmatrix} 1 & -1 & \varepsilon \\ \varepsilon & 1 & -1 \\ -1 & \varepsilon & 1 \end{bmatrix}^{-1} = \frac{1}{\varepsilon(3+\varepsilon^2)} \begin{bmatrix} (1+\varepsilon) & (1+\varepsilon^2) & (1-\varepsilon) \\ (1-\varepsilon) & (1+\varepsilon) & (1+\varepsilon^2) \\ (1+\varepsilon^2) & (1-\varepsilon) & (1+\varepsilon) \end{bmatrix}$$

$$= \frac{1}{(3+\varepsilon^2)} \begin{bmatrix} 1 & \varepsilon & -1 \\ -1 & 1 & \varepsilon \\ \varepsilon & -1 & 1 \end{bmatrix} + \frac{1}{\varepsilon(3+\varepsilon^2)} \begin{bmatrix} 1 & 1 & 1 \\ 1 & 1 & 1 \\ 1 & 1 & 1 \end{bmatrix} \tag{J.3.2}$$

In the limit when $\varepsilon \to 0$, the first matrix is finite and the second becomes infinitely large. Thus, we have:

$$\begin{bmatrix} \tilde{V}_{ae} \\ \tilde{V}_{be} \\ \tilde{V}_{ce} \end{bmatrix} = \frac{1}{3} \begin{bmatrix} 1 & 0 & -1 \\ -1 & 1 & 0 \\ 0 & -1 & 1 \end{bmatrix} \begin{bmatrix} \tilde{V}_{ab} \\ \tilde{V}_{bc} \\ \tilde{V}_{ca} \end{bmatrix} + \lim_{\varepsilon \to 0} \frac{1}{3\varepsilon} \begin{bmatrix} 1 & 1 & 1 \\ 1 & 1 & 1 \\ 1 & 1 & 1 \end{bmatrix} \begin{bmatrix} \tilde{V}_{ab} \\ \tilde{V}_{bc} \\ \tilde{V}_{ca} \end{bmatrix} \tag{J.3.3}$$

According to Equation J.1.2 we have $\tilde{V}_{ab} + \tilde{V}_{bc} + \tilde{V}_{ca} = 0$ under all conditions. Despite this fact, the second term in Equation J.3.3 remains indeterminate. If we write:

$$\Delta \tilde{V} = \lim_{\varepsilon \to 0} \frac{1}{3\varepsilon} (\tilde{V}_{ab} + \tilde{V}_{bc} + \tilde{V}_{ca}) \qquad (\text{J.3.4})$$

then:

$$\begin{bmatrix} \tilde{V}_{ae} - \Delta \tilde{V} \\ \tilde{V}_{be} - \Delta \tilde{V} \\ \tilde{V}_{ce} - \Delta \tilde{V} \end{bmatrix} = \frac{1}{3} \begin{bmatrix} 1 & 0 & -1 \\ -1 & 1 & 0 \\ 0 & -1 & 1 \end{bmatrix} \begin{bmatrix} \tilde{V}_{ab} \\ \tilde{V}_{bc} \\ \tilde{V}_{ca} \end{bmatrix} \qquad (\text{J.3.5})$$

$$\tilde{V}_{ae} + \tilde{V}_{be} + \tilde{V}_{ce} - 3\Delta \tilde{V} = 0 \qquad (\text{J.3.6})$$

In a similar manner we determine:

$$\begin{bmatrix} \tilde{I}_{ab} - \Delta \tilde{I} \\ \tilde{I}_{bc} - \Delta \tilde{I} \\ \tilde{I}_{ca} - \Delta \tilde{I} \end{bmatrix} = \frac{1}{3} \begin{bmatrix} 1 & -1 & 0 \\ 0 & 1 & -1 \\ -1 & 0 & 1 \end{bmatrix} \begin{bmatrix} \tilde{I}_{a} \\ \tilde{I}_{b} \\ \tilde{I}_{c} \end{bmatrix} \qquad (\text{J.3.7})$$

with:

$$\Delta \tilde{I} = \lim_{\varepsilon \to 0} \frac{1}{3\varepsilon} (\tilde{I}_{a} + \tilde{I}_{b} + \tilde{I}_{c}) \qquad (\text{J.3.8})$$

$$\tilde{I}_{ab} + \tilde{I}_{bc} + \tilde{I}_{ca} - 3\Delta \tilde{I} = 0 \qquad (\text{J.3.9})$$

J.4 Neutral Voltage

Let the geometrical neutral of the voltage triangle be identified by gn so that the voltages with respect to the geometrical neutral may be identified as $\tilde{V}_{agn}$, $\tilde{V}_{bgn}$, and $\tilde{V}_{cgn}$. From purely geometric considerations we know that:

$$\tilde{V}_{agn} + \tilde{V}_{bgn} + \tilde{V}_{cgn} = 0 \qquad (\text{J.4.1})$$

Then we may write:

$$(\tilde{V}_{agn} + \Delta \tilde{V}) + (\tilde{V}_{bgn} + \Delta \tilde{V}) + (\tilde{V}_{cgn} + \Delta \tilde{V}) - 3\Delta \tilde{V} = 0 \qquad (\text{J.4.2})$$

Comparing Equation J.4.2 with Equation J.3.6 we may say that $\tilde{V}_{ae}$, $\tilde{V}_{be}$, and $\tilde{V}_{ce}$ are equal to the voltages with respect to the geometric neutral respectively, plus a common neutral shift equal to $\Delta \tilde{V}$. If only line-to-line voltages are known, $\Delta \tilde{V}$ is indeterminate. We may write:

$$\begin{bmatrix} \tilde{V}_{a}' \\ \tilde{V}_{b}' \\ \tilde{V}_{c}' \end{bmatrix} = \frac{1}{3} \begin{bmatrix} 1 & 0 & -1 \\ -1 & 1 & 0 \\ 0 & -1 & 1 \end{bmatrix} \begin{bmatrix} \tilde{V}_{ab} \\ \tilde{V}_{bc} \\ \tilde{V}_{ca} \end{bmatrix} \qquad (\text{J.4.3})$$

$$\tilde{V}_{a}' + \tilde{V}_{b}' + \tilde{V}_{c}' = 0 \qquad (\text{J.4.4})$$

Because we can identify the geometrical neutral of the voltage triangle, we recognize that $\tilde{V}_{a}' = \tilde{V}_{agn}$, $\tilde{V}_{b}' = \tilde{V}_{bgn}$, and $\tilde{V}_{c}' = \tilde{V}_{cgn}$.

J.5 Circulating Current

In a similar manner we determine that:

$$\begin{bmatrix} \tilde{I}'_{ab} \\ \tilde{I}'_{bc} \\ \tilde{I}'_{ca} \end{bmatrix} = \frac{1}{3} \begin{bmatrix} 1 & -1 & 0 \\ 0 & 1 & -1 \\ -1 & 0 & 1 \end{bmatrix} \begin{bmatrix} \tilde{I}_a \\ \tilde{I}_b \\ \tilde{I}_c \end{bmatrix} \tag{J.5.1}$$

with

$$\tilde{I}'_{ab} + \tilde{I}'_{bc} + \tilde{I}'_{ca} = 0 \tag{J.5.2}$$

The true values for $\tilde{I}_{ab}$, $\tilde{I}_{bc}$, and $\tilde{I}_{ca}$ are equal to $\tilde{I}'_{ab}$, $\tilde{I}'_{bc}$, and $\tilde{I}'_{ca}$ respectively, plus a common circulating current equal to $\Delta\tilde{I}$. If only line currents are known, $\Delta\tilde{I}$ is indeterminate.

J.6 Alternate Mathematical Analysis

A comparison of Equations J.1.2 and J.2.2 shows that:

$$[T_i] = [T_v]^t \tag{J.6.1}$$

As a matter of curiosity we consider:

$$[T_i]\,[T_v] = \begin{bmatrix} 1 & 0 & -1 \\ -1 & 1 & 0 \\ 0 & -1 & 1 \end{bmatrix} \begin{bmatrix} 1 & -1 & 0 \\ 0 & 1 & -1 \\ -1 & 0 & 1 \end{bmatrix} = \begin{bmatrix} 2 & -1 & -1 \\ -1 & 2 & -1 \\ -1 & -1 & 2 \end{bmatrix} \tag{J.6.2}$$

which shows that $[T_i]$ and $[T_v]$ are not true inverses of one another. With

$$[U] = \begin{bmatrix} 1 & 0 & 0 \\ 0 & 1 & 0 \\ 0 & 0 & 1 \end{bmatrix} \tag{J.6.3}$$

$$[\xi] = \begin{bmatrix} 1 & 1 & 1 \\ 1 & 1 & 1 \\ 1 & 1 & 1 \end{bmatrix} \tag{J.6.4}$$

we may write Equation J.6.2 in symbolic matrix format as:

$$[T_i]\,[T_v] = 3\,[U] - [\xi] \tag{J.6.5}$$

We consider now Equation J.1.3 and write:

$$[T_i]\left[\tilde{V}_{delta}\right] = [T_i]\,[T_v]\left[\tilde{V}_{wye}\right] = [3\,[U] - [\xi]]\left[\tilde{V}_{wye}\right] \tag{J.6.6}$$

Thus, given $\left[\tilde{V}_{wye}\right]$, which may include a neutral shift, we correctly determine $\left[\tilde{V}_{delta}\right]$ according to Equation J.1.2 . Then Equation J.6.6 returns $\left[\tilde{V}_{wye}\right]$ minus the neutral shift.

Similarly, on the basis of Equation J.2.3 we may write:

$$[T_v]\left[\tilde{I}_{wye}\right] = [T_v]\,[T_i]\left[\tilde{I}_{delta}\right] = [3\,[U] - [\xi]]\left[\tilde{I}_{delta}\right] \tag{J.6.7}$$

Thus, given a set of $\left[\tilde{I}_{delta}\right]$, which may include a circulation current component, we may correctly determine $\left[\tilde{I}_{wye}\right]$ according to Equation J.2.2. Then Equation J.6.7 returns $\left[\tilde{I}_{delta}\right]$ minus the circulating current.

J.7   Practical Considerations

In Figure J.3.1, when the network elements are linear, there can be no circulating current in the delta since the sum of the delta voltages is always zero. Similarly, there can be no neutral shift across a neutral grounding impedance since the sum of the wye (line) currents is necessarily zero. If the wye voltages are balanced, there can also be no neutral shift due to the source voltages. Under these conditions, Equations J.6.6 and J.6.7 will indeed return the true results. If the source voltages are not balanced, the neutral shift in the source voltages cannot be determined on the basis of a knowledge of the delta voltages alone. If the delta represents a transformer winding, then the nonlinear magnetizing reactance may give rise to circulating triplen harmonic currents in the delta. Additionally, in the real world there will be capacitances between the windings and earth that may lead to a neutral shift. These matters require a more detailed analysis than that contained in this appendix.

Appendix K

Analysis of Electromagnetic Circuits

This presentation assumes that the reader is familiar with the analysis of magnetic circuits as presented in textbooks on electric machinery. In this appendix we present a brief review of a few of the principles that are of particular importance in this book.

K.1 Closed Iron-Core Magnetic Circuits

Consider the current-carrying toroid shown in Figure K.1.1, with a closed iron-core ring of rectangular cross section. On the basis of Ampere's circuital law, an H field exists only within the iron-core. However, the strength of the B field within the iron is considerably greater than that for the air-core solenoid due to the high relative magnetic permeability of iron.

Example K.1.1

For the closed iron-core toroid shown in Figure K.1.1, we have a magnetic field only within the iron-core:

$$B = \frac{\mu_i N i}{2\pi r} \qquad [Wb/m^2]$$

Then:

$$\phi = \int_{r_i}^{r_o} B w \, dr = \frac{\mu_i N i w}{2\pi r} \ln \frac{r_o}{r_i} \qquad [Wb]$$

If we employ the convergent power series:

$$0.5 \ln \frac{r_o}{r_i} = \left(\frac{r_o - r_i}{r_o + r_i} \right) + \frac{1}{3} \left(\frac{r_o - r_i}{r_o + r_i} \right)^3 + \cdots$$

and let the second term in the series be 1% of the first term, so that we may neglect all higher-order terms:

$$\frac{1}{3} \left(\frac{r_o - r_i}{r_o + r_i} \right)^3 = 0.01 \left(\frac{r_o - r_i}{r_o + r_i} \right)$$

then:

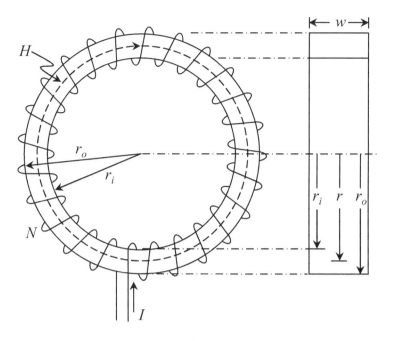

Figure K.1.1: Closed Iron-Core Toroid with Fully Distributed Turns

$$\left(\frac{r_o - r_i}{r_o + r_i}\right)^2 = 0.03$$

and:

$$\left(\frac{r_o - r_i}{r_o + r_i}\right) = 0.1732$$

and:

$$\frac{r_i}{r_o} = 0.7$$

Thus, if $r_i/r_o > 0.7$, we may write:

$$0.5 \ln \frac{r_o}{r_i} \approx \left(\frac{r_o - r_i}{r_o + r_i}\right)$$

with only about 1% error. Then we may write:

$$\phi = \frac{\mu_i\, Niw}{\pi}\left(\frac{r_o - r_i}{r_o + r_i}\right) \qquad [Wb]$$

Now:

$$\frac{r_o + r_i}{2} = r_{mean} \qquad [m]$$

and:

$$w\,(r_o - r_i) = A \qquad [m^2]$$

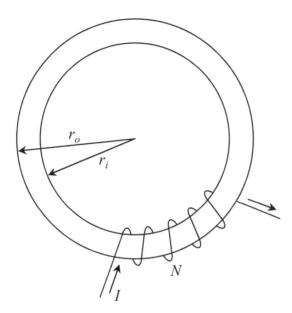

Figure K.1.2: Closed Iron-Core Toroid with Partially Distributed Turns

so that we may write:

$$\phi = \frac{Ni\mu_i A}{2\pi r_{mean}} \qquad [Wb]$$

Thus, in order to calculate the flux in the toroid, we may use the mean value for H, the mean length of the closed flux line, and the actual cross sectional area of the toroid, with less than about 1% error for $r_i/r_o > 0.7$.

Consider now the closed iron-core magnetic circuit shown in Figure K.1.2, where the winding is concentrated over only a portion of the iron-core. Now it becomes somewhat awkward to begin with Ampere's circuital law. Instead, we recognize that for a given NI, almost all of the magnetic flux will flow along the iron core because of the high relative magnetic permeability of iron. Again the B field will remain (almost) entirely within the core and, in turn, loci of constant H will again be (almost) concentric circles.

From another point of view, for a given amount of required magnetic flux within the iron toroid, we require only $\mu_0/\mu_i = 1/5000$ of the ampere turns required for an air toroid. We would therefore expect very little flux outside the iron core.

K.2 Magnetic Ohm's Law

According to the analysis in Example K.1.1, we concluded that for closed iron-core magnetic circuits, where we may assume the magnetic flux to be constrained to flow entirely within the iron-core circuit, we may evaluate the magnetic flux as follows:

$$\phi = \frac{Ni\mu_i A}{\ell} \qquad [Wb] \qquad\qquad (K.2.1)$$

where A is the cross sectional area through which the flux flows, and l is the length of the closed flux path. Of course, as shown in Example K.1.1, this is not rigorously correct, but assumes a uniform flux density B across the cross section of the core. For most practical electromagnetic machines, the error is negligibly small.

We define:

$$\Re = \frac{\ell}{\mu_i A} \qquad [H^{-1}] \qquad (K.2.2)$$

$$\wp = \frac{\mu_i A}{\ell} \qquad [H] \qquad (K.2.3)$$

where $\Re$ is called the magnetic *reluctance*, and $\wp$ is called the magnetic *permeance*. Then we may write Equation K.2.1 in the form:

$$\phi = (Ni)\,\wp \qquad [Wb] \qquad (K.2.4)$$

or:

$$MMF = Ni = \Re\phi \qquad [A] \qquad (K.2.5)$$

We observe the striking similarity to Ohm's law for electric circuits:

$$EMF = RI \qquad [V] \qquad (K.2.6)$$

This similarity can be demonstrated formally as follows. By Ampere's circuital law, and following a path of integration from A to B along path ℓ_1 and returning to A along path ℓ_2 in Figure K.2.1, we have:

$$\int_A^B \vec{H} \cdot d\vec{\ell}_1 + \int_A^B \vec{H} \cdot d\vec{\ell}_2 = \oint \vec{H} \cdot d\vec{\ell} = 0 \qquad [A] \qquad (K.2.7)$$

since this closed path does not encircle any net current. Therefore:

$$\int_A^B \vec{H} \cdot d\vec{\ell}_1 = \int_A^B \vec{H} \cdot d\vec{\ell}_2 \qquad [A] \qquad (K.2.8)$$

Consequently, the open-path integral between two points in a magnetic circuit will have the same value no matter which path the integration is performed over, so long as there is no net current enclosed between the two chosen paths. It is convenient to introduce the practical concept of *magnetic scalar potential* or *magnetomotive force*, which we formally define by:

$$\vec{H} = -grad\,(MMF) = -\nabla\,(MMF) \qquad [A/m] \quad (K.2.9)$$

so that:

$$\int_A^B \vec{H} \cdot d\vec{\ell} = (MMF_A - MMF_B) \qquad [A] \qquad (K.2.10)$$

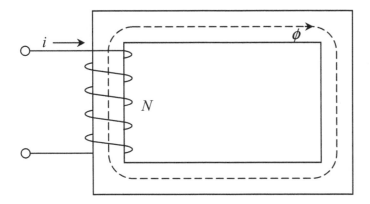

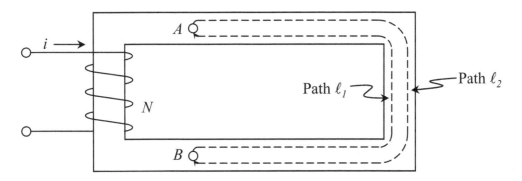

Figure K.2.1: Closed Iron-Core Magnetic Circuit

For most practical magnetic circuits where the flux is constrained to flow along an easily defined path with uniform flux density, we can avoid the formality of Maxwell's equations and employ Equation K.2.5 instead. In view of the dimensions associated with magnetic permeance $\wp$ [H], we may anticipate that this geometric factor will be a significant part of the self and mutual inductance coefficients of windings, which will be introduced in a later section.

K.3   Fringing Flux and Leakage Flux Concepts

Ampere's circuital law remains the basis for relating the strength of magnetic fields to an exciting current. The $MMF = \Re\phi$ law is simply a practical algebraic formulation of Ampere's law and is considerably easier to employ in most practical magnetic circuit configurations where the flux density is uniform over a specified constrained path. This is always the case in magnetic circuits where the flux flows more or less completely in iron and where the permeability of iron is assumed to be infinitely large.

The iron-core reactor with air gap does not rigorously fall into this category even if the permeability of the iron is assumed to be infinitely large. For example, in Figure K.3.1(c), the total MMF acting upon the magnetic circuit is $MMF_t = Ni$. If the iron is assumed to have infinite permeability, then the reluctance of the iron portion of the circuit is zero and the total MMF_t appears across the air gap g. However, this same MMF_t also appears between the points 1 and 2 across the window of the core. Even though the length of the path between points 1 and 2 is considerably greater than that across the gap g, invariably the cross sectional area available for the window is greater than that for the gap g. Thus, there does exist some flux across the window of the reactor magnetically in parallel with the flux across the gap g.

Simply for the sake of identifying the various components, we define *main flux* as that portion of the total flux linking the winding that crosses the gap g, and the remainder of the flux as *leakage flux*. Ordinarily, the magnitude of the leakage flux (in this type of configuration) cannot be easily evaluated directly since the flux density will generally not be uniform, but usually can be deduced from voltage and current measurements. Therefore, the total flux can be determined by the application of the MMF law by simply taking an equivalent reluctance path for the flux in the window parallel to the reluctance of the gap g.

Consider now the configuration shown in Figure K.3.1(b), where the exciting winding has been placed directly over the air gap. For infinitely permeable iron, the total $MMF_t = Ni$ is consumed by $MMF_g = \Re_g\phi$ so that the MMF acting between points 3 and 4 is zero. Consequently there can be no leakage flux across the window of the core. Thus, placement of the exciting current winding can have a significant influence upon the performance characteristics of magnetic circuits.

Figure K.3.1(d) shows the nature of fringing flux at the air gap. The flux lines are not parallel across the air gap but assume a pattern that can be determined from a magnetic field plot assuming the iron surfaces to be magnetic equipotential surfaces. In order that the flux fringing can be appropriately taken into account while still taking advantage of the simple form of $MMF = \Re\phi$, field plots have been made for many practical gap configurations and correction factors established. For most cases where the cross sectional area of the core is given by $A_c = wh$, the effective cross sectional area of the air gap can be taken as:

$$A_g = (w + g)(h + g) \qquad [m^2] \tag{K.3.1}$$

where g is the actual length of the air gap.

Again the placement of the exciting winding has an influence upon the degree of flux fringing. In view of the fact that the MMF inside a (long) solenoidal winding is essentially parallel to the solenoid, locating the winding as shown in Figure K.3.1(b) will tend to reduce the effect of flux fringing by forcing the flux lines across the gap to remain more nearly parallel.

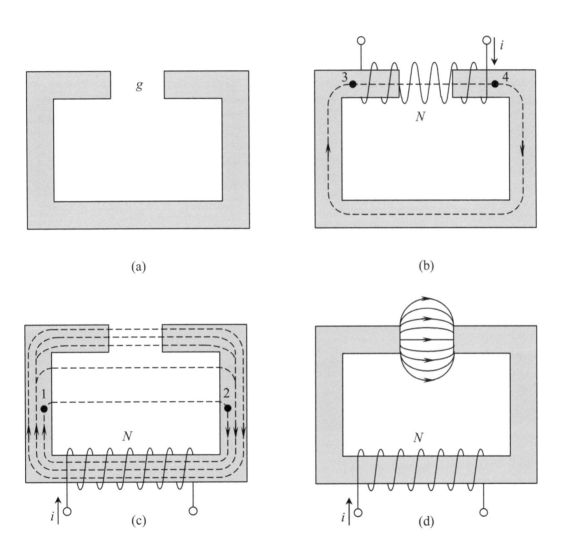

Figure K.3.1: (a) Iron-Core with Air Gap; (b) Exciting Winding over Air Gap; (c) Exciting Winding Remote from Air Gap; and (d) Fringing at Air Gap

K.4   Characteristics of Ferromagnetic Materials

It has been shown that it is highly desirable to employ materials with a high relative permeability to guide magnetic flux into a region where high energy densities can be achieved. Of the pure elements only iron, nickel, and cobalt exhibit high relative magnetic permeabilities at normal temperatures. Steel alloys, containing combinations of cobalt, nickel, aluminum, tungsten, and silicon also have high relative magnetic permeabilities and all are known as ferromagnetic materials. All other materials are called *nonmagnetic* and have magnetic permeabilities very nearly that of air (free space).

For practical engineering purposes, fortunately, it is not necessary to delve into the physical characteristics of materials to explain why some materials are ferromagnetic and others are not. We do need to concern ourselves with some external characteristics of magnetic materials, however.

When a closed iron-core magnetic circuit, such as that shown in Figure K.1.1, is energized with a pure sinusoidal voltage:

$$v(t) = \sqrt{2}\, V_{rms} \cos \omega t \tag{K.4.1}$$

and ignoring all losses (watts), we have according to Faraday's law:

$$\lambda(t) = \int_0^r v(t)\, dt = \left(\frac{\sqrt{2}}{\omega}\right) V_{rms} \sin \omega t$$

$$= \sqrt{2}\left(N^2 \wp\, I_{rms}\right) \sin \omega t \tag{K.4.2}$$

so that:

$$\frac{V_{rms}}{I_{rms}} = \omega N^2 \wp \tag{K.4.3}$$

For:

$$\wp = \frac{\mu A}{\ell} \tag{K.4.4}$$

$$V_{rms} = \omega N B_{rms} A \tag{K.4.5}$$

$$I_{rms} = \frac{H_{rms}\, \ell}{N} \tag{K.4.6}$$

we may write:

$$\left(\frac{V_{rms}}{I_{rms}}\right)\left(\frac{\ell}{\omega N^2 A}\right) = k\left(\frac{V_{rms}}{I_{rms}}\right) = \frac{B_{rms}}{H_{rms}} = \mu \tag{K.4.7}$$

If we apply voltages V_{rms} of increasing magnitudes, and measure I_{rms}, and plot B_{rms} vs. H_{rms} we obtain an *apparent AC magnetization characteristic* as shown in Figure K.4.1. The nonlinear relationship between B and H is due to the fact that the magnetic permeability of the iron is not constant but depends upon the flux density in the iron. At low flux densities the iron is *unsaturated* and has a very high permeability. At high flux densities the iron is *saturated* and the permeability approaches that of air.

Now consider an iron-core magnetic circuit with air gap, shown in Figure K.3.1, with:

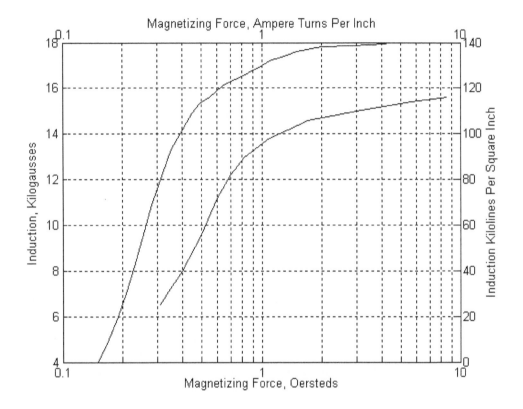

Left Curve: kG vs. Oersteads   Right Curve: kL/in^2 vs. At / in
1 Oerstead = 79.58 A / m   1 kiloGauss = 0.1 Wb / m^2

*Data from USS Transformer 66 – 29 Gage, Cold Reduced Coils
or Sheets. Replotted from the original USS specification.
Test Conditions: Lengthwise sample annealed at 1450^0 F, tested
at 60 Hz with double lap joint.

Figure K.4.1: Apparent AC Magnetization Characteristic

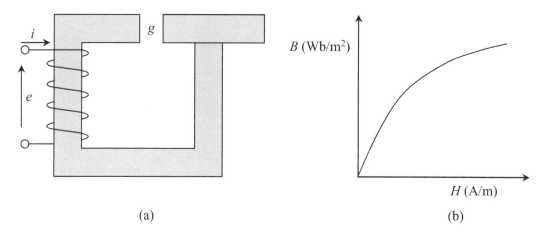

Figure K.5.1: Ferromagnetic Core and Normal Magnetization Curve

$$\Re_i = \frac{\ell_i}{\mu_i \, A_i} \qquad\qquad \Re_g = \frac{\ell_g}{\mu_0 \, A_g} \qquad [H^{-1}] \tag{K.4.8}$$

Then:

$$MMF = NI = (\Re_i + \Re_g)\,\phi \qquad [A] \tag{K.4.9}$$

$$\phi = \frac{NI}{\Re_i + \Re_g} \qquad [Wb] \tag{K.4.10}$$

If the voltage (flux) is known, it is a simple matter to determine the permeability to use for $\Re_i$ from Figure K.4.1. Then Equation K.4.9 may be evaluated directly for the total MMF required.

On the other hand, if the total MMF is known, we do not know the flux nor the MMF_i drop in the iron, so that we cannot directly establish the permeability to use for $\Re_i$ in Equation K.4.9. A trial-and-error or iterative procedure would be required in this case. An alternative approach would be to employ the graphical analysis described in the next section.

K.5   Nonlinear Ferromagnetic Core Structures – Graphical Analysis

Because of the nonlinear characteristics of ferromagnetic materials, the evaluation of the electrical performance characteristics of stationary magnetic circuits may become somewhat more complex than that described in Section K.2 for ideal linear circuits.

For the ferromagnetic core with air gap, illustrated in Figure K.5.1, there is little added complication to the analysis when voltage e is held constant. Then, according to Faraday's law:

$$\lambda = N\phi = \int e\,dt \qquad [V \cdot s] \tag{K.5.1}$$

the flux-linkages also maintain a constant peak value regardless of the circuit nonlinearity. We simply add the MMF drops in the iron and air portions to obtain the total ampere-turn excitation required:

$$Ni = MMF_{iron} + MMF_{gap} = MMF_{iron} + \frac{g\phi}{\mu_0 A_g} \qquad [A] \qquad \text{(K.5.2)}$$

In order to calculate the MMF drop in the iron in the same manner as that for the air gap, an appropriate value for μ must be taken. This estimation can be circumvented if the $B - H$ relationship for the iron, shown in Figure K.5.1(b) is used directly. The flux density in the iron-core is determined as:

$$B_i = \frac{\phi}{A_i} \qquad [Wb/m^2] \qquad \text{(K.5.3)}$$

where A_i is the effective cross sectional area of the iron-core. For reasons that will be taken up shortly, iron-cores are generally assembled as layers of thin steel laminations, each lamination being either enameled on one side or both, or simply have some other insulated coating. After stacking the core, there remains an unavoidable nonmagnetic space between laminations so that the cross sectional area of the core must be calculated on the basis of:

$$A_i = F_s wd \qquad [m^2] \qquad \text{(K.5.4)}$$

where w and d represent the actual measured width and depth of the stacked core, and F_s is a stacking factor (somewhat less than unity) to take into account the nonmagnetic space between laminations.

With the calculated value of B_i, the value of H_i is read directly from Figure K.5.1(b). This must be multiplied by the effective length of the flux path in iron. Again, at the corners where the separate laminations or punchings form a joint (butt or miter), small unavoidable air gaps exist and an experience factor must be included to correct the effective length of the flux path.

Somewhat more complicated is the situation where the exciting ampere-turns Ni are held constant and it is required to predict the magnitude of flux for the magnetic circuit shown in Figure K.5.1(a). The difficulty arises because we cannot predict in advance how the MMF drops will be proportioned between the nonlinear core and the linear air gaps. A graphical analysis resolves the problem quite simply. Since:

$$MMF_{total} = MMF_i + MMF_g \qquad [A] \qquad \text{(K.5.5)}$$

we may write:

$$MMF_i = MMF_{total} - MMF_g = MMF_{total} - \Re_g \phi \qquad [A] \qquad \text{(K.5.6)}$$

Now MMF_i is plotted as ϕ vs. MMF in Figure K.5.2, correctly representing the nonlinearity of the iron circuit, and the right-hand side of Equation K.5.6 is plotted as a linear relationship. The intersection of the two plots represents the equality of both sides of the equation so that the flux ϕ_{net} is read directly from Figure K.5.2.

It is of interest to note that a change in excitation Ni produces an air gap line parallel to the one shown in Figure K.5.2. Thus, in the previous example, where e (hence ϕ) was constant in amplitude, an air gap line could be found intersecting the MMF_i curve at ϕ, and the required Ni could thus be determined by this graphical method as well.

Attention should be called to the fact that the curve in Figure K.5.1(b) is different in nature than the one shown in Figure K.5.2. The former represents a $B - H$ curve for the steel alone (usually obtained from steel manufacturers), while the latter is an empirical curve obtained from tests on actual cores, relating ϕ and Ni and taking into account the influence of air gaps at the joints and the stacking factor.

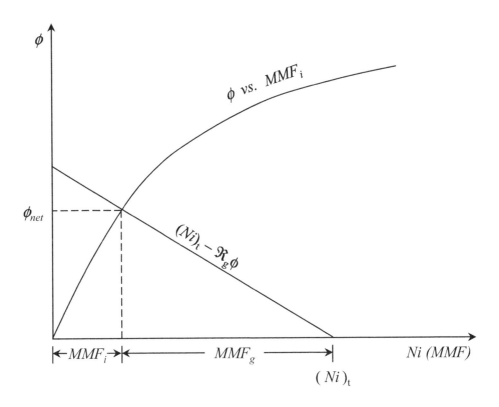

Figure K.5.2: Graphical Magnetic Circuit Analysis

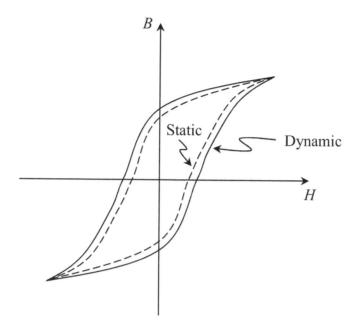

Figure K.6.1: Static and Dynamic Hysteresis Loops

K.6 Core Losses – Exciting and Magnetizing Current

In Figure K.4.1, we showed the rms relationship between B and H for an iron-core magnetic circuit for VARs only. Figure K.6.1 illustrates the complete instantaneous relationship between B and H for ferromagnetic materials with a sinusoidal variation in B. It is apparent that there is not a one-to-one correspondence between B and H as we would expect for a material with a constant μ, but rather the value of B, for a given value of H, depends upon whether H is increasing or decreasing. This lag of magnetic flux density behind magnetic field intensity is called *hysteresis*.

During one complete cycle, the energy supplied to the magnetic circuit is greater than that returned to the source. The area under the hysteresis loop represents real energy loss, attributable to the stress and strain within the material undergone during the reversals in magnetic field, and appears as heat generated within the material. Since there is one complete loop per cycle of electrical frequency, hysteresis loss is directly proportional to this frequency.

The area under the hysteresis loop increases with increase in B_{max} and H_{max}. Moreover, the shape of the hysteresis loop also changes with B_{max}. The hysteresis loss can be estimated on the basis of the empirical formula proposed by Charles Proteus Steinmetz in 1892:

$$P_h = k_h f B_{max}^n \quad [W] \qquad (K.6.1)$$

where k_h is a constant of the particular ferromagnetic material and geometry of the core, f is the electrical frequency of the exciting current, B_{max} is the peak value of magnetic flux density in the ferromagnetic material, and n is called the *Steinmetz exponent*, which varies from 1.5 to 2.5 for different materials, with $n = 1.6$ taken as a typical value.

Indicated on the outermost loop in Figure K.6.1 is B_r at $H = 0$, which is called the *residual flux density*, and H_c at $B = 0$, which is called the *coercive force*. The dotted curve through the positive tips of successively larger hysteresis loops is called the *normal magnetization curve*. It is a representation of the magnetic characteristics of the material without consideration of the real loss component.

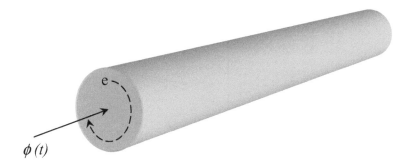

Figure K.6.2: Eddy Current Losses in Iron

There appears another energy loss in iron-cores caused by induced voltages according to Faraday's law and their consequent $I^2 R$ losses in the iron itself. In Figure K.6.2, we show a cylindrical section of a solid iron-core that carries a time-varying magnetic flux, $\phi(t)$. For a sinusoidally varying flux we have, for example:

$$\phi(t) = \phi_{max} \sin \omega t = NAB_{max} \sin \omega t \qquad [Wb] \qquad \text{(K.6.2)}$$

and by Faraday's law, eddy current voltages are induced:

$$e(t) = \frac{d\phi(t)}{dt} = \omega NAB_{max} \cos \omega t \propto fB_{max} \qquad [V] \qquad \text{(K.6.3)}$$

which, in turn, produce eddy currents:

$$i(t) = \frac{e(t)}{R} \propto fB_{max} \qquad [A] \qquad \text{(K.6.4)}$$

with a real power loss:

$$p_e = k_e \, f^2 \, B_{max}^2 \qquad [W] \qquad \text{(K.6.5)}$$

Thus, core losses are composed of eddy current losses defined by Equation K.6.5, and hysteresis losses defined by Equation K.6.1.

The effective resistance R in Equation K.6.4 depends not only upon the electrical resistivity of the ferromagnetic material, but also upon the shape of the cross sectional area of the iron-core. Because skin effect will cause a nonuniform flux density across the cross section of the core, a rigorous solution involves the use of Bessel functions.

For a steel lamination, illustrated in Figure K.6.3, with a thickness t, it can be shown that:

$$p_e = \frac{\pi^2 \, B_{max}^2 \, f^2 \, t^2}{6\rho} \qquad [W/m^3] \qquad \text{(K.6.6)}$$

for a sinusoidal variation in flux density and with ρ being the electrical resistivity of the lamination. It is apparent that the thinner the lamination the smaller will be the eddy current losses. However, core assembly costs increase with a decrease in lamination thickness, so that practical transformer steel laminations have a thickness of 0.3556 mm (0.014 inches).

The hysteresis loop shown in Figure K.6.1, containing only hysteresis loss, is called a dc or *static hysteresis loop*, obtained at very low frequency alternating current so that eddy currents are negligible. At power frequencies, the additional eddy current losses will affect the $B - H$ loop as

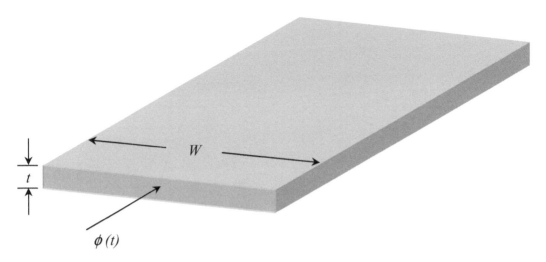

Figure K.6.3: Eddy Currents in Core Lamination

shown in Figure K.6.1. The $B - H$ loop obtained under power frequency conditions is called the *dynamic hysteresis loop* and contains both eddy current and hysteresis losses.

It is of practical interest to see what influence this magnetic nonlinearity has upon the electrical performance characteristics of ferromagnetic cores. Figure K.6.4(a) illustrates the dynamic hysteresis loop obtained for *sinusoidal flux* conditions. The exciting current (or MMF drop) is shown to be severely distorted indicating not only significant 3d, 5th, 7th, etc., higher harmonic components, but apparently also a significant fundamental frequency component responsible for causing an unsymmetrical waveshape. Figure K.6.4(b) illustrates the corresponding normal magnetization curve, which represents the variation of flux with *magnetizing* current (no losses). The magnetizing current exhibits significant 3d, 5th, 7th, etc., harmonic components, but the waveshape is symmetrical. If we write:

$$i_{mag} = a_1 \sin \omega t + a_3 \sin 3\omega t + a_5 \sin 5\omega t + \cdots \qquad (K.6.7)$$

it would appear that:

$$i_{exc} = i_{mag} + b_1 \cos \omega t \qquad (K.6.8)$$

thereby accounting for the dissymmetry in i_{exc}, as well as properly allowing for the 90 degree phase shift between the real loss component and the inductive magnetizing current. Figure K.6.5 illustrates the appropriate phasor relationship between total exciting current i_{exc}, the magnetizing current i_{mag}, and the total loss component i_{e+h}.

The above rationale can be substantiated based upon the following considerations. The watts that must be supplied to cover p_h and p_e, at *pure sinusoidal applied voltage*, must come from the fundamental frequency in-phase component of the exciting current since:

$$\int_0^{2\pi} v \cos \omega t\, i \cos k\omega t\, d\omega t = 0 \qquad (k \neq 0) \qquad (K.6.9)$$

After subtracting this fundamental frequency in-phase component from the exciting current, there must remain only magnetizing current. In practical situations, the distorted exciting current in conjunction with source impedance will result in some distortion of the voltage actually impressed upon a winding. Under controlled laboratory conditions, the above performance characteristics can be verified with satisfactory accuracy.

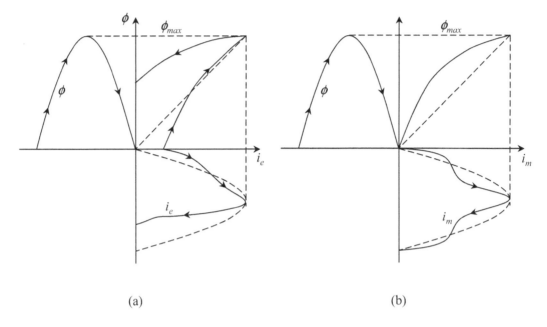

Figure K.6.4: Exciting Current and Magnetizing Current

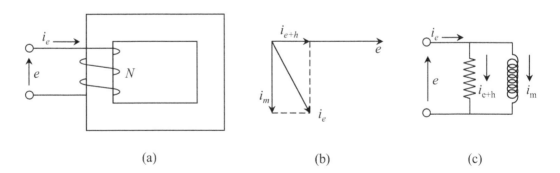

Figure K.6.5: Phasor Relationships among Exciting Current Components

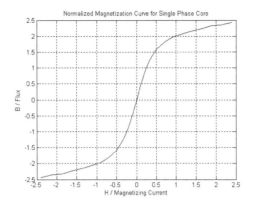

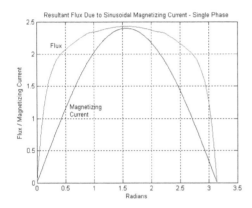
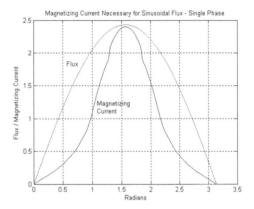

Figure K.6.6: Calculated Flux and Magnetization Current with Normal Magnetization Curve

Figures K.6.6 and K.6.7 show calculated flux and magnetizing currents under the conditions that each is purely sinusoidal for the cases when a normal magnetization curve is used and when a magnetization characteristic with a hysteresis loop is used respectively.

K.7 Two-Winding Magnetic Circuits – Mutual Flux and Leakage Flux

The analysis of two-winding magnetic circuits requires an extension of the principles already developed. Consider the two-winding magnetic circuit shown in Figure K.7.1. Let the windings be energized with alternating currents, which we shall represent in terms of rms (effective) values and in phasor notation. We have:

$$\begin{bmatrix} N_1 \tilde{I}_1 \\ N_2 \tilde{I}_2 \end{bmatrix} = \begin{bmatrix} (\Re_1 + \Re3) & -\Re_3 \\ -\Re_3 & (\Re_2 + \Re_3) \end{bmatrix} \begin{bmatrix} \tilde{\phi}_1 \\ \tilde{\phi}_2 \end{bmatrix} \tag{K.7.1}$$

$$\begin{bmatrix} \tilde{\phi}_1 \\ \tilde{\phi}_2 \end{bmatrix} = \left(\frac{1}{\nabla} \right) \begin{bmatrix} (\Re_2 + \Re3) & \Re_3 \\ \Re_3 & (\Re_1 + \Re_3) \end{bmatrix} \begin{bmatrix} N_1 \tilde{I}_1 \\ N_2 \tilde{I}_2 \end{bmatrix} \tag{K.7.2}$$

with:

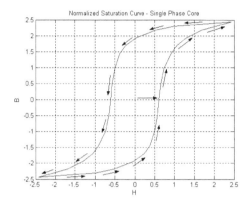

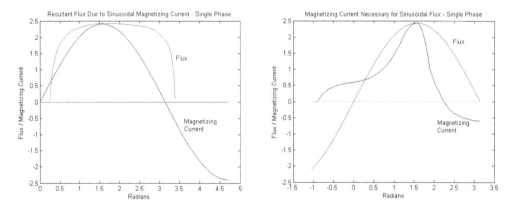

Figure K.6.7: Calculated Flux and Magnetization Current with Hysteresis Loop

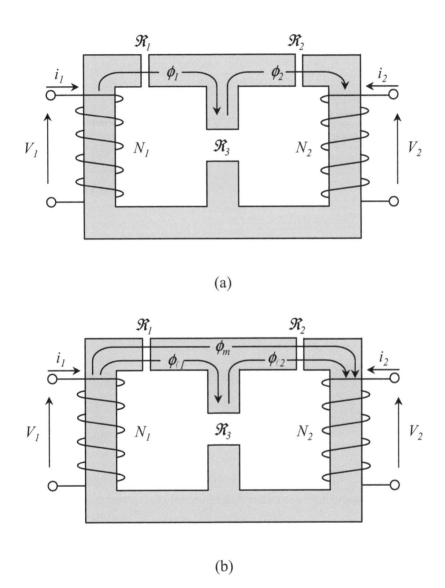

(a)

(b)

Figure K.7.1: Two-Winding Magnetic Circuit

$$\nabla = \Re_1 \Re_2 + \Re_3 (\Re_1 + \Re_2) \tag{K.7.3}$$

Therefore:

$$\tilde{\phi}_1 = \left(\frac{\Re_2}{\nabla}\right) N_1 \tilde{I}_1 + \left(\frac{\Re_3}{\nabla}\right) \left(N_1 \tilde{I}_1 + N_2 \tilde{I}_2\right) \tag{K.7.4}$$

$$\tilde{\phi}_2 = \left(\frac{\Re_3}{\nabla}\right) \left(N_1 \tilde{I}_1 + N_2 \tilde{I}_2\right) + \left(\frac{\Re_1}{\nabla}\right) N_2 \tilde{I}_2 \tag{K.7.5}$$

We observe that the flux linking the first winding is composed of one component that depends only upon the ampere turns in the first winding acting upon a permeance:

$$\wp_{\ell 1} = \frac{\Re_2}{\nabla} = \frac{1}{\Re_1 + \Re_3 \left(1 + \dfrac{\Re_1}{\Re_2}\right)} \tag{K.7.6}$$

and similarly for the second winding:

$$\wp_{\ell 2} = \frac{\Re_1}{\nabla} = \frac{1}{\Re_2 + \Re_3 \left(1 + \dfrac{\Re_1}{\Re_2}\right)} \tag{K.7.7}$$

and a second component, which depends upon the ampere turns of both windings acting together upon a permeance:

$$\wp_m = \frac{\Re_3}{\nabla} = \frac{1}{\Re_1 + \Re_2 + \dfrac{\Re_1 \Re_2}{\Re_3}} \tag{K.7.8}$$

Therefore, we may write:

$$\begin{bmatrix} \tilde{\phi}_{\ell 1} + \tilde{\phi}_m \\ \tilde{\phi}_{\ell 2} + \tilde{\phi}_m \end{bmatrix} = \begin{bmatrix} \tilde{\phi}_1 \\ \tilde{\phi}_2 \end{bmatrix} = \begin{bmatrix} (\wp_{\ell 1} + \wp_m) & \wp_m \\ \wp_m & (\wp_{\ell 2} + \wp_m) \end{bmatrix} \begin{bmatrix} N_1 \tilde{I}_1 \\ N_2 \tilde{I}_2 \end{bmatrix} \tag{K.7.9}$$

Figure K.7.1(b) shows the two-winding magnetic circuit in terms of the flux components.

We shall now change our definition for leakage flux from that given in Section F.1 for a singly-excited magnetic circuit. In a two-winding magnetic circuit, we shall call *mutual flux* that *component* of total flux that links both windings, whether this flux is in iron or air or both. We shall call leakage flux associated with one winding as that *component* of total flux that links only one winding, whether this flux is in iron or air or both.

More specifically, we shall call $\tilde{\phi}_{\ell 1}$ the *leakage flux* component of $\tilde{\phi}_1$, and $\tilde{\phi}_{\ell 2}$ the *leakage flux* component of $\tilde{\phi}_2$, each linking only their respective windings. We shall call $\tilde{\phi}_m$ the *mutual flux* component, which links both windings.

We emphasize the *component* aspect of these mutual and leakage fluxes in the mathematical sense as contrasted to the true net flux, as measurable with flux probes, which actually links both windings. In Figure K.7.2(a), where the respective ampere turns are in *opposition*, the *total* flux linking the winding N_1 is:

$$\tilde{\phi}_1 = \tilde{\phi}_m + \tilde{\phi}_{\ell 1} \tag{K.7.10}$$

and the *total* flux linking winding N_2 is:

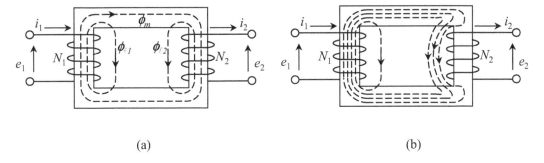

(a)   (b)

Figure K.7.2: Two-Winding Magnetic Core

$$\tilde{\phi}_2 = \tilde{\phi}_m + \tilde{\phi}_{\ell 2} \tag{K.7.11}$$

When the secondary winding is short-circuited ($E_2 = 0$), a field plot obtained by actually measuring the flux around the core would result in Figure K.7.2(b) where it would appear (on the surface) that there existed no mutual flux component nor leakage flux component associated with winding N_1, but only a leakage flux associated with winding N_2. This is *not* in keeping with our definition of *leakage* and *mutual* flux *components*. The actual net flux distribution is the result of the superposition of leakage and mutual flux components contributed by each of the exciting windings. Thus, the leakage flux and mutual flux concepts are mathematical conveniences that allow us to identify (at least in principle) these separate components and which permit us to employ the principle of superposition (in linear cases) to obtain the net results directly and simply.

The leakage flux components are determined, for all practical purposes, by the permeance of the air gap between the windings, which is independent of any nonlinear saturation effects. The mutual flux component is determined, for all practical purposes, by the permeance of the flux path around the core, which may exhibit nonlinear saturation effects. Since, for practical machines, these permeances can be identified in a reasonable manner, we can appreciate why these flux components have been defined. It is also possible to evaluate these parameters from electrical tests.

K.8   Two-Winding Magnetic Circuits – Stored Magnetic Energy

We may write Equation K.7.9 in terms of flux-linkages as follows:

$$\begin{bmatrix} \dfrac{\tilde{\Lambda}_1}{N_1} \\ \dfrac{\tilde{\Lambda}_2}{N_2} \end{bmatrix} = \begin{bmatrix} \wp_{11} & \wp_{12} \\ \wp_{21} & \wp_{22} \end{bmatrix} \begin{bmatrix} N_1 \tilde{I}_1 \\ N_2 \tilde{I}_2 \end{bmatrix} \tag{K.8.1}$$

Phasor power is defined by:

$$\tilde{S} = P + jQ = \tilde{V}\tilde{I}^* \quad [VA] \tag{K.8.2}$$

so that we may define *phasor energy* by:

$$\tilde{W} = \tilde{\Lambda}\tilde{I}^* \quad [J] \tag{K.8.3}$$

In a purely magnetic circuit, the current lags the voltage by 90 degrees, and the current is in phase with $\tilde{\Lambda}$, so that:

$$W_{mag} = \tilde{\Lambda}\tilde{I}^* = \Lambda I \qquad [J] \tag{K.8.4}$$

Now the electrical energy input to the magnetic network (ignoring losses) is:

$$dW_{el-in} = I_1 d\Lambda_1 + I_2 d\Lambda_2$$

$$= (N_1 I_1)\, d\left(\frac{\Lambda_1}{N_1}\right) + (N_2 I_2)\, d\left(\frac{\Lambda_2}{N_2}\right)$$

$$= \begin{bmatrix} N_1 I_1 & N_2 I_2 \end{bmatrix} \begin{bmatrix} d\left(\dfrac{\Lambda_1}{N_1}\right) \\[2ex] d\left(\dfrac{\Lambda_2}{N_2}\right) \end{bmatrix} \qquad [J] \tag{K.8.5}$$

$$\tag{K.8.6}$$

and for a stationary magnetic circuit in which $[\wp]$ is constant, we have:

$$dW_{el-in} = \begin{bmatrix} N_1 I_1 & N_2 I_2 \end{bmatrix} \begin{bmatrix} \wp_{11} & \wp_{12} \\ \wp_{21} & \wp_{22} \end{bmatrix} \begin{bmatrix} d\,(N_1 I_1) \\ d\,(N_2 I_2) \end{bmatrix}$$

$$= \begin{bmatrix} 1 & 1 \end{bmatrix} \begin{bmatrix} N_1^2 \wp_{11} I_1 dI_1 & N_1 N_2 \wp_{12} I_1 dI_2 \\ N_2 N_1 \wp_{21} I_2 dI_1 & N_2^2 \wp_{22} I_2 dI_2 \end{bmatrix} \begin{bmatrix} 1 \\ 1 \end{bmatrix}$$

$$= \frac{1}{2}\left[N_1^2\, \wp_{11} d\left(I_1^2\right) + N_2^2\, \wp_{22} d\left(I_2^2\right) + 2 N_1 N_2\, \wp_{12} d\left(I_1 I_2\right) \right] \qquad [J] \tag{K.8.7}$$

Because Equation K.8.7 is a perfect differential equation, it can be integrated directly, and for stationary magnetic circuits (neglecting losses) the electrical energy input is equal to the energy stored in the magnetic field so that:

$$W_{mag-stored} = \frac{1}{2}\left[(N_1 I_1)^2\, \wp_{11} + (N_2 I_2)^2\, \wp_{22} + 2\,(N_1 I_1)\,(N_2 I_2)\, \wp_{12} \right] \qquad [J] \tag{K.8.8}$$

which can also be written in the generalized form:

$$W_{mag-stored} = \frac{1}{2}\sum_{j=1}^{n}\sum_{k=1}^{n} (N_j I_j)\,(N_k I_k)\, \wp_{jk} \qquad [J] \tag{K.8.9}$$

Example K.8.1

In Figure K.7.1 let:

$$\Re_1 = \Re_2 = 0.005 \qquad \Re_3 = 10.0$$

$$\begin{bmatrix} N_1 \tilde{I}_1 \\ N_2 \tilde{I}_2 \end{bmatrix} = \begin{bmatrix} 10.005 & -10.0 \\ -10.0 & 10.005 \end{bmatrix} \begin{bmatrix} \dfrac{\tilde{\Lambda}_1}{N_1} \\ \dfrac{\tilde{\Lambda}_2}{N_2} \end{bmatrix}$$

Now let:

$$\frac{\tilde{\Lambda}_1}{N_1} = 1.00 \qquad \frac{\tilde{\Lambda}_2}{N_2} = 0.90$$

so that:

$$N_1 \tilde{I}_1 = 1.005 \qquad N_2 \tilde{I}_2 = -0.9955$$

Then:

$$\begin{bmatrix} \dfrac{\tilde{\Lambda}_1}{N_1} \\ \dfrac{\tilde{\Lambda}_2}{N_2} \end{bmatrix} = \begin{bmatrix} 100.025 & 99.975 \\ 99.975 & 100.025 \end{bmatrix} \begin{bmatrix} N_1 \tilde{I}_1 \\ N_2 \tilde{I}_2 \end{bmatrix}$$

so:

$$W_{mag-stored} = \frac{1}{2} \left[(N_1 I_1)^2 \, \wp_{11} + (N_2 I_2)^2 \, \wp_{22} + 2 \, (N_1 I_1) \, (N_2 I_2) \, \wp_{12} \right]$$
$$= 50.513875 + 49.5634 - 100.022738 = 0.054525 \qquad [J]$$

We may also write:

$$W_{mag-stored} = \frac{\phi_1^2 \, \Re_1}{2} + \frac{\phi_2^2 \, \Re_2}{2} + \frac{(\phi_1 - \phi_2)^2 \, \Re_3}{2}$$
$$= 0.0025 + 0.002025 + 0.05 = 0.054525 \qquad [J]$$

where we observe that about 92% of the magnetic energy is stored in the leakage flux path $\Re_3$, which we could, therefore, also have determined on the basis of:

$$\wp_{\ell 1} = \wp_{11} - \wp_m \qquad \wp_{\ell 2} = \wp_{22} - \wp_m$$

$$\frac{(N_1 I_1)^2 \, \wp_{\ell 1}}{2} + \frac{(N_2 I_2)^2 \, \wp_{\ell 2}}{2} = 0.050014 \qquad [J]$$

We observe, from the results of Example K.8.1 that for nearly equal flux-linkages of the first and second windings, the net ampere turns acting on the mutual flux path:

$$N_1 \tilde{I}_1 + N_2 \tilde{I}_2 = 0.0095 \quad [A] \tag{K.8.10}$$

is very small even though the ampere turns in each winding are comparatively large and of opposite sign. This accounts for the fact that most of the magnetic energy is stored in the leakage flux path.

K.9 Inductance Concept

Thus far, we have studied the performance of the magnetic circuit from a fundamental physical point of view. A thorough understanding of this aspect is necessary for the application of new ideas and materials to the design of electromechanical machines. For the design of singly-excited magnetic circuits, this approach is in itself practical and often to be preferred.

For doubly-excited magnetic circuits it has been found expedient to analyze the performance by means of equivalent electrical networks instead. The transition from the physical approach to the electrical network approach already begun in Section K.4, where we introduced the concepts of mutual and leakage flux, mutual and leakage permeance, and where we pointed out that these parameters are more easily evaluated by means of electrical tests than by field analysis.

We may rewrite Equation K.8.1 as follows:

$$\begin{bmatrix} \tilde{\Lambda}_1 \\ \tilde{\Lambda}_2 \end{bmatrix} = \begin{bmatrix} N_1^2 \, wp_{11} & N_1 N_2 \wp_{12} \\ N_2 N_1 \, \wp_{21} & N_2^2 \, \wp_{22} \end{bmatrix} \begin{bmatrix} \tilde{I}_1 \\ \tilde{I}_2 \end{bmatrix} = \begin{bmatrix} L_{11} & L_{12} \\ L_{21} & L_{22} \end{bmatrix} \begin{bmatrix} \tilde{I}_1 \\ \tilde{I}_2 \end{bmatrix} \tag{K.9.1}$$

We define *self-inductance* as:

$$L_{jj} = N_{jj}^2 \, \wp_{jj} \quad [H] \tag{K.9.2}$$

where $\wp_{jj}$ is the permeance to the total flux linking N_{jj}, and *mutual inductance* as:

$$L_{jk} = L_{kj} = N_j N_k \, \wp_{jk} \quad [H] \tag{K.9.3}$$

where $\wp_{jk}$ is the permeance to the mutual flux linking N_{jj} and N_{kk}.

We may generalize Equation K.9.1 for many windings on a common core by employing symbolic matrix notation:

$$[\tilde{\Lambda}] = [L] \, [\tilde{I}] \tag{K.9.4}$$

so that, on the basis of Equation K.8.5 we have:

$$W_{mag-stored} = \frac{1}{2} \sum_{j=1}^{n} \sum_{k=1}^{n} L_{jk} I_j I_k \quad [J] \tag{K.9.5}$$

On the basis of Faraday's law:

$$[v] = [L] \, \frac{d\,[i]}{dt} = \frac{d\,[L]}{dt} \, [i] \quad [V] \tag{K.9.6}$$

whereby the second term on the right-hand side of Equation K.9.5 is zero for stationary windings, and nonzero when the mutual inductance between windings varies with time (as in the case of rotating machines).

If the currents are sinusoidal and of common frequency, then the voltages are also sinusoidal and of common frequency. In terms of phasor notation and rms (effective) values we therefore have:

$$[\tilde{v}] = (j\omega)\,[L]\,[\tilde{I}] = j\,[X]\,[\tilde{I}] \tag{K.9.7}$$

For the two-winding magnetic circuit in Figure K.7.1, energized with sinusoidal current we have:

$$\begin{bmatrix} \tilde{V}_1 \\ \tilde{V}_2 \end{bmatrix} = j \begin{bmatrix} X_{11} & X_{12} \\ X_{21} & X_{22} \end{bmatrix} \begin{bmatrix} \tilde{I}_1 \\ \tilde{I}_2 \end{bmatrix} \tag{K.9.8}$$

On the basis of Equation K.7.9, we have:

$$X_{12} = \omega N_1 N_2 \, \wp_m \tag{K.9.9}$$

$$X_{11} - X_{12} = \omega N_1^2 \, \wp_{11} + \omega N_1 \,(N_1 - N_2)\, \wp_m \tag{K.9.10}$$

$$X_{22} - X_{12} = \omega N_2^2 \, \wp_{22} - \omega N_2 \,(N_1 - N_2)\, \wp_m \tag{K.9.11}$$

We observe that, in this form, the defined reactances are not nearly as directly related to the mutual and leakage fluxes as the designer of magnetic circuits would prefer. In Appendix I there are presented other forms of equivalent networks for this type of magnetic circuit that may be more appealing to the electric power engineer.

Appendix L

List of Symbols and Contexts

Symbol	Property	Units
A	Area	Square Meters $[m^2]$
B	Flux Density	Webers per Square Meter $[Wb/m^2]$
	Susceptance	Siemens $[S]$
C	Capacitance	Farads $[F]$
		Farads per meter $[F/m]$
D	Electric Flux	Coulombs per Square Meter $[C/m^2]$
d	Distance	Meters $[m]$
E	Electric Field Intensity	Volts per Meter $[V/m]$
ε	Permittivity	Farads per Meter $[F/m]$
f	Frequency	Hertz $[Hz]$
F	Force	Newton $[N]$
ϕ		
	Magnetic Flux	Webers $[Wb]$
	Electrostatic Potential	Volts $[V]$
Φ	Magnetic Flux (Used only in SS Harmonic Cases)	Webers $[Wb]$
G	Conductance	Siemens $[S]$
GMR	Geometric Mean Radius	Meters $[m]$
GMD	Geometric Mean Distance	Meters $[m]$

Symbol	Property	Units
H	Magnetic Field Intensity	Amp per Meter $[A/m]$
I,i	Current	Amperes $[A]$
J	Current Density	Amps per Dimension $[A/m, m^2, m^3]$
L	Inductance	Henrys $[H]$ Henrys per Meter $[H/m]$
λ	Flux-Linkage	Volt-Second $[V \cdot s]$
Λ	Flux-Linkage (Used only in SS Harmonic Cases)	Volt-Second $[V \cdot s]$ Volt-Second per Meter $[V \cdot s/m]$
ℓ	Path Length	Meters $[m]$
MMF	Magnetomotive Force	Amperes $[A]$
N	Turns	Dimensionless
P		
	Real Power	Watts $[W]$
	Potential Coefficient	1/Farads $[F^{-1}]$
$\wp$	Permeance	Henry $[H]$
Q	Reactive Power	Volt-Amp Reactive $[VAR]$
q,Q	Charge	Coulomb $[C]$
r	Radius	Meter $[m]$
R	Resistance	Ohms $[\Omega]$
ρ		
	Resistivity	Ohm-Meter $[\Omega \cdot m]$
	Charge Density	Coulombs per Dimension $[C/m, m^2, m^3]$
$\Re$	Reluctance	1/Henry $[H^{-1}]$
S	Complex Power	Volt-Amp $[VA]$
σ	Conductivity	1/Ohm-Meter $[1/\Omega \cdot m]$
V,v	Voltage (Electric Potential)	Volts $[V]$
v	Velocity	Meters per Second $[m/s]$
w	Work, Energy Density	Joules per Cubic Meter $[J/m^3]$
W	Work, Energy	Joules $[J]$

Symbol	Property	Units
ω	Radian Frequency	Radians per Second $[rad/s]$
X	Reactance	Ohms $[\Omega]$
		Ohms per Meter $[\Omega/m]$
		Ohm-Meters $[\Omega \cdot m]$
Y,y	Admittance	Siemens $(1/\Omega)$ $[S]$
Z,z	Impedance	Ohm $[\Omega]$

Context	Property
$\bar{F}$	Per Unit Quantity
$\tilde{F}$	Phasor Quantity
$\bar{\tilde{F}}$	Per Unitized Phasor Quantity
lowercase	Transient
UPPERCASE	Steady State
$\vec{F}$	Vector Quantity
$[F]$	Matrix, Row, or Column Vector
$[F]^{-1}$	Matrix Inverse
$[F]^{T}$	Matrix Transpose
$\tilde{F}^{*}$	Complex Conjugate
F_s	Sending-End Quantity
F_r	Receiving-End Quantity
F_1	Symmetrical Component Positive Sequence Quantity
F_2	Symmetrical Component Negative Sequence Quantity
F_0	Zero Sequence Quantity (Used with Several Transformations)

Context	Property
F_α	Clarke Component
	α Axis Quantity
F_β	Clarke Component
	β Axis Quantity
F_d	Park's Transformation
	D Axis Quantity
F_q	Park's Transformation
	Q Axis Quantity

Index